土壤地下水污染协同防治理论与技术丛书

Theories and Methods for Visualization of Enzyme Activities in Contaminated Soil and Groundwater
（水土污染酶活性可视化表征理论与方法）

蒲生彦　刘世宾　著

科学出版社

北　京

内 容 简 介

本书对土壤和地下水环境中多种酶的活性、功能和效应等进行介绍，以期为酶学在土壤和地下水协同修复方面的应用提供参考和依据；并结合多年在水土污染协同修复方面的研究概述修复过程对酶的功能和活性的影响，根据研究结果总结酶对环境变化（如污染、修复、气候变化、施肥）的响应，详细介绍可表征水土环境中酶活性时空分布特征的原位酶谱技术的基本原理、实验方法、研究与应用等，定量分析热点区域（即根际、碎屑）酶活性的时空变化过程及其主要控制因素，系统阐述原位酶谱法的基本原理、操作过程、应用与研究进展等。

本书可供从事土壤地下水环境保护工作从业人员和土壤学、地下水科学与工程、环境科学与工程、农业资源与环境相关领域专业技术人员及高校师生参考。

图书在版编目（CIP）数据

水土污染酶活性可视化表征理论与方法 = Theories and Methods for Visualization of Enzyme Activities in Contaminated Soil and Groundwater：英文 / 蒲生彦，刘世宾著. — 北京：科学出版社，2024.6
（土壤地下水污染协同防治理论与技术丛书）
ISBN 978-7-03-077113-1

Ⅰ. ①水… Ⅱ. ①蒲… ②刘… Ⅲ. ①酶—应用—地下水污染—修复—研究—英文 ②酶—应用—土壤污染—修复—研究—英文 Ⅳ. ①X523 ②X53

中国国家版本馆 CIP 数据核字 (2023) 第 229292 号

责任编辑：罗　莉　李小锐 / 责任校对：彭　映
责任印制：罗　科 / 封面设计：墨创文化

科学出版社 出版
北京东黄城根北街 16 号
邮政编码：100717
http://www.sciencep.com
四川煤田地质制图印务有限责任公司印刷
科学出版社发行　各地新华书店经销
*
2024 年 6 月第　一　版　开本：720 × 1000　1/16
2024 年 6 月第一次印刷　印张：27
字数：545 000
定价：298.00 元
（如有印装质量问题，我社负责调换）

About the editors

Shengyan PU, Ph.D, Professor
E-mail address: pushengyan13@cdut.edu.cn, pushengyan@gmail.com
Address: State Key Laboratory of Geohazard Prevention and Geoenvironment Protection (Chengdu University of Technology), Dongsanlu 1#, Erxianqiao, Chengdu 610059, Sichuan, P.R. China.

Shengyan PU is a professor in the College of Ecology and Environment, Chengdu University of Technology. He obtained his Ph.D degree from the Graduate School of Environmental Studies, Nagoya University, and completed his post-doctoral researches at the Hong Kong Polytechnic University and the Chinese Research Academy of Environmental Sciences, respectively. His research focuses on the development of key theories, technologies and equipment for collaborative remediation of contaminated soil and groundwater. In particular, his research interests include: ① Environmental interfacial behavior of pollutants of multiphase, multi-process and their multi-field coupling mechanism; ②The geochemical genesis of soil combined pollution and the behavioral mechanism of active components; ③The development of green remediation materials and exploitation of theories and technologies of collaborative remediation of cross-media pollution; ④The application of collaborative control technologies for contaminated soil & groundwater and localization of key equipments. He is the principal scientist of the National Key R&D Program of China and is selected into the Hong Kong Scholars Program. Prof. Pu has handled over 30 research projects granted by national, provincial and ministerial levels. He has published over 140 research papers in reputed journals, obtained 7 software copyrights, applied 26 invention patents, and authored 7 textbooks/monographs. He was awarded the Gold Prize of the Young Scientist Award of Chinese Society for Environmental Sciences, the second prize (2019) and the first prize (2020) of the National Environmental Protection Science and Technology Award, the prize of Sichuan Province Scientific and Technological Progress, the 14th Sichuan Youth Science and Technology Award, and the Sichuan Outstanding Environmental Scientist. He also serves as the deputy director of Research

Center for Soil and Groundwater Environmental Remediation and Risk Control Engineering Technology in Sichuan; the principal investigator of Soil and Groundwater Collaborative Remediation Innovation Research Team at State Key Laboratory of Geohazard Prevention and Geoenvironment Protection; the deputy director member of Young Scientists Branch of Chinese Society for Environmental Sciences; the deputy director of Ecological and Environmental Modeling Professional Committee of Chinese Society for Environmental Sciences, and a member of National Environmental Benchmarking Expert Committee.

Shibin LIU, Ph.D, Associate Professor
E-mail address: liushibin17@cdut.edu.cn
Address: School of Ecology and Environment, Chengdu University of Technology, 1# Dongsanlu, Erxianqiao, Chengdu 610059, Sichuan, P. R. China.

Shibin LIU is an Associate Professor at Chengdu University of Technology. He obtained his master's degree from Northwest University of Agriculture and Forestry and his Ph.D degree from the Department of Soil Science of Temperate Ecosystems, University of Göttingen. Dr. Liu's research focuses on the mechanisms of soil degradation and restoration. His main research areas include: ①Fate and environmental behaviors of soil contaminants (e.g., heavy metals, microplastics and organic pollutants) and mechanisms of their remediation; ②Influence of organic fertilizer and its pyrolyzed biochar on soil biochemical properties and plant growth. Various methods (i.e. zymography, isotope labelling, 16 sRNA sequencing and so on) are involved in these studies. He has published 27 academic articles and presided or precipitated over 10 projects. Dr. Liu also authored 2 textbooks/monographs. He also serves as the reviewer for *Soil Biology & Biochemistry, Geoderma, Biogeosciences, Science of the Total Environment, Biology and Fertility of Soils, Land Degradation and Development, Agriculture Ecosystems and Environment, European Journal of Soil Science, Ecotoxicology and Environmental Safety, Journal of Environmental Management, Rhizosphere, European Journal of Soil Biology, Scientific Reports*, and so on.

Preface

The concept of enzymes or more appropriately natural catalysts is an area of interest to researchers for over a century. Environmental enzymologists recognize that the measured activity may be a composite of reactions taking place in different locations and at different rates. Enzymes, excreted by both plants and microbes, are early indicators of environmental quality and the main mediators of organic matter decomposition, as they depolymerize organic compounds and generate soluble oligomers and monomers that can be transported into the cells. Enzymes also play an important role in the management of environment by detoxifying or transforming harmful substances into useful products. A variety of enzymes have been isolated from bacteria, fungi, and plants having a wider application in degradation and/or transformation of toxic environmental pollutants. They have been used, alone or in combination, for the decontamination of water and soil contaminated with organic (pesticides, polyaromatic hydrocarbons, polycyclic biphenyls, etc.) and inorganic (heavy metals and radionuclides) pollutants.

This book renders overviews of the role of enzymes in response to environmental contamination and remediation, and climate change. Its relationships with major environmental factors were also comprehensively investigated. Furthermore, novel approaches to quantify enzyme activities were also jointly presented. In summary, this book satisfies the urgent need to update the information on progress of enzymes for the scientific community and industrial personnel for the progress of knowledge and application of technological know-how for the benefit of mankind.

This book consists of six distinctive chapters and gives an overview of the functions, activities and analysis methods of enzymes in the soil and water environments (Chapter 1). Besides, the response of enzymes to environmental changes [e.g., climate change (Chapter 2), fertilization (Chapter 3), contamination (Chapter 4), remediation (Chapter 5)] was also summarized based on experimental results. Spatial and temporal distribution of enzyme activities was also assessed using in-situ zymography (Chapters 3, 6 and 5). Furthermore, variation of enzyme activities in hotspots (Chapter 6 i.e., rhizosphere, detritusphere) and the controlling factors were also summarized. This book has the following three key features.

(1) Detailed information of in-situ zymography technique that can visualize enzyme activities in soil and water environments.

(2) Quantified distribution of enzyme activities in response to nano-metal oxides contamination. The use of enzymes is also applied to evaluate remediating effect of various carbon-based biomass materials on heavy metal contamination.

(3) Sufficient responses of enzymes to environmental changes (e.g., contamination, remediation, climate change and fertilization).

This book is of interest to all scientists and professionals in soil enzymology, as well as those working in environmental sciences, including biogeochemists, hydrologists, biologists, hydrogeologists and so on. In addition, this book may provide a comprehensive reference for chemists and biotechnologists in thoroughly understanding the functions of enzymes and its great significance during environmental remediation. As enzyme application becomes more common both now and into the future, environmental organizations may utilize the work to explore new and more effective ways of behaving. Moreover, this book can be used by students in the above-mentioned fields as a guide for their design of research experiments.

Shengyan Pu

Acknowledgements

We would like to start by thanking Science Press for giving us the wonderful opportunity of editing the book, *Theories and Methods for Visualization of Enzyme Activities in Contaminated Soil and Groundwater*. We appreciate the trust, patience and knowledge they demonstrated throughout the whole process. We wish to thank all of the past and present members of the Innovation Research Group on Collaborative Remediation of Contaminated Soil & Groundwater for their persistence, dedication and intellectual contributions. We thank the editors and reviewers for their constructive comments on this book. We are also very grateful for two specialists' contribution to English polishment.

This work was supported by the National Key Research and Development Program of China (2020YFC1808300), the National Natural Science Foundation of China (U22A20591, 42077185), the Sichuan Science and Technology Program for Distinguished Young Scholars (2022ZYD0040, 2022JDJQ0010), the Natural Science Foundation of Sichuan Province (2023NSFSC0135), and the Research Fund of State Key Laboratory of Geohazard Prevention and Geoenvironment Protection (SKLGP2020Z002, SKLGP2021Z020).

Abbreviations

The following are the abbreviations that are used in this book.

Abbreviations	In English	In Chinese
Acp	acid phosphatase	酸性磷酸酶
AK	available potassium	可利用态钾
Akp	alkaline phosphatase	碱性磷酸酶
AMC	7-amino-4-methylcoumarin	7-氨基-4-甲基香豆素
AN	available nitrogen	可利用态氮
AP	available phosphorus	可利用态磷
ATP	adenosine triphosphate	腺嘌呤核苷三磷酸
BC	biochar	生物炭材料
BCR	the Community Bureau of Reference	标准物质局
B-glu/β-glu	β-glucosidase	β-葡糖苷酶
BN	biochar slow-release nitrogen fertilizer	生物炭缓释氮肥
C	carbon	碳
Ca-Fe-B	calcium-based magmetic biochar	钙基磁性生物炭
Deh/Dehydro	dehydrogenase	脱氢酶
DNA	deoxyribonucleic acid	脱氧核糖核酸
DOC	dissolved organic carbon	可溶性有机碳
DTPA	diethylene triamine pentaacetic acid	二乙基三胺五乙酸
EDS	Energy Dispersive Spectroscopy	能量色散谱
ENOPs	Engineered nanometal oxide particles	工程纳米金属氧化物
Fe-B	magmetic biochar	磁性生物炭
FT-IR	Fourier Transform infrared spectroscopy	傅里叶变换红外光谱仪
GT	green tea	绿茶
GTBC	green tea biochar	绿茶生物炭
LAP	leucine-peptidase	亮氨酸肽酶
MAP	mean annual precipitation	年均降水量
MAT	mean annual temperature	年均温度
MBC	microbial biomass carbon	微生物生物量碳
MBN	microbial biomass nitrogen	微生物生物量氮
MUF	4-methylumbelliferone	4-甲基伞形酮
N	nitrogen	氮
Nag	chitinase	几丁质酶
NAG	N-acetyl-β-glucosaminidase	β-N-乙酰基氨基葡糖苷酶
NH_4^+	ammonium	铵根离子
NO_3^-	nitrate	硝酸根离子

Abbreviations	In English	In Chinese
NPs	nano-particles	纳米颗粒
nZVI	nano zero-valent iron	纳米零价铁
nZVI@GTBC	nano zero-valent iron incorporated green tea biochar	纳米零价铁负载绿茶生物炭
OTU	operational taxonomic unit	运算分类单元
P	phosphorus	磷
PBAT	physiologically based extraction test	基于生理学的提取试验
PSD	particle size distribution	颗粒尺寸分布
SEM	scanning electron microscope	电子扫描显微镜
SEP	sequential extraction procedure	连续提取法
SOC	soil organic carbon	土壤有机碳
SOM	soil organic matter	土壤有机质
SPAD	soil and plant analyzer development	土壤、作物分析仪器开发
Sul	sulfatase	硫酸酯酶
TEM	transmission electron microscope	透射电子显微镜
TK	total potassium	总钾
TN	total nitrogen	总氮
TOC	total organic carbon	总有机碳
TP	total phosphorus	总磷
UR	urea	尿素
Ure	urease	脲酶
VSM	vibrating sample magnetometer	振动样品磁强计
XPS	X-ray photoelectron spectroscopy	X 射线光电子能谱
XRD	X-ray diffraction technique	X 射线衍射仪

Contents

CHAPTER 1

Enzymes and their functions in soil and groundwater

1.1 Introduction

Soil enzymology refers to the study of the activity and relevant features of soil enzymes, which is a borderline interdisciplinary subject between soil biology and biochemistry (Burns and Dick, 2001). As one part of the soil ecosystem, enzyme functions as biological catalysts of the ecosystem and fuels for metabolism of soil organic matter. Enzymes also have close relationships with soil physiochemical properties, types, fertilization, cultivation and other agricultural actions, as well as play a significant role in material circulation and energetic transformation in the soil. Their activity is very sensitive to changes caused by external factors, such as environmental pollution. Therefore, they also act as early indicators and sensitivity indexes for alterations of soil ecosystem (Badiane et al., 2001). They participate in the material circles and act as both an executor of the transformation of soil organic matter and an active library of plant nutrients. As indicated by the 60 enzymes already identified in soils, enzyme activity is related to many soil physiochemical properties. Catalytic function of enzymes exerts a great impact on the circulation and migration of elements [e.g., carbon (C), nitrogen (N), phosphate (P) and sulfur (S)].

Soil physiochemical properties (e.g., nutrients, structure, etc.) have long been used as indicators of soil quality and fertility. Alongside the intensifying climate changes, growing population, reinforced land exploitation and utilization, it becomes increasingly important to assess and monitor soil environmental quality for sustainable use of land resources and prevention of soil degradation. Traditional physiochemical properties can no longer meet the demands of studies on soil quality and health, as well as process and potential of soil recovery. In particular, under natural and anthropogenic disturbances, the degradation of terrestrial ecosystem has strongly reduced the vegetation productivity and input of soil organic matter, accelerated decomposition of soil organic matter and degradation of soil ecosystems. Therefore, it is of great realistic significance to analyze and discuss the structure, function and sustainable utilization of soil ecosystems using activity of soil enzymes as one of the main early-warning indicators as it could relatively and comprehensively reflect the changes in soil environment, quality and fertility, as well as judge soil degradation under stressed environment conditions.

Since Wood firstly detected the activity of catalase in the soil in 1898 (Wood, 1991), the study on soil enzymes has gone through a long period of establishment and development. In 1910, Fermi detected the proteinase and other soil enzymes. Later, deaminase, phosphatase and urease were also detected. Soil enzymology was founded before 1950s, when scholars detected over 40 soil enzymes in various soils, and advanced the analysis methods and theories on activity of soil enzyme. At that time, the study on soil enzymes became a rising borderline interdisciplinary subject between soil biology and biochemistry (Burns and Dick, 2001). In the rapid developing period of soil enzymology from the 1950s to the mid-1980s, breakthroughs in detection technologies and methods of soil enzymes spurred remarkable progress in soil biological chemistry and biology. New soil enzymes were also detected. In the mid-1980s, around 60 soil enzymes were detected and the theoretical framework for soil enzymology was basically erected. Activity of soil enzymes was recommended widely among pedologists as an indicator of soil fertility. The mid-1980s and later marked a period when soil enzymology penetrated with forestry, ecology, agriculture and environmental science. The study of soil enzymology has gone beyond the scope of classical pedology. Measurement of enzyme activity has become an essential item during investigations of almost all terrestrial ecosystems. Since enzyme activity is strongly related with soil biological properties, physiochemical properties and environmental conditions (Dick et al., 1996), the response of enzyme activity to environmental perturbations, the functional importance of rhizosphere soil enzymes, the research techniques of soil enzymology and the role of enzymes as indicators of soil quality have become the main focuses of the research (Dick, 1994, 1997).

The research of soil enzymes covers the sources of soil enzymes, relationship between soil enzymes and soil formation, interrelation between soil enzymes and soil components, soil enzyme assay, and their roles in nutrient cycling, pollutant transformation and degradation, soil quality evaluation, etc.

1.2 Types and sources

1.2.1 Types

By the distribution of enzymes in soil, Burns classified soil enzymes into two categories. The first category refers to the biological enzymes associated with free proliferating cells, including intracellular enzymes distributed in the cytoplasm, enzymes in the peripheral space, and enzymes on the cell surface. The second category refers to the abiotic enzymes or the nonliving ones, including the excreta of living cells during their growth or division, enzymes associated with cell debris and dead cells, as well as enzymes that leak from living or dissolved cells into soil solutions with their original functional sites remaining inside and outside the cells. Abiotic enzymes, as a stable type, exist in two forms by being absorbed to the interior/exterior of clay particles, and complexed with humus colloids through adsorption, trapping and copolymerization in the process of humus formation.

To effectively study and apply various enzymes, the International Enzyme Committee (IEC) proposed a classification system in 1961 to divide the known enzymes into six major categories according to the types and functions of their catalytic reaction: oxidoreductases, hydrolases, transferases, lyases, ligases and isomerases. Among them, the first four are chiefly involved in the study on the activity of soil enzymes.

Oxidoreductases: They mainly include dehydrogenases, polyphenol oxidases, catalases, nitrate reductases and sulfate reductases. Oxidoreductase is a well-studied enzyme. Since reactions catalyzed by them are mostly related to energy acquisition or release, they play an important role in the material and energy transfer in soils. The study on soil oxidoreductase is conducive to working out the problems concerning soil genesis and soil fertility since it participates in the synthesis of soil humus components, and the soil formation process.

Hydrolases: They mainly include sucrases, amylases, ureases, proteases, lipases, phosphatases, cellulolytic enzymes, β-glucosidase and fluorescein diacetate enzymes. Hydrolases can hydrolyze macromolecular substances, such as polysaccharides and proteins, to form simple micromolecular substances that can be easily absorbed by the plants which play an important role in carbon and nitrogen cycling in soil ecosystems. For instance, higher plants have urease, which can enzymatically promote the hydrolysis of peptide bonds in organic matter molecules. The soil urease activity shows a highly significant positive correlation with the total nitrogen and a highly significant negative correlation with the nitrate nitrogen, fast-acting phosphorus and fast-acting potassium. Therefore, soil urease activity can be used to characterize the nitrogen status of the soil.

Transferases and lyases: They mainly include aminotransferases, fructosan sucrases, transglycosidases, aspartate decarboxylases and glutamate decarboxylases. Both transferases and lyases play an equally important role in the transformation of soil substances. Transferases are involved not only in the metabolism of proteins, nucleic acids and lipids, but also in the synthesis, as well as transformation of hormones and antimicrobials; by far lyase activity in soil has been little studied.

Of soil enzymes, ureases, phosphatases, nitrate reductases, transferases, catalases and polyphenol oxidases have been mainly studied. Soil ureases are significantly or highly significantly correlated with soil organic matter, total nitrogen and total phosphorus traits. Phosphatases are closely related to phosphorus conversion, functioning as an important indicator of soil management system concentration and content of soil organic matters. Knowles argues that nitrate reductase is an important enzyme in the denitrification process and catalyzes the reduction of nitrate to nitrite reductase; its activity under anaerobic conditions is stronger than under aerobic conditions. Transferases can catalyze the hydrolysis of sucrose to glucose.

1.2.2　Sources

The source of soil enzyme has always been the focus of scientists, as it plays an important role in revealing the function of soil enzymes and the mechanism of material circulation of ecosystems. In early studies, people considered that soil enzymes came from soil microorganisms. With the deepening of the research, the results show that soil enzymes primarily come from three channels: microorganisms, soil animals, plant roots and plant residues.

Many microorganisms can produce extracellular enzymes. It's discovered in the study on *Aspergillus* oryzae that various enzymes are released to the medium in a certain order. Glycoenzymes and phosphatases are firstly released, followed by proteinases, lipases, and finally catalases. Some are released at the very beginning of the growth period and others later, when the weight of hyphae is lowered. Interestingly, however, catalases found in medium are typical intracellular enzymes, in free status. Scientists have also launched extensive studies on the release of extracellular enzymes by microorganisms in synthesized and natural media. Phaff has done brilliantly in this regard. Many bacteria and fungi can release amylases, cellulases, pectinases and proteinases. For instance, the chitinase is produced extracellularly in Streptomyces; *Fusarium* can release phosphatases and lipases.

Roots can also release some enzymes. Wood (1898) firstly reported on the soil extracellular enzymes, specifying that plant roots can secrete oxidases. Knudson and Smith (1919) discovered amylases secreted by plant roots. Rogers (1942) found phosphatases in maize and tomato root secretions and soil nucleases in roots. Admittedly, a good deal of information accumulated by plant physiologists indicates that plant roots do secrete some enzymes into the rhizosphere soil. However, technical limitation makes it difficult to distinguish the contribution of plants and microorganisms to enzyme activity in the rhizosphere soil.

Plant residues are also able to release enzymes into the soil during decomposition or remain partially active in the tissues of decomposing plant cells. Indirect evidences suggest that the enzymes extracted from all possible sources of plant debris are quickly broken down but are able to promote the growth of microorganisms in the soil. Other studies show that when fresh maize root tissue is added to the soil, the soil shows phosphatase activity, which rapidly decreases, and then presents proteinase activity. Adding steam-sterilized roots can enhance the activity of acid phosphatases, which means that plant residues can stimulate the synthesis of acid phosphatases of microorganism in comparison with direct provision of soil acid phosphatases.

The amount of enzymes provided by the soil fauna to the soil has rarely been studied. Kiss (1957) has studied the effect of earthworms on invertase activity, and indicated that earthworm excreta plays an important role in enhancing the invertase activity in grassland and cultivated soil, especially in the surface soil. In grassland and cultivated soil, especially in the surface soil, the excretion of maggots plays an important role in the activity of soil sucrases, while that of ants is a less important role. The research results of Kozlov further support the conclusion of Kiss, i.e., soil fauna may to a certain degree dictate the content of soil enzymes.

1.3 Kinetics, functions and influencing factors
1.3.1 Enzyme kinetics
The reaction speed can be measured quantitatively in enzyme-catalyzed reaction. Enzyme kinetics refers to the action mechanism of an enzyme which can be inferred by the effects of varying the reaction conditions on the reaction speed. Enzyme kinetics deals with the speed of enzyme-catalyzed

reaction, impact of various factors on the reaction speed and reaction process from reactants to products. For the system of a certain enzymatic concentration, the correlation between the speed of enzyme-catalyzed reaction (V) and the substrate concentration (S) is described as the curve in Figure 1.1. When the substrate concentration is very low, the enzyme is far from being saturated. The speed of enzyme-catalyzed reaction depends on the rate at which the enzyme reacts with the substrate molecule to form the enzyme-substrate complex, which is proportional to the substrate concentration, as a first-order reaction. At a high concentration of substrate with enzyme molecule saturated by the substrate, it is a zero order reaction where the reaction rate, independent of the substrate concentration, reaches its maximum. At a medium concentration of substrate with enzyme molecules partially saturated by the substrate, it is a mixed-order reaction between zero order and first order. With the three phases of the enzyme-catalyzed reaction taken as a whole, the following relational expression can be used to describe the kinetic features of the enzyme-catalyzed reaction:

$$V = \frac{V_{max}[S]}{[S] + K_m} \tag{1.1}$$

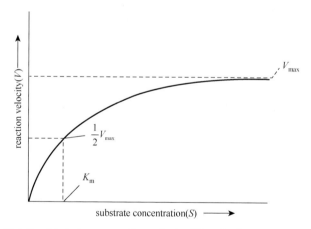

Figure 1.1 Relationship between reaction velocity (V) and substrate concentration (S).

This is the famous Michaelis-Menten formula, proposed by Michaelis and Menten in 1913. It quantitatively describes the relationship between enzyme-catalyzed reaction speed (V) and substrate concentration (S), where V_{max} is the maximum reaction speed, K_m is Michaelis constant, whose physical meaning is the substrate concentration when the enzyme-catalyzed reaction speed reaches half of the maximum; thus in the same unit as S.

Michaelis constant K_m is a dissociation constant characterizing the complex formed by the combination of enzyme and substrate. For a specific enzyme-substrate system, it is an eigenvalue of enzyme-catalyzed reaction, which is independent of enzyme concentration, but subject to environmental conditions (pH and temperature) and other external factors such as activators and inhibitors. The smaller the K_m value, the stronger the affinity of the enzyme to the substrate. If an enzyme can act on several substrates simultaneously, the specificity of the enzyme and the condition of the binding groups can be evaluated by comparing the K_m values between the enzyme and different substrates.

Michaelis-Menten formula also features other conversion forms; they are,

Lineweaver-Burk formula:

$$\frac{1}{V} = \frac{K_m}{V_{max}} \times \frac{1}{S} + \frac{1}{V_{max}} \tag{1.2}$$

Eadie-Hofstee formula:

$$V = -K_m \times \frac{V}{S} + V_{max} \tag{1.3}$$

Hanes formula:

$$\frac{S}{V} = \frac{1}{V_{max}} \times S + \frac{K_m}{V_{max}} \tag{1.4}$$

These equations turn the two variables of reaction rate and substrate concentration into a linear relationship, and the two enzymatic constants V_{max} and K_m can be easily obtained by graphing. In fact, it is no longer necessary to make such a conversion, because many computer softwares can calculate V_{max} and K_m values directly from the Michaelis-Menten curve equation.

Enzyme kinetics is a common method used in the research of soil enzymes. A single enzyme may have different enzyme kinetic characteristics in different soils. The characteristics of enzyme activity and the impact of the soil environment, whether the enzyme is inhibited or activated, can be reflected by measuring the maximum reaction rate and kinetic constant of a particular enzyme in a given soil.

1.3.2 Functions of enzymes

1.3.2.1 Assessment of soil fertility

In 1850s, Hofmam et al. firsty proposed to use the sucrase activity as an indicator to evaluate the soil fertility, as it comprehensively reflects the changes in climate, cultivation, soil reclamation and soil nature. In the 1880s, some suggested to predict the soil fertility with the comprehensive activity of soil enzymes. Compared with the evaluation based on activity of a certain enzyme, this could better reflect the release of nutrients during the metabolism of organic matters and the relative effectiveness of inorganic nutrients. According to the positive relationship between the enzyme activity and the soil fertility, Chinese scholars made a classified statistical analysis of various soil enzymes and divided China's soil into four levels of fertility. Under the action of various enzymes, organic matters in the soil can release specific plant nutrients. Thus the enzyme activity is related to not only the soil fertility but also the effectiveness of plant nutrients. For instance, under the action of fungi, some carbohydrases participate in the decomposition of the litters to increase the content of available phosphorus and available nitrogen in the soil. The activity of acid phosphatase and various forms of soil phosphorus and the activity of acid/alkaline phosphatase and the reduction of organic phosphorus in the rhizosphere of wheat and triticale are positively correlated. Moreover, the activity of acid phosphatase is positively correlated with the wheat yield. Speir's study on the soil of Tonga indicates that sulfatase can be used as an indicator for diagnosing sulfur nutrition in soil. In addition, the conversion of nitrogen in soil is related to the activity of protease and urease. The sucrase activity can reflect the conversion of intensity of carbon and respiration in soil. Catalase is closely related to the conversion rate of organic matters in soil and cellulase is related to the cellulose content in soil. It is thus believed that enzymatic methods can be used to evaluate the effectiveness of other microelements in soil. The soil pH can be evaluated via soil enzymes for their sensitivity to pH. For example, there are principally acid phosphatases but barely alkaline phosphatases in the acid soil. Peshakov further broadened the above conception regarding pH by defining the optimal amount of nitrogen, phosphorus and potassium fertilizer as the amount that triggers the maximum activity of urease, sucrase, catalase and phosphatase.

1.3.2.2　*Diagnosis of soil pollution*

Since soil enzymes can be used to evaluate the impact of several pollutants on the relatively "healthy" soil, active studies on the changes in enzyme activity caused by acid rain, heavy metal, pesticide, other industrial and agricultural chemicals were available. The study by Jarvis found that simulated acid rain at pH 2.7 inhibited the activity of dehydrogenase, protease, phosphatase and aryl sulfatase in the soil. Ohlinger's study concluded that acid rain significantly attenuated the enzyme activity in the transformed soil, but had little effect on the enzyme activity in the mound soil. Mathu also conducted a study on the way in which heavy metals and exotic biological agents affect the enzyme activity and found that these two affected in two different ways (Liu et al., 2017). In general, the concentrations of heavy metals causing biological impacts, especially mercury, silver, cobalt and copper, are lower than those of heterologous biological agents. Also, adding trace amount of copper to organic soils can inhibit the activity of carbohydrase, leading to mineralization and loss of organic matters in soil. Nevertheless, many metals, if in low concentrations, can be components of the soil enzymes, functioning to maintain the optimal enzyme activity. The effect of pesticide application on enzyme activity varies depending on the type, concentration, application time, and dosage of the pesticide. Insecticides (monocrotophos, quinalphos, and cypermethrin alone or in combination) significantly increase the dehydratase activity but decrease it at high concentrations (Gundi et al., 2005). Dehydrogenase activity is weakened by high content of ethephon residues, but applying organic fertilizer to ethephon-treated soil stimulates the soil dehydrogenase activity. The activity of acid phosphatase, alkaline phosphatase, urease, catalase and convertase is only inhibited by primary and secondary pesticide applications (clotulanib), with urease, catalase and acid phosphatase being affected the most (Long et al., 2006).

Usually, detecting the enzyme activity could reveal the relevant pollution of heavy metals and other industrial pollutants in soil. Among them, the activity of sucrase can be used to give a preliminary rating of soil contamination caused by heavy metals and boron. Ascorbate oxidase, nitrate and nitrite reductase can be used as diagnostic indexes for soil contamination. The results show that soil enzymes involved in the degradation of organic compounds may be applied in the treatment of contaminated soil. Repeated use of pesticides of one or more similar chemical structures may lead to rapid degradation of the pesticides and total failure in the end. In soil extracts, it accelerates the degradation of pesticides. For the failure mechanism of soil enzymes, Dick's study concluded that when partially purified free-state enzymes were added to the soil, then the enzymes bound to the soil particles, thereby masking the active site of the enzymes (Dick, 1997). The inhibition of enzymes by soluble salts in soil and the biodegradability of the enzymes also contributed to the inactivation of enzymes. Therefore, to apply the enzyme in treatment of soil pollution, the mechanism of soil enzymology must be studied deeply and corresponding measures must be taken.

Many studies have been reported on the effects of heavy metal pollution on enzyme activity. Recently, a study showed that Hg caused a decrease in the activity of soil urease and convertase. Morenoa et al. (2001) studied the impact of Cd on the activity of soil urease and dehydrogenase and the content of ATP. The results showed that Cd had a significant inhibitory effect on enzyme activity, and a weaker effect on enzymes with a higher clay grain content. Speri et al. (1992) studied the toxic effects of Cr(VI) and As(V) on soil organisms, and the results also showed that the enzyme activity decreased with the increase of heavy metal ion content in the soil. These studies indicated that enzyme activity can be used to monitor the pollution status of heavy metal contaminated areas, because soil enzymes are proteins that are highly sensitive to heavy metal contamination (e.g., Hg, Cd, As, Cr, Pb and Zn). In addition, the effects of chemical fertilizers, herbicides, sterilizers and acid mist on the enzyme activity are used as an important bioactivity indicator of soil contamination. It's one of the current focuses of soil science to use the enzyme activity as a bioactivity indicator of soil contamination.

1.3.2.3 Bioactivity indicator of soil quality

Soil microorganisms and their activities are often used as early sensitivity indicators of the soil stress process or ecological restoration process in natural and agricultural ecosystems. Especially since the late 1980s, it has been one of the focuses of soil science to use the enzyme activity as a bioactivity indicator of soil contamination. The change in land-use mode imposes a rather evident impact on enzyme activity. For example, the transformation of forest ecosystems to agricultural ecosystems has resulted in soil degradation, reducing the content of β-glucoglycase, phosphomonolipase, soil organic matter and soil microbial biomass. The activity of invertase, β-glucomannase, phosphatase and urease for rotational crop planting is generally higher than that for singular crop planting, which is normally used as a biological evaluation indicator of the cultivation mode's impact on soil quality. Fertilizer application and crop residues can affect the enzyme activity by improving the soil hydrothermal conditions and microbiota. For example, returning wheat and corn straw to field can improve the activity of dextranase, acid and alkaline phosphatase, phosphate monolipase, pyrophosphatase, and aryl sulfate lipase in soil. Additional application of organic fertilizer could increase the number of microorganism and enzyme activity in the orchard soil. Additional application of organic fertilizer and microbial fertilizer could improve the physicochemical properties of the soil and the microbiota, as well as boost the activity of invertase, phosphatase, dextranase, peroxidase and urease. Therefore, using enzyme activity as an evaluation index of soil quality represents a main direction and focus of soil enzymology research.

One could be taken as an indicator of soil quality, if its testing value can rationally indicate the soil quality at a specific time point instead of comparison with other soils in the control area. However, today's soil testing methods still have many shortcomings and distinctive regional characteristics, making it hard to attain the said effects. Nonetheless, people still tried to take the enzyme activity as an empirical indicator of soil biology and fertility, and then made some constructive trials. Stefanic et al. (1984) proposed a biological index of fertility (BIF), which is calculated with the following formula:

$$BIF=(DH+kCA)/2 \tag{1.5}$$

where DH is the dehydrogenase activity, CA is the peroxidase activity, and k is the scaling factor.

Beck (1984) proposed the index of enzyme number (EAN) as follows:

$$EAN=0.2(DH+CA/10+AP/40+PR/2+AM/20) \tag{1.6}$$

where DH is the dehydrogenase activity, CA is the activity of catalase, AP is the alkaline phosphatase activity, PR is the protease activity, and AM is the amylase activity.

Beck (1984) studied enzyme activity in different utilization modes, and pointed out that the EAN value of tillage soil fell between 1 and 4, while that of grassland and forest soil between 4 and 8. Perucci (1992) assumed a hydrolysis coefficient (HC), equal to the total amount of diacetate fluorescein added to the soil divided by the amount of diacetate decomposed by the soil, within a variation range of 0−1. Studies over the past 3 years have shown that the HC value for the control group was 0.142, while the variation ranges of HC upon addition of 3×10^7g and 9×10^7g of municipal sludge per hectare of soil per year were 0.218−0.367 and 0.245−0.442, respectively. Using these soil samples, Perucci (1992) calculated the BIF and EAN values, and found no significant correlation between BIF and enzyme activity or biomass C, but a high correlation between EAN and HC and biomass C ($r = 0.84$ and 0.92, respectively). The advantage of HC is that it is expressed with only one indicator, while EAN needs to test activity of five soil enzymes.

1.3.2.4 Maintaining a normal biological process

Soil enzymes play an important role in soil microbe's life process and several important reactions necessary for soil structural stability, organic waste decomposition, organic matter formation and nutrient cycling (Dick et al., 1994). They are both the implementers of soil organic matter transformation and the active reservoir of plant nutrients. These enzymes are continuously synthesized, accumulated, inactivated and/or broken down in the soil, playing an important role in maintaining the soil health, especially in nutrient cycling (Marinari et al., 2013). The diversity of enzyme functions can reflect the diversity of soil ecological functions, as expressed by the diversity of enzyme activity. The ratio of enzyme activity associated with the C, N and P cycles can reflect to some extent the relative importance of nutrient elements in the soil for plant and microbial growth (Caldwell, 2005). Soil enzymes catalyze the mineralization of complex organic matters containing carbon, nitrogen and phosphorus into small inorganic molecules, which are converted into forms that can be absorbed and used by plants to promote the uptake of carbon, nitrogen, phosphorus, sulfur and other elements. The cycle of such elements is an important link in the natural material cycle and plays a pivotal role in the metabolism of the soil ecosystem. For example, sucrase can convert sucrose to small molecules such as glucose and fructose, thus increasing soluble nutrients in the soil; urease can catalyze the hydrolysis of urea to NH_4^+-N, CO_2 and H_2O, thereby providing nitrogen source for plants and microorganisms; protease can catalyze the conversion of protein—like nitrogen and amino acids. All soils contain a set of enzymes (Mclaren, 1975) that determine the soil metabolic processes, which in turn are related to the physical, chemical, microbial and biochemical properties of the soil so as to maintain the functional diversity and stability of the soil.

1.3.2.5 Soil enzymes and global changes

A variety of soil enzymes play important roles in the soil carbon cycle, such as the hydrolases that break down sugars and oxidoreductases that oxidize phenolic substances, as they largely regulate the soil greenhouse gas emissions and lay a profound impact on global climate change. Freeman et al. (2004) compared the activity of enzymes in aerobic and anaerobic soils and found that only the activity of phenol oxidase increased under aerobic conditions, seven times more than that under anaerobic conditions, and the increase in phenol oxidase activity resulted in a 27% decrease in the concentration of soil phenolic compounds. Another experiment they designed showed that the activity of sulfatase and β-glucosidase in soil treated with phenol-containing water (2.4mg/L) was significantly weaker than that in soil treated with nonphenolic water. The results of the field survey and statistical collection showed that the increase in the activity of soil phenol oxidase was always accompanied by an increase in CO_2 release. The results suggested that the activity of hydrolases, which plays an important role in peat decomposition, is maintained at very low levels due to the inactivation of phenol oxidase in soil under anaerobic conditions. Based on this groundbreaking discovery, they suggest that phenol oxidase may be an important switch for the huge carbon pool (455Gt) of wetlands on Earth. The suggestion was also proved in turf ecosystems. The study by Yao et al. (2006) showed that the activity of phenol oxidase in alkaline turf soil was more than twice that of acidic soil, the activity of hydrolytic enzymes such as cellulase, chitinase, and β-glucosidase was lower in alkaline soil, and the activity of phenol oxidase was positively correlated with the difference between the content of soluble phenol and soluble carbon in alkaline and acidic soil. This indicates that the high concentration of soluble phenolic compounds in alkaline soil inhibited the activity of hydrolase, thereby resulting in limited degradation of soluble organic carbon and accumulation in soil.

1.3.3　Influencing factors of enzyme activities

1.3.3.1　Soil microbes

Since the 1960s, attempts had been made to establish a relationship between soil microorganisms and enzyme activity, but no consistent conclusion was drawn. Scarcely any evidence suggested solid correlation between them. Not until the advancement of soil enzyme and microbetesting techniques, the close relationship between enzyme activity and soil microorganisms has been gradually revealed. Soil microorganisms, as the core of soil ecosystems, are directly or indirectly involved in regulating the soil nutrient cycling, energy flow, organic matter conversion, fertility formation, pollutant degradation and environmental purification (Petersen et al., 2012; Kuzyakov and Xu, 2013). Different types and compositions of soil microorganisms would cause qualitative and quantitative differences in the enzyme activity, especially in soil urease activity. Soil urease is a kind of hydrolytic enzyme that decomposes nitrogenous organic matters. As a direct source of nitrogen nutrition in plants, it ubiquitously exists in fungi. Urease activity was found to be very significantly and significantly positively correlated with the number of diazotroph and fungi in the *Stipa baicalensis* steppe, which was associated with the characteristics of soil urease itself. In 2002, the study by Groffman et al. showed that the enzyme activity was significantly correlated with the soil microbial activity, microbial biomass and soil microbial quantity. The soil microbial activity was directly related to the activity of sucrase, urease, phosphatase and catalase. Guo et al. (1999) studied the enzyme activity of alkalized meadow soil with different grassland vegetation to discover that the activity of cellulase, urease and phosphatase intensified with the increase of soil microbial biomass. Soil microbial biomass carbon is the total amount of tecarbon in all living microorganisms in soil, usually accounting for 40%−50% of the dry microbial matters. It is an important microbiological index and a sensitive indicator of soil nutrients. Soil microbial biomass nitrogen is the main part of the active pool of soil available nitrogen, of which the basic content reflect the soil fertility and soil nitrogen supply capacity. Shen et al. found through pot experiments that the microbial biomass carbon and nitrogen of loamy brown soil were significantly or very significantly correlated with the activity of soil oxido-reductase (catalase) and hydrolase (sucrase and urease). Some studies in northeastern Australia found that overgrazing led to a decrease in microbial biomass carbon, and a significant decline in peptidase and amidase activities as well, and the results suggested that the content of microbial biomass carbon dictated the enzyme activity.

Previous studies showed the relationship between the enzyme activity in soil of half-drought regions and the physiochemical properties and microorganisms, and the results showed that the soil microbial content was significantly correlated with the sucrase, urease and alkaline phosphatase. Taylor et al. (2002) used various techniques to compare the relationships between microbial population and enzyme activity in topsoil and subsoil, and found that the increase in soil microbial population resulted in the enhancement of phosphoryl monolipase and dehydrogenase activity. Sucrase activity has a direct dependence on the soil microbial population and soil respiration intensity. However, some studies have reached different conclusions: for example, the relationship between the soil microbial population of rocky coastal protection forests, the enzyme activity and physiochemical properties, and found that the correlation between the soil microbial population and enzyme activity was not very close; the effect of organic fertilizer on biochemical properties of soil between Moso bamboos and in the root zones, found no correlation between soil microorganisms and enzyme activity. In studies on the growth of the poplar-acacacia mixed woods, the interrelationship between soil enzymes and fertility, Zhao et al.

(2009) found that the cellulose and amylase were positively correlated with the number of fungi and bacteria, while the sucrase activity was irrelevant to fungi and bacteria. At low latitudes, urease was positively correlated with both bacteria and fungi; at high latitudes, it was only positively correlated with the number of fungi. Phosphatase showed a correlation with the number of fungi only at low latitudes. There are various conclusions on the correlation between enzyme activity and soil microorganisms, as there are many kinds of microorganisms which act in different modes, or as soil enzymes come from not only microorganisms but also other animals or plants, or as the interaction between enzyme activity and microorganisms is also dictated by soil physiochemical properties, soil type, climate, fertilization, planting system and other factors.

Overall, the study on the relationship between the soil microorganisms and the source and activity of soil enzymes is of great significance for the research and development of soil enzymology. Using advanced research techniques to study the correlation between microorganisms and enzyme activity, to reveal the source, nature and function of soil enzymes is one of the focuses for future research of soil enzymology.

1.3.3.2 *Soil organic matters and nutrients*

The content of organic matters in soil, despite of a few thousandths to a few hundredths, has a great impact on the physiochemical properties of the soil. Soil enzymes can be adsorbed on the organic matter. An array of soil enzymes, such as urease, diphenol oxidase, protease and hydrolytic enzymes, have been extracted from the soil in the form of "enzyme-humic substance compound", in which the enzymes can still keep active. In some cases, they even have great resistance against decomposition and thermal stability. In general, the content of total nitrogen and phosphorus in soil is proportional to the organic matter content. Thus the content of nitrogen and phosphorus in soil is related to the enzyme activity. Organic matters, total nitrogen and total phosphorus in soil become the main factors dictating the activity of urease, acid phosphatase and invertase through direct and indirect effects. The enzyme activity is related to the distribution profile of organic matters, the former descending with the deepening of the latter. The activity of catalase, protease, phosphatase and urease in soil is very significantly correlated ($P<0.01$) or significantly correlated ($P<0.05$) with the organic matters (organic carbon) in soil, and significantly correlated with total nitrogen; the catalase, invertase, protease phosphatase and urease are relatively significantly correlated, significantly correlated or very significantly correlated with the fast-acting nitrogen and phosphorus; the urease is very significantly correlated with the total phosphorus. Yang Yuanping's study (2002) on activity of phosphatase in Bijie region also showed that the activity of phosphatase was closely related to the total nitrogen, organic matter, fast-acting phosphorus and hydrolysable nitrogen. The study by Fan et al. (2014) showed that the activity of urease, alkaline phosphatase and protease in soil improved with the increase of content of organic carbon in soil, and the relationship between the sucrase and peroxidase activity and the organic carbon varied depending on the type of fertilizer applied and the planting method. Wang et al. (2009) measured the urease activity of main arable soil in Guizhou in a relatively systematic and comprehensive manner. Their regression analysis showed that the soil urease activity was mainly influenced by the organic matters, nitrogen, phosphorus and potassium in soil, with the biggest impact brought by the basic content of ammonium in soil. Trace elements function as activators and inhibitors of plants, microorganisms and enzymes. The content of trace elements in soil may be an important ecological factor that indicates the enzyme activity. The impact of trace elements on enzyme activity depends on the nature of the soil and the specific properties of different enzymes to trace elements; trace elements that activate some enzymes may inhibit others. While one amount of a trace element activates the enzymes, another amount of it inhibits them. The study by Li et al. (2019) showed that the Zn and Mn laid the greatest impact, that is, the largest promotion, on the soil protease activity. Zinc has a negative effect on urease and catalase to a certain extent, that is a certain inhibitory effect, while manganese has a positive effect on them, that is a promoting effect.

1.3.3.3 Soil aggregates and clay particles

Soil aggregates are stable structures composed of tiny mineral particles, used as indicators of the physiochemical properties and nutrients of the soil. They can be divided into large aggregates (sized>250μm) and micro-aggregates (sized 50–250μm). Aggregates with a diameter of 0.5–3mm are one of the deciding factors for soil fertility. Aggregates of different particle sizes feature different enzyme activity. Smaller aggregates have higher enzyme activity than larger ones. The stability of aggregates is also related to enzyme activity. For example, the urease activity is significantly negatively correlated with the stability of soil aggregates and soil bulk density, while the invertase activity is significantly positively correlated with the stability of soil aggregates. Zhou et al. (2017) proposed that the urease activity in black soil and brown soil was mainly concentrated on micro-aggregates, as equivalent to the clay part of soil grain size. The larger the particle size, possibly the lower the urease activity. Carbohydrase is chiefly absorbed on powdered sand particles. The amount of enzyme absorbed by the clay particles and sand particles depends on the mineral composition of such soil particles.

Clay particles are the most active mineral component in the soil due to their fine grain size, large surface area and specific characteristics of certain mineral structures. Clay particle is the reason why the soil features so many physiochemical properties, which can form compound colloid together with the soil humus. Only a small fraction of soil enzymes exist in soil solution, and most of them are clinged by soil. Absorbed substances such as humus can be bound to the clay particle minerals via cation exchange reactions. The absorption of enzyme by clay particles is subject to environmental conditions such as acidity and temperature. The lower the pH of the soil (lower than the isoelectric point of enzyme protein), the more enzymes absorbed by the clay particles. Since urease is a weakly acidic enzyme, the adsorption of urease in weakly acidic media is greater than that in weakly basic media; at 20—60℃, the adsorption of urease in each soil clay particle decreases with the rise of temperature. Sharing same distribution pattern of urease, the neutral phosphatase and proteinase in soil are chiefly absorbed on the colloids and clay particles; the soil sulfatase activity is related to the total surface area of clay particles; the phosphatase activity is overwhelmingly adsorbed by clay particles. Different clay particles absorb different amounts of enzymes. The study by Feng et al. (2008) found that all clay particles absorb urease, but each absorbs a specific amount. For the same clay particle, original soil absorbs more than soil with organic substance removed. For different clay particles, the absorption is as follows: original building soil > original black kiln soil > original loess soil > black kiln soil with organic substances removed > loess soil with organic substances removed > original yellow brown soil > building soil with organic substances removed > yellow brown soil with organic substances removed.

1.3.3.4 Soil pH

The soil pH can change the stability of the base point of enzyme-catalyzed reaction and enzyme absorbed by the soil. Some enzyme-catalyzed reactions are very sensitive to pH changes, which can only be carried out within a narrow pH range. Some studies found that the two optimal pH value ranges for soil enzymes were 6.5—7.0 and 8.8—9.0, the optimal pH value ranges for soil phosphatase were 4.0—5.0, 6.0—7.0, 8.0—10.0, known as acid, neutral and alkaline phosphatase, respectively. When the pH value is below 5.0, the catalase and dehydrogenase are almost completely deactivated. The invertase and urease are less affected by acidity, but positively correlated with the soil humus content. In addition, some studies indicated that the impact of pH on the status of urease's sulfhydryl, amino carboxyl and other components and protein configuration (tertiary structure) would also lead to the change of enzyme activity. Other studies showed that low acidity first left a certain activation effect on urease and neutral phosphatase, then an inhibition effect. When the concentration of [H^+] ions is $0-55 mmol \cdot kg^{-1}$, exogenous acid has an obvious activation effect on the activity of invertase and acid phosphatase.

1.3.3.5 Water, air and heat conditions in soil

Water, air and heat conditions have an apparent impact on the enzyme activity. On the one hand, they have a significant correlation with the activity and type of soil microorganisms. As a result, they must have a massive impact on the enzyme activity. On the other hand, different moisture conditions, air composition and moisture status also directly affect the existence and activity of soil enzymes. In general, enzyme activity is higher when soil moisture is greater. Some scientistes studied the enzyme activity in regions with continuous wet and dry seasons, then noted that enzyme activity improves significantly when the dry season ends and the wet season begins; enzyme activity decreases when water content in the soil declines. The soil temperature directly affects the population and quantity of microorganisms that release enzymes. Therefore soil temperature makes one of the influencing factors of enzyme activity. Studies showed that the enzyme activity improved significantly when the temperature rose from $10\,^{\circ}\!C$ to $60\,^{\circ}\!C/70\,^{\circ}\!C$. However, with the further rise of temperature, the urease rapidly passivated. If heated at $150\,^{\circ}\!C$ for 24h or $115\,^{\circ}\!C$ for 15h, soil enzymes would be completely deactivated.

Since CO_2 and O_2 in soil are subject to the activity of soil microorganisms, the air in soil has direct impact on the enzyme activity. Overrein pointed out that oxygen was related to urease activity; the sucrase, amylase, cellulase, urease, phosphatase and sulfase except hemicellulase are all positively correlated with the oxygen intake of soil.

It can be seen that the moisture, air and heat in soil have an obvious impact on the enzyme activity. Just like in studies on microorganisms where moisture, air and heat conditions must be properly set, the control and selection of optimal moisture, air and heat conditions must be considered in studies on certain soil enzymes. The analysis and setting of moisture, air and heat conditions are more emphasized in the studies on enzymatic remediation of polluted soil, biochemical disposal and manufacture of organic fertilizers.

1.3.3.6 Plants

Impact of plants on the enzyme activity chiefly manifests as the action on the rhizosphere microflora by the root exudates. Due to the active interaction between root exudates and the active activities of rhizosphere microorganisms, the soil enzymatic process in plant rhizosphere is much stronger than that outside the rhizosphere. The urease and catalase in rhizosphere soil in *Eucalyptus grandis* and oil-tea camellia forests are more active than those in non-rhizosphere soil, showing a significant or very significant difference. The enzyme activity in young Chinese fir forests shows a rule—rhizosphere < non-rhizosphere. With growth of the trees, the enzyme activity of rhizosphere soil is obviously higher than that of non-rhizosphere soil. Liu et al. (2017) found through indoor pot experiment that the ryegrass roots increased the activity of polyphenol oxidase and dehydrogenase in paddy soil, thus increasing the degradation rate of P-benzo pyrene in plants. Kandeler studied the activity of invertase, ligninolytic enzyme, and protease at the soil-litter interface to find that the ecological interface had the highest enzyme activity (Kandeler et al., 1999). Now that rhizosphere soil plays an important role in the root-soil interface ecosystem, the study of rhizosphere soil enzymes is of great significance for exploring the effects of plants on soil ecosystem processes.

The enzyme activity is also subject to the vegetation types and plant communities. The enzyme activity is under direct impact of the vegetation litters and indirect impact of the moisture/heat, physiochemical properties, and microbiota of the soil changed by vegetation variation. Jin Suying (1996) studied the enzyme activity of different vegetation types, and the results showed that the protease activity was the highest in paddy soil, and the activity of catalase in beet soil was higher than that in wheat soil. He Bin (2002) studied the enzyme activity of different mangrove communities in Yingluo Port, Guangxi. The study results showed that the activity of soil sucrase, protease, urease and acid phosphatase in the rhizophora stylosa community was higher than those in the Bruguiera gymnorrhiza community.

Soil enzyme activity is closely related to the growth process and yield of plants. Enzymatic changes are most active during the period of most vigorous growth of crops. The study by Verstraete and Voets (1977) showed that the activity of phosphatase, invertase, urease and β-glucosidase was positively correlated with the crop yield. Yang Lijuan (2000) found that the activity of phosphatase, catalase, invertase and urease in the soil of vegetable garden was significantly or very significantly positively correlated with the cucumber yield.

The enzyme activity is correlated with the seasonal variation of plant growth. The study by Zhang et al. (1999) showed that the hydrolase activity was highest in summer and autumn, lowest in winter, and rose again in spring. Zhang et al. (1990) studied the seasonal changes of enzyme activity in red soil of different types of mixed forests. It was the general trend that the enzyme activity was higher in spring, highest in summer, slightly lower in autumn, and lowest in winter, but different mixed forests varied in this regard. With the continuous updating and improvement of research methods, the research on the effects of plants on soil enzymes will also be advanced.

1.3.3.7 *Fertilization and other agricultural management measures*

Soil management systems that can maintain or improve the soil quality are sustainable. Soil quality assessment is a basic evaluation method for soil or agricultural management practices. Soil enzyme activity can more rapidly reflect the effects of agricultural management measures, such as fertilization, on the soil quality and fertility than organic matters. Soil enzymes can be used as an early-warning indicator of changes in the soil ecosystem. Fertilization can improve the physical and chemical properties, hydrothermal conditions and microbiota of the soil, thereby affecting the enzyme activity. The application of organic fertilizer and chemical fertilizer could obviously affect the enzyme activity of soil. Dick's study of fallow wheat fields showed that long-term application of farm manure could improve the microbial and enzyme activity in soil (Dick, 1988). Returning of maize and wheat straw could increase the activity of sucrase and alkaline phosphatase in soil. The application of green fertilizer could enhance the activity of dehydrogenase in soil. The application of cattle and pig manure could increase the activity of urease in soil. The application of a large amount of organic fertilizer alone for a long term could apparently elevate the activity of urease and alkaline phosphatase in brown soil. The study by Nayak et al. (2007) in Katak, India found that the urease activity of the soil with combined application of organic and inorganic fertilizers was higher than that of the soil with no fertilizer applied. Wang et al. (2009) studied soil enzyme activity under different fertilization methods for a long time and found that applying chemical fertilizer could significantly increase the activity of invertase and alkaline phosphatase. Zhang et al. (2006) studied the soil enzyme activity in red paddy soils under different long-term fertilization treatments, and the results showed that the combined application of nitrogen, phosphorus and potassium fertilizers could enhance the soil invertase activity. Other studies indicated that fertilization can reduce enzyme activity. For example, Prietzel (2001) studied the effect of $(NH_4)_2SO_4$ on soil aryl sulfatase activity in the mountain and forest region in southwestern Germany, and the results showed that aryl sulfatase activity decreased under the application of $(NH_4)_2SO_4$. The results of these studies varied to the differences in soil types, climatic characteristics, precipitation, fertilization methods and amounts, tillage practices.

Sludge features a complex composition, possibly containing toxic and harmful substances, such as heavy metals, pathogenic bacteria and parasites (eggs). Meanwhile, the sludge also contains a large amount of nitrogen, phosphorus, potassium, trace elements, organic matters and other nutrients. It is a trend to compost the sludge and use it as fertilizer resources. Some studies showed that the application of sludge and grass-sewage mixture could improve the activity of sucrase, urease and neutral phosphatase in soil. The activity of urease, protease and invertase in turfgrass soil increased significantly with the increase of sludge content in compound fertilizer with the same amount of nitrogen, phosphorus and potassium applied. Wastewater sludge can improve the dehydrogenase and alkaline phosphatase activity in soil. However, some studies suggested that the dehydrogenase activity was significantly reduced by sludge application, possibly due to the inhibition of heavy metals in the sludge. The distribution and activity of soil enzymes could be affected by the soil tillage practices. The activity of sucrase, urease and acid/alkaline phosphatase in soil of monoculture was lower than that in soil of rotational cultivation. Badiane et al. (2001) used the enzyme activity to monitor the soil quality in the semiarid tropics. The results showed that the vegetation types, tillage methods and years all would affect the enzyme activity, but different enzymes had different responses to such factors.

1.4 Methods for visualizing enzyme activities: in situ zymography

1.4.1 Introduction

Since soil enzymes are highly sensitive to changes in physical and chemical properties and environmental factors of the soil, their species diversity and activity can not only reflect the health and fertility level of soil but also be sensitive indicators to characterize the status and changes of plants and microorganisms in the soil microenvironment (Aon et al., 2001; Caldwell, 2005). Studies on soil enzyme activity have a long history, and the most commonly used traditional methods are fluorescence and spectrophotometry. These methods require destructive sampling and have disadvantages such as difficulties in sample transportation and preservation, cumbersome pretreatment process and long determination cycle, leading to low precision and large errors in results of enzyme activity determination (Jiang et al., 2021). Therefore, an advanced visualization technique is urgently needed to fine tune the spatial and temporal distribution of soil enzyme activity (Kuzyakov and Razavi, 2019).

In situ zymography, firstly introduced in 1962, is a technique for studying hydrolytic enzymes on the basis of substrate degradation. As an emerging technique for visualizing and quantifying soil enzyme activity in two dimensions, it has been rapidly developed in agriculture and ecology in recent years due to its nondestructive, high-resolution and repeatable measurement characteristics (Spohn et al., 2013). In situ zymography is a substrate-dependent enzyme activity visualization technique, which is based on the oxidative decomposition of a specific fluorescent substrate synthesized from a fluorescent dye and the corresponding substrate by an active enzyme in the soil, resulting in a fluorescent signal that is visualized on a nylon membrane (Razavi et al., 2019). The synthesized fluorescent substrate is initially located on the membrane, and the nylon membrane saturated with the fluorescent substrate is closely adhered to the soil surface. After the synthesized fluorescent substrate is saturated with fluorescent substrate, the synthesized fluorescent substrate is decomposed when it encounters specific enzymes on the soil surface, thus activating the fluorescence (Figure 1.2, Razavi et al., 2019). The result of the in situ zymography is a two-dimensional image taken by developing under a

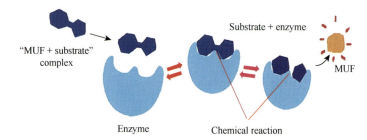

Figure 1.2　Principle of fluorescent assay.

ultraviolet (UV) lamp and using a camera set up with suitable equipment, known as in situ enzyme spectroscopy (Razavi et al., 2019).

Development of zymography over five decades was mostly focused on the analysis of proteases and their inhibitors in various matrices and media besides soil (Choi et al., 2009; Hughes and Herr, 2010; Pan et al., 2011), for example, to gain insights into tumor formation (Kleiner and Stetler-Stevenson, 1994; Wilkesman and Kurz, 2009). Kurzbaum et al. (2010) proposed a novel approach to visualize dehydrogenase activity of plant roots by use of tetrazolium violet dye without destructive steps, allowing repeated observations of growing plants and the impact of inhibitors such as sodium azide and cycloheximide. However, this approach was not tested in soil specimens. Visualization of enzyme activities developed rapidly once fluorescently labeled substrates became widely applied in environmental samples. During the first attempt at visualization of enzyme activity in the soil matrix, the fluorescently labeled substrate was dissolved in agarose solution that was then directly poured onto the sample (Baldrian and Vetrovsky, 2012). The approach was successful in visualizing the spatial distribution of enzyme activity in soils and biological specimens, such as fungal cell colonies. However, due to the diffusion of the substrate in agar gel, the resolution of this enzyme mapping method was low. The same limitation was visible following the standard zymography assays for the detection of protease and amylase activity in electrophoresis gels (Spohn et al., 2013). The revolutionary optimization of the method started by integrating dissolved fluorescently labeled substrates in membrane filters instead of gels (Kuzyakov and Razavi, 2019; Razavi et al., 2016; Sanaullah, et al., 2016; Spohn et al., 2013).

Soil zymography techniques can be utilized for hydrolases or oxidases acting on any biological substrate such as proteins and peptides, oligosaccharides and polysaccharides, lipids and sugars (Spohn and Kuzyakov, 2013; Voothuluru et al., 2018). To date, soil zymography has been adapted for various applications such as studying the impact of plant species (Razavi et al., 2016), root morphology (Ma et al., 2018), pathogens (Razavi et al., 2017), abiotic controls like temperature (Ge et al., 2017), drought (Ahmadi et al., 2018; Guhr et al., 2015), nutrient availability (Giles et al., 2018; Heitkoetter and Marschner, 2018; Wei et al., 2019a) and heavy metal pollution (Duan et al. 2018) on the activity of different enzymes in various spheres such as the rhizosphere (Sanaullah et al., 2016; Spohn et al., 2013), detritusphere (Liu et al., 2017; Ma et al., 2017; Spohn and Kuzyakov, 2014; Wei et al., 2019b), and biopores (Hoang et al., 2016b; Razavi et al., 2017), in both lab and field studies (Razavi et al., 2017). Guber et al. (2021)

developed time-lapse zymography methodology to avoid the errors induced by signal nonlinearity and diffusion losses of the product in the activity calculations. This increases the accuracy of traditional in situ zymography. Khosrozadeh et al. (2022) attempted to extend time-lapse zymography to oxidoreductases by using Amplex Red, which formed the brightly fluorescent product resorufin when oxidized. Benefiting from all of these developments, we can now test a larger array of hypotheses related to enzyme-based processes and their roles in biogeochemical cycling. Besides its potential application, the simple sample preparation procedure and relatively worldwide accessibility of all necessary chemicals and equipment have made soil zymography as one of the most influential imaging techniques in soil.

1.4.2 Materials and methods

1.4.2.1 Substrates corresponding to different hydrolytic enzymes

Table 1.1 presents a summary of selected enzymes, their main ecological functions and substrate proxies.

Table 1.1 Summary of selected enzymes, their main ecological functions and substrate proxies.

Enzyme	Synthetic substrate[*]	Enzyme function
β-glucosidase	MUF-β-D-glucopyranoside	Releases glucose from cellulose
Cellobiohydrolase	MUF-β-D-cellobioside	Releases disaccharides from cellulose
Xylanase	MUF-β-D-xylopyranoside	Releases xylose from hemicellulose
Chitinase	MUF-N-acetyl-β-D-glucosaminide	Releases N-acetyl glucosamine from chitin
Phosphatase	MUF-phosphate	Releases attached phosphate groups
Leucine-aminopeptidase	L-Leucine-7-amido-AMC	Hydrolysis of the peptide bonds
Tyrosine-aminopeptidase	L-Tyrosine-7-amido-AMC	Hydrolysis of the peptide bonds

Note: [*] For each synthetic substrate, the fluorescent dye 4-methylumbelliferone (MUF) and 7-amino-4-methylcoumarin (AMC) are presented.

1.4.2.2 Polyamide membrane

Polyamide membrane (pore size: 0.45μm) should be cut into pieces of biopore sizes and shapes (e.g., 5mm×10mm) (Hoang et al., 2016a). Thereafter, membrane should be soaked in prepared substrate solutions.

1.4.2.3 Experiment under lab condition (Rhizobox preparation)

(1) A transparent (or opaque) plastic box (size variable depending on

the objectives) is used for the experiment, with a removable front panel that could be easily opened without affecting the earthworm habitats or root.

(2) Before filling the boxes with soil, a sand layer is placed on bottom of the pots for drainage to prevent water saturation (in case of earthworm).

(3) Regulate water content and keep it stable at 30% of soil dry weight during whole experiment.

(4) Depending on the objectives, place number of mature earthworms (three or more) by different length (5—10cm) in each pot.

(5) Prepare similar extra box without earthworm as a control box. Control box will comfort your further comparison between biopores and bulk soil.

(6) If the effect of plant roots is also going to be considered in the study, then plant seedling in the soil simultaneously with earthworms.

(7) Keep the boxes under a stabled temperature of (18 ± 1)℃ and a daily light cycle of 16h, with light intensity of 300μmol·m^{-2}·s^{-1}.

(8) After 2 weeks that earthworms formed biopores, samples are ready for zymography (Figure 1.3).

Figure 1.3 An example of the rhizobox and biopores formed by earthworms during 2 weeks. Dashed lines indicate some of the biopores made by earthworm.

(9) Place the rhizoboxes horizontally with one side open (like a door) and then slowly and continuously pour soil into the rhizoboxes through a 2mm sieve to achieve a uniform soil packing and avoid soil layering. The open side should be closed, turn samples vertically and gently shake them to achieve a stable soil packing.

(10) Cover the rhizoboxes to avoid algae growth.

(11) The opening side should be covered by plastic cap to avoid soil disturbance at the opening time.

(12) Irrigate your rhizoboxes (without earthworm) from the bottom.

1.4.2.4 Experiment under field condition: Biopore-window system (rhizotrons installation)

(1) The biopore-window consists of a transparent acrylic sheet (3—5mm thick).

(2) The window can be 30cm×30cm or 20cm×20cm.

(3) It is recommended to install rhizotrons in field simultaneously with plant sowing (if effect of plant roots would be considered in the study).

(4) Install rhizotrons at an angle of 90 degrees to the soil surface, and fix it in place by 2 vertical steel rods and backfill it with soil to remove air gaps.

(5) Place some earthworms (2 or 3) behind the rhizotron to increase the number of burrows (if earthworm is point of interest).

(6) Backfill rhizotron with the soil (Figure 1.4).

(7) After 2 or 3 weeks, samples are ready for zymography (Figure 1.5).

Figure 1.4 Scheme of biopore-window under field conditions. Rhizotron should be fixed, while the dug-out soil should be placed back.

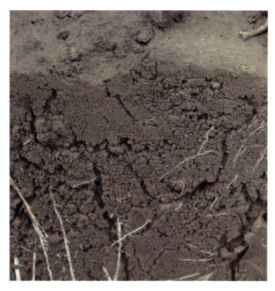

Figure 1.5　An example of the biopores formed by earthworms under field condition within 4 weeks after biopore-window establishment.

1.4.3　Procedures
1.4.3.1　Solution preparation
1. Hydrolytic enzyme

(1) $0.1 mol \cdot L^{-1}$ MES buffer: weigh 20.673g of MES ($C_6H_{13}NO_4SNa_{0.5}$) salt in one 1000mL volumetric flask and make the volume by adding sterile water.

(2) $0.05 mol \cdot L^{-1}$ Trizma buffer: weigh 0.985g of TRIZMA-Base (α- α-α —Tris (hydroxymethyl)-methylamin A.C.S. $0.016 mol \cdot L^{-1}$) and 2.66g of TRIZMA-HCl (Tris (hydroxymethyl) aminomethane hydrochloride buffer, $0.036 mol \cdot L^{-1}$) salts together in 500mL volumetric flask and make the volume by adding sterile water.

(3) $10 mmol \cdot L^{-1}$ MUF stock solution: in a 100mL volumetric flask, 0.1762g of 4-methylumbelliferone and 50mL of methanol are added. After the 4-methylumbelliferone has completely dissolved in methanol, final volume of 100mL is made with MES buffer.

$5 mmol \cdot L^{-1}$ AMC stock solution: weigh 0.0875g of 7-amino-4-methylcoumarin into a 100mL volumetric flask and add 50mL of methanol. After complete dissolution of 7-amino-4-methylcoumarin in methanol, make the final volume of 100mL with TREZMA buffer.

(4) 12mmol·L^{-1} Substrate working-solution: weigh 6/10000 of the molecular weight of each substrate in 25mL sterile centrifuge tubes and dissolve it in 500μL of dimethyl sulfoxide (DMSO) (C$_2$H$_6$SO). Then add 19.5mL of sterile buffer (MES buffer for MUF substrates and TRIZMA buffer for AMC substrates) and shake it properly to make it homogeneous solution.

2. Oxidoreductase enzyme

(1) 50mmol·L^{-1} Trizma-buffer: weigh 0.985g of TRIZMA-Base (α- α- α -Tris (hydroxymethyl)-methylamin A.C.S. 0.016mol·L^{-1}) and 2.66g of TRIZMA-HCl [Tris (hydroxymethyl) aminomethane hydrochloride buffer, 0.036mol·L^{-1}] salts together in 500mL volumetric flask and make the volume by adding sterile water.

(2) 1mmol·L^{-1} Resorufin stock solution: dissolve 0.1065g of resorufin in 50mmol·L^{-1} Trizma buffer (Trihydroxymethyl aminomethane), then diluting in 50mL volumetric flask and make the volume by adding Trizma buffer.

(3) 2mmol·L^{-1} substrate working-solution: dissolve 25mg of Amplex Red reagent (10-acetyl-10H-phenoxazine-3,7-diol; ADHP) in 1mL of DMSO then diluting in 50mL volumetric flask and make the volume by adding sterile water. The stock solution was further diluted in 50mmol·L^{-1} Trizma buffer (pH 7.4) to obtain a 2mmol·L^{-1} working solution of the substrate. Inert gas (N$_2$) was bubbled through the working solution, in a glass vial for 5min, keeping the lid tightly closed thereafter.

1.4.3.2 Standard curve establishment

Calibration standards were prepared to convert the brightness of the image pixels to resorufin content equivalents, as follows. The 1mmol·L^{-1} resorufin stock solution was diluted in the Trizma buffer to obtain 0.2mmol·L^{-1}, 0.4mmol·L^{-1}, 0.6mmol·L^{-1}, 0.8mmol·L^{-1} and 1mmol·L^{-1} calibration solutions. To protect the resorufin solutions from degradation by light, the standard solutions were prepared in a dark room and all flasks were covered by aluminum foil. A 10μL portion of each calibration and validation solutions was added to membranes, which were then covered by a transparent glass sheet to mimic the zymography settings. The membranes with standards were photographed under UV light using the same camera settings as zymograms (Figure 1.6). The amount of resorufin on an area basis (nmoL·cm^2) was calculated from resorufin solution volume soaked by the membrane and its size. The calibration relates enzyme activities to the gray value of zymogram fluorescence.

The establishment of standard curve of hydrolase is the same as that of oxidoreductase. The corresponding fluorescent substance is 4-methylumbelliferone

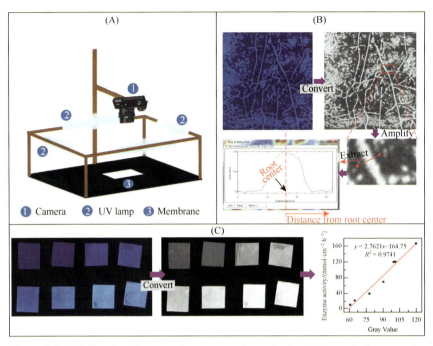

Figure 1.6 (A) Sketch of the setup for zymography analysis. (B) Example for extraction of scatter-plot of distance from root center and gray values. (C) Establishment of calibration line. Notes: in (A), to simplify the setup, the electric wires and light-proof box were not drawn. In (B), the extraction of scatter-plot of distance from bulk soil hotspot center and gray values was not drawn, but the procedure was the same.

(MUF) or 4-methylcoumarin (AMC), the stock solution is $10 mmol·L^{-1}$ MUF or $5 mmol·L^{-1}$ AMC, the buffer solution is $0.1 mmol·L^{-1}$ MES solution or $0.05 mol·L^{-1}$ Trizma solution. Membranes ($2cm×2cm$) were uniformly saturated during 30min either with MUF solutions at increasing concentrations of 0, $0.01 mmol·L^{-1}$, $0.2 mmol·L^{-1}$, $0.5 mmol·L^{-1}$, $1 mmol·L^{-1}$, $2 mmol·L^{-1}$, $4 mmol·L^{-1}$, $6 mmol·L^{-1}$, $10 mmol·L^{-1}$ or with AMC solutions at increasing concentrations of 0, $0.01 mmol·L^{-1}$, $0.02 mmol·L^{-1}$, $0.05 mmol·L^{-1}$, $0.1 mmol·L^{-1}$, $0.2 mmol·L^{-1}$, $0.5 mmol·L^{-1}$, $1 mmol·L^{-1}$, $3 mmol·L^{-1}$, $5 mmol·L^{-1}$.

1.4.3.3 Soil zymography

In this section, visualization of enzyme activities in the rhizosphere is taken as an example to demonstrate the zymography process. Polyamide filter membranes with adimension of $130mm×130mm$ (length×width) and a pore size of $0.45μm$ were cut into the size suitable for rhizotrons. The membranes were saturated in the substrate solution for each enzyme. The detachable front plates of rhizotrons were removed, and the saturated membranes were applied directly to the soil surface. Soil zymography was performed for each enzyme separately on the same rhizotrons. During these measurements, there was no overlying water in the rhizotrons, but the soils remained saturated. Each membrane contains different enzymes. Substrates were deployed on the soil surface for 1h. After that, the membranes were carefully peeled off from the soil surface and any attached soil particles on the membranes were gently removed using tweezers and transferred to a dark room and dried for 10min, before being taken photographed under UV light with a wavelength of 368nm.

1.4.3.4 Image processing and analysis

The analyzed image is added to the ImageJ software. The "Split Channels" tool is used to extract the red channel signal, and then convert the image to an 8-bit color grayscale image. Before extracting the gray value data, the "Subtract Background" tool is used to subtract the background gray value to ensure the accuracy of the data. By comparing with the actual photos of plant root distribution in soil profile, several suitable roots are selected as characteristic roots for treatment. The feature area with "Rectangle" "Straight" "Freehand selection" etc. are selected, and then the "Plot Profile" tool is used to obtain the gray value distribution of the feature area. The data is then extracted to excel for subsequent processing, and the space-time distribution characteristics of enzyme activity is obtained according to the standard curve. The main analysis indexes include the expansion range of plant rhizosphere enzyme activity, the average enzyme activity size of root surface/root tip, the distribution of enzyme activity in different dimensions of rhizosphere and the average enzyme activity in non rhizosphere hot spot soil.

1.4.4　Combination with other visualization techniques

In situ zymography represents a novel technique to enable the quantification of micro-scale distribution of enzyme activities in soil and groundwater environments. In recent years, development of other techniques was also paid great attention, which helps to visualize more important factors involving in various processes. Lots of studies were also performed to combine these techniques to deepen our understanding of these biochemical processes (Ma et al., 2019; Li et al., 2019). For instance, Ma et al. (2021) investigated the spatial distribution of phosphatase activity and pH in the rhizosphere in response to P availability between legumes and cereals and along individual roots to understand plant P-acquisition strategies (Figure 1.7). They found that before the formation of cluster roots, the main strategy of phosphorus uptake by the root system is to increase phosphatase activity, acidify the interroot soil and expand the interroot range around the main root; whereas after the formation of roots, the root system adjusts the key mechanism of phosphorus uptake, mainly by increasing the phosphatase activity hotspots around the roots. After the formation of cluster roots, the root system will adjust the key mechanism of phosphorus uptake, mainly by

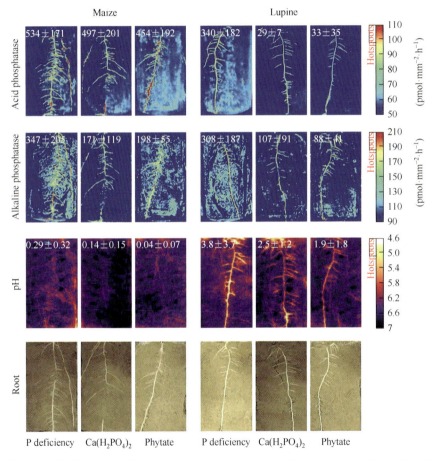

Figure 1.7 Examples of spatial distribution of alkaline phosphatase activities, pH and roots in the rhizosphere of maize and lupine. Reprinted with permission from Ma et al. 2021.

31

increasing the area of phosphatase activity hotspots around the cluster roots and expanding the effective interroot range. In this section, brief introduction of planar optode and diffusive gradients in thin-film (DGT) technique were demonstrated.

1.4.4.1 Planar optode technique

The conventional study of soil physicochemical properties often relies on the ex situ sampling and determination, failing to reflect the heterogeneity of soil environment because of the destructive sampling processes. The process disturbs the original structure and ignores the heterogeneity of soil microsites. In recent years, planar optode with high spatiotemporal resolution has gained more attention in investigating soil biogeochemical processes. Based on the luminescence principle, the specific luminescent indicator that is sensitive to different solutes is embedded into the matrix to make a sensing membrane, then its two-dimensional emission signals are recorded using digital imaging technology. Therefore, high resolution and sensitive dynamic characteristics of solutes can be continuously obtained by changing the indicator luminescence signal with relatively simple experimental configurations. So far, planar optode has been successfully deployed to monitor the dynamics of O_2, pH (Figure 1.7), CO_2, NH_4^+ and other important physicochemical parameters in soil.

The planar photopolar technique is based on the principle of luminescence sensing. A luminescent indicator is embedded in a matrix to form a sensing film. An imaging device is used to record the two-dimensional characteristic emission spectrum of its excitation. The concentration of the analyte is then quantified based on a pixel-calibrated transfer function. The planar optode consists of a luminescent sensing system and a signal collection system. The luminescence sensing system consists of an optical planar sensor and an excitation source. The excitation source provides excitation energy for the luminescent indicator, which is usually a high-power LED (light emitting diode) or a halogen xenon lamp. The signal collection system consists of a charge-coupled device or a complementary metal-oxide semiconductor connected to a computer terminal. Similar with zymography, image analysis software (e.g., Image J) can be used for data processing to acquire the information on the two-dimensional distribution of the substances (e.g., O_2, pH, CO_2, NH_4^+).

The sensing film is a key component of the planar optical pole, which determines the sensitivity, selectivity, dynamic range, response time and stability of the optical planar sensor. The production method of the sensing film is similar to the traditional coating process: first, the luminescent indicator, embedding agent, reference dye and some light scattering auxiliary particles are dissolved in organic solvent in a certain ratio to form a homogeneous mixture of the master batch. The master batch is then uniformly coated on the surface of the supporting substrate by knife coating, spin coating or spraying, etc. The sensing film with a thickness of less than 10μm is formed after the solvent evaporates. The luminescent indicator is the key component of the sensing film and the core of the research to extend the application of planar photopolar technology. Since most of the solutes in soil are colorless or do not fluoresce themselves, they cannot be measured directly by optical methods. The luminescent indicator is used to convert the concentration of the solute to be measured in soil into a measurable light signal based on its luminescent properties and specific sensitivity to the analyte to be measured. In general, an increase in the concentration of the luminescent indicator can enhance the light signal, but when the concentration is too high, the self-quenching of the fluorescence emitted will occur. Therefore the energy absorption of the luminescent indicator is enhanced by adding light scattering particles such as nanoscale TiO_2, diamond, or gold to the matrix solution. It should be noted that TiO_2 can also act as a photocatalyst, so it could reduce the stability of the optical planar sensor to some extent. Another common method is to add antenna dyes (such as coumarin) to the matrix solution. Coumarin can effectively absorb the excitation light and transfer the energy to the light-emitting indicator to enhance the emitted light signal, and this method has been applied to O_2, pH and other planar photopolar sensors. In practical applications, the sensing membrane is highly stable and none of the individual chemical components would diffuse or leak into the soil environment.

1.4.4.2 *Diffusive gradients in thin-film technique*

DGT offers significant advantages in the in situ and high-resolution characterization of nutrients and contaminants in their active state. The technique was invented in 1994 by Davison and Zhang (1994) at Lancaster University, UK. The device consists of a filter membrane, a diffusion membrane, an adsorption membrane and a plastic housing to hold the three membranes. The DGT technique has the advantages of quantification, trace and high spatial resolution characterization, and has been applied to the monitoring of more than 100 elements or compounds such as nutrients, metals, metalloids, radionuclides and organic pollutants in various environmental media. The greatest advantage of the DGT technique is that can be used to collect two-dimensional high-resolution (submillimeter) distribution information of elements in environmental media and at microinterfaces, providing a powerful methodological support for characterizing the microscopic distribution/aggregation of elements and the transport characteristics across media/interfaces. The adsorption membrane can be recovered by using a flat plate DGT device or a bilayer structure consisting of only the adsorption membrane and the diffusion layer (which can include only the filter membrane) after being placed in the ambient medium or at the microinterface for a period of time (usually hours to days). The concentration of elements accumulated on the membrane can be quantified and two-dimension distribution of these elements can be identified.

However, not all adsorbent membranes can be used for high-resolution analysis. When conducting high-resolution studies with submillimeter precision, LA-ICP-MS is used to analyze adsorbent membranes. To ensure the accuracy and precision of the analysis, the adsorbent membranes should meet the following standards: the adsorbent particles should be distributed uniformly on the adsorbent membrane; the adsorbent material particles on the adsorbent membrane are small enough ($\leqslant 10\mu m$). This helps to obtain information on the one- and two-dimensional concentration distributions of multiple solutes, which has fueled the mechanistic studies of biogeochemical processes of nutrients and contaminants in soils and sediments. For instance, Zhang et al. (2021) revealed the P uptake processes and mechanisms of *Vallisneria natans* during two vegetation periods (i.e., week 3 and week 6) were revealed using 3 noninvasive 2D imaging techniques: planar optode, DGT and zymography (Figure 1.8). The increase in phosphatase activity, O_2 concentration and root-induced acidification was identified together in the rhizosphere of root segments and tips. They also confirmed that *Vallisneria natans* roots acquired P from Fe plaque via organic acid complexation of Fe(III). The emergence of high-resolution sorbent membranes, the diversification of sorbent membrane analysis techniques and the low-cost of analysis, as well as the combined use of other imaging techniques further highlight the advantages of DGT in the study of microinterface processes.

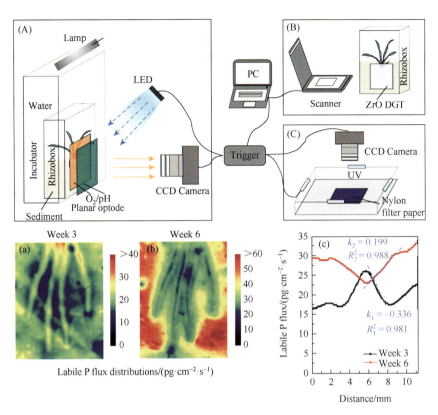

Figure 1.8 Experimental setup in Zhang et al. (2021) and two-dimensional distribution of DGT-labile P with roots of *Vallisneria natans*. Credit: Reprinted with permission from Zhang et al. 2021.

1.5 Enzymes and their relations with environmental researches

There are a number of reports in literature that describe the use of enzymes in soil bioremediation. Examples include the use of horseradish peroxidase (HRP) in the detoxification of industrial wastewater (Maloney et al., 1986), the use of HRP and hydrogen peroxide in the removal of phenol from coal conversion aqueous effluents (Klibanov et al., 1983), as well as the application of enzymes to detoxify pesticide-contaminated soils and waters (Nannipieri and Bollag, 1991). Another notable report is the application of urease on a wide range of compounds and its ability to render harmless some of the "unnatural" chemicals in the environment (Emsley, 1992). The use of enzymes for waste management can be classified into four categories: (1) effluent treatment and detoxification; (2) bioindicators for pollution monitoring; (3) biosensors; (4) sustainable energy resources.

1.5.1 Effluent treatment and detoxification

Compared with the traditional physiochemical processes and biological treatment processes, enzymatic treatment of organic wastewater has the following advantages (Wang and Wen, 2001): (1) It can treat organic compounds that are difficult to degrade; (2) it's applicable to wastewater of both high concentration and low concentration; (3) there is a wide range of pH, temperature and salinity during operation; (4) the processing will not be slowed down by the aggregation of biological substances, making it simple and easy to control the treatment process; (5) it is not easily inhibited by biotoxic substances; (6) the reactor with immobilized enzyme has strong impact resistance.

1.5.1.1 Treatment of phenolic wastewater

Aromatic compounds, including phenols and aromatic amines, are among the pollutants to be first controlled. Such pollutants could be found in the wastewater of petrochemical plants, resin factories, plastic factories and dye factories. Many enzymes have been used for treatment of wastewater of this type.

1. Peroxidase

Peroxidases are a type of oxido-reductases produced by microorganisms or plants. They can catalyze many reactions, but all of them require the presence of peroxides, such as hydrogen peroxide, to activate them. At present, the more studied and applied peroxidases include HRP, lignin peroxidase (LiP), etc. (Stoner, 1994).

a. Horseradish peroxidase. HRP is one of the most used enzymes in the enzymatic treatment of wastewater. Hydrogen peroxide, if any, could catalyze the oxidation of a variety of toxic aromatic compounds, including phenols, aniline, benzidine and related isomers, to produce water-insoluble precipitates. HRP is particularly suitable for wastewater treatment also because of its ability to maintain activity over a wide range of pH and temperature. Many applications of HRP are intended for the treatment of phenol-containing contaminants, including aniline, hydroxyquinolines and carcinogenic aromatic compounds. Moreover, HRP can precipitate with some hard-to-remove pollutants to form polymers, thus increasing the difficulty of removal. This phenomenon has important practical significance in the treatment of wastewater containing multiple pollutants. For example, PCBS can precipitate together with phenols in the solution.

b. Lignin peroxidase. LiP is a part of the *Phanerochacte chrysosporium*. LiP can process many refractory aromatic compounds and oxidize a variety of polycyclic aromatic hydrocarbons and phenols. The optimum conditions for removal of phenols are as follows: high concentration of enzyme, pH greater than 4.0 and a certain amount of hydrogen peroxide. Immobilization of LiP on porous ceramics does not affect its activity. It does well in degrading aromatic substances.

c. Enzymes of plant sources. Peroxidases extracted from tomatoes are used to polymerize phenolic compounds. Roots of some plants can also be used to remove pollutants. For the 2, 4-dichlorophenol of a concentration up to $850mg \cdot L^{-1}$, plant peroxidase features a removal rate similar to that of pure HRP. The removal rate is subject to the pH value of the reaction mixture system, the particle size of plant raw material, the amount of raw material, and whether hydrogen peroxide has been involved.

2. Polyphenol oxidase

Polyphenol oxidases mark another type of oxido-reductases that catalyze the oxidation of phenols. They can be divided into two groups: tyrosinase and laccase. They both require the involvement of oxygen molecules, but not the coenzymes.

a. Tyrosinase. Also known as phenolase or catecholase, they can catalyze two sequential reactions: (1) the oxidation-reduction reaction of a single-molecule phenol with an oxygen molecule to form catechol; (2) dehydrogenation of catechol to form benzoquinone, which is very unstable and forms water-insoluble products through enzymatic polymerization, and water-insoluble products can be removed by filtration. Tyrosinase has been successfully used to precipitate and remove phenols $(0.01-1.00g \cdot L^{-1})$ from wastewater. Tyrosinase, immobilized with chitin, is used to treat the phenol containing wastewater, recording a removal rate of 100% within 2 hours. The tyrosinase is immobilized so as not to be washed away by water and inactivated due to reaction with benzoquinone. The immobilized tyrosinase remains effective even after be used 10 times.

b. Lactase. Produced by some fungi, the laccase can remove toxic phenols through polymerization reactions. Moreover, its nonselectivity allows it to simultaneously reduce the content of many phenols. The detoxification feature of laccase is related to the substance being treated, the source of the enzyme and some environmental factors.

1.5.1.2 *Treatment of papermaking wastewater*

1. Peroxidase and laccase

HRP and LiP have been used to decolorize the papermaking wastewater.

Their immobilized state features better treatment effect than their free state. The action mechanism of LiP is to degrade the lignin through catalytic oxidation of benzene ring units into cationic groups that can be automatically degraded. Laccase can remove chlorophenols and chlorinated lignin from bleaching wastewater through precipitation.

2. Enzymes that break down cellulose

This type of enzyme is mainly used in pulp-making and de-inking operations for pollution treatment. The high content of cellulose from wastewater treatment in pulp and paper making processes can be used to produce ethanol and other energy materials. The enzyme used is a mixture of cellobiose hydratase, cellulase and β-glucose-enzyme. The low content of fibrous waste produced during the de-inking operation can be converted into fermentable sugars. The enzyme used cannot be inhibited in the presence of high concentrations of ink.

1.5.1.3   Treatment of cyanide wastewater

1. Cyanide enzyme

Cyanidase converts cyanide to ammonia and formate, through a one-step process. It is a Gram-negative bacterium that can produce cyanidase, of high affinity and stability, and process cyanide at a mass concentration below $0.02 \text{mg} \cdot \text{L}^{-1}$. The kinetic properties of cyanase conform to the Michaelis-Menten equation. The activity of cyanase is neither affected by the common cations in wastewater such as Fe^{2+}, Zn^{2+} and Ni^{2+}, nor by organic matters such as acetic acid and formamide. The fittest pH value is 7.8—8.3. The diffusion-type plate membrane reactor, suitable for cyanase, has the advantages of preventing the enzyme from being washed by macromolecules and keeping the substrate used to fix the enzyme from being destroyed. The cyanide reacts with the enzyme inside the membrane through the semi-permeable membrane, with the reaction product penetrating back into the solution (Sohal and Srivastava, 1994).

2. Cyanide hydrase

Cyanide hydrase can hydrolyze the cyanide into formamide. These enzymes can be obtained through a variety of fungi. If immobilized, they are more stable and easier for the treatment of cyanide-containing wastewater.

1.5.1.4   Treatment of food processing wastewater

The wastewater of the food processing industry is easy to decompose, or convert into feed or other products of economic value. Enzymes can be used in the treatment of food processing wastewater to purify wastewater and obtain high value-added products.

1. Protease

Protease, as a kind of hydrolytic enzyme, has been widely used in the treatment of fish and meat processing wastewater. Protease can hydrolyze the protein in wastewater and obtain the recyclable solution or the nutritious feed. During the protein hydrolysis process, protease is absorbed by solid proteins in water. When it unbinds the polypeptide chains on the surface of the protein, the tighter core is gradually dissolved in water.

An alkaline enzyme (EC unknown), derived from *Bacillus subtilis* and can be used for feather treatment in poultry slaughterhouses, could be a feed ingredient of high protein content through NaOH pretreatment, mechanical crushing and enzymatic hydrolysis.

2. Amylase

Amylase, as a kind of polysaccharide hydrolase, catalyzes the transformation of polysaccharide into monosaccharide and ferments at the same time. Amylase can be used in the treatment of starch-containing wastewater, converting organic matters in the wastewater arising from rice processing into alcohol. Amylase can also shorten the time of wastewater treatment by the activated sludge method.

1.5.1.5 *Application of microbial lipase*

Microbial lipases (glycerolipid hydrolase) can catalyze a range of reactions, including hydrolysis, alcoholysis, acidolysis, esterification and ammonolysis. A large number of microorganisms can be used to produce lipase, among which Candida, Pseudomonas and Rhizopus are important sources of enzymes. The application of lipase in bioremediation and waste treatment of contaminated environment marks an emerging field. Oil spills from oil extraction and refining, and lipid-containing wastes from lipid processing can be effectively treated with lipase from different sources. Lipase is widely used in wastewater treatment. Dauberhe and Boehnke worked out a technique that uses a mixture of enzymes, including lipase, to turn dehydrated sludge into biogas. A Japanese patent is reported to treat wastewater by growing lipophilic microorganisms directly in wastewater. Lipase has been widely used in bioremediation of contaminated environment. Using the lipase from *Aspergillus oryzae* to produce cystine in waste hair proves a promising application prospect. Using yeast to produce single-cell proteins from industrial wastewater shows another attractive prospect for lipase application in waste management.

Enzymes are currently charged a high price. Thus, they are unsuitable for the treatment of pollutants of high concentrations, but suitable for the treatment of pollutants of low concentrations and high toxicity. Enzyme-based treatment embraces a broad application prospect. Further studies must be conducted on the characteristics and stabilization of enzyme reaction byproducts, the treatment of reaction residues and the treatment costs.

1.5.2 Bioindicators for pollution monitoring

Soil environmental quality indexes mainly include the soil environmental capacity, total heavy metal content, availability of heavy metal elements, residual amount of organic pollutants (including pesticides), soil pH and soil texture. All these are physical and chemical indexes. Studies have found that compared with physiochemical indexes of the soil, the biological and biochemical indexes are more sensitive to changes in the soil environment, which can be used to rapidly evaluate the changes in soil environment quality in a short period of time (Nannipieri et al., 1990; Yakovchenko et al., 1996). Among the biological components of the soil, microorganisms and enzymes play an important role in the material as well as energy conversion process. The changes of soil microbial communities can quickly reflect the impact of foreign matters on the soil environmental quality (Visser and Parkinson, 1992). But to obtain the biological characteristics of soil, molecular biology technology and professional microbial technology need to be used, with expensive reagents and sophisticated experimental instruments equipped. This undoubtedly increases the difficulty of large-scale soil research. Therefore, in recent years, scholars have begun studying the indication of soil enzymes on soil environmental quality (Ayten et al., 2011).

In fact, all the biochemical reactions in soil are catalyzed by enzymes. As one of the essential properties of soil, the enzyme activity reflects the intensity and direction of various biochemical processes in soil. Dick and Tabatabai (1992) pointed out in 1992 that "Soil enzymes may contribute to describing and predicting the function, quality and interactions of different ecosystems." Dick (1997) summarized the significance of soil enzymes in maintaining soil function, arguing that soil enzymes can decompose foreign organic matters, promote the transformation of soil organic matters, decompose organic matters into mineral elements available to plants, participate in the N_2 fixation, degrade harmful foreign matters (artificial compounds such as pesticides and industrial wastes), participate in nitrification and denitrification processes. The study by Bandick and Dick

(1999) further pointed out that the soil enzymes can be used to symbolize the dynamics of soil quality, because ①they can catalyze the nutrient cycling process in soil, including the cycling of C (β-glycosidase, and β-galactose glucoside enzyme), N (urease), P (phosphatase) and S (sulphur esterase) in soil; ②associated with the soil biology, they can be easily measured; ③they react rapidly to the changes upon the soil pollution. At present, studies taking soil enzymes as an indicator of soil environmental quality chiefly focus on the polluted, disturbed and agricultural soils.

1.5.2.1 *Indication of agricultural pollutants by soil enzymes*

When agricultural pollutants such as fertilizers and pesticides enter the soil, they will affect the biological properties of the soil, and change the activity of soil enzymes accordingly. Since the range of change is related to the type, concentration, duration and application amount of pollutants, many scholars take soil enzymes as an indicator of agricultural pollution (Ajwa et al., 1999). For example, Gianfreda et al. (2005) found that compared with normal cultivated land, the activity of aryl sulfatase, β-glucoside, phosphatase, urease, dehydrogenase and fluorescein diacetate hydrolase in land moderately or severely polluted by organic matters were decreased or inactivated. However, other studies found that when the agricultural pollutants applied were of low concentration, the enzyme activity remained unchanged or improved (Ajwa et al., 1999; Russo and Biles, 2003).

In general, when excessive fertilizer is applied, enzyme activity will decrease. For example, Dick et al. (1997) found that chemical fertilizers would inhibit the phosphatase activity. Qian et al. (2007) found that the application of excessive nitrogen fertilizer in paddy soil would inhibit the urease activity. Li et al. (2019) found that nitrogen, phosphorus and potassium could significantly reduce the activity of catalase. Liu et al. (2017) found that with calcium added to the original application of nitrogen, phosphorus and potassium fertilizer, the activity of dehydrogenase and aryl sulfoesterase in soil was reduced, while the activity of urease, luciferin hydrolase and acid phosphatase in soil was inhibited. Graham and Haynes (2005) found that the activity of dehydrogenase, alkaline phosphatase, aryl sulfatase and histidinase in sugarcane soil decreased with the increase of fertilizer application. However, some studies believed that the application of chemical fertilizer can boost or leave no impact on the activity of some soil enzymes. Studies of Ajwa et al. (1999) and Acosta-Martinez et al. (1999) showed that with the increase of nitrogen application, the activity of β-glucosidase, acid phosphatase and dehydrogenase in soil also increased.

Liu et al. (2017) found that the activity of urease, invertase and alkaline phosphatase in soil with sole application of nitrogen fertilizer increased. Russo and Biles (2003) found that the soil β-galactosidase activity increased or remained unchanged when a mixture of nitrogen, phosphorus and potassium fertilizers was applied to pepper fields. It was also found that the increase of fertilizer application can increase the activity of acid phosphatase, invertase and protease (Graham and Haynes, 2005).

The application of pesticides (insecticides, herbicides and fungicides) will affect the growth and activity of soil microorganisms, thereby affecting the enzyme activity. A large number of studies have shown that the action mechanism of enzyme on pesticides is to regulate, modify or control the gene of synthase through inhibition or induction. The products and change of total number of intracellular enzyme and extracellular enzyme can be controlled by different concentrations of pesticides in soil (Cervelli et al., 1978; Gianfreda and Rao, 2008; Kaare et al., 2001). However, the application of enzyme activity as an evaluation index of pesticide pollution varies greatly to the soil and pesticide types (Sannino and Gianfreda, 2001; Speri et al., 1992).

(1) Existing studies indicated that pesticides had certain effects on the urease, dehydrogenase, cellulase, amylase, protease, phosphatase, phosphomonoesterase, catalase and fluorescein diacetate hydrolase in soil. For example, except endosulfan that can activate the dehydrogenase, most of the others have inhibitory effect on the dehydrogenase activity (Beulke and Malkomes, 2001; Kalam et al., 2004; Yao et al., 2006; Jastrzebska, 2011; Kalyani et al., 2010; Defo et al., 2011). Rain et al. (2008) found that urease and dehydrogenase were most sensitive to chlorpyrifos, while the cellulase, amylase and protease gradually produced resistance after the application of chlorpyrifos. Some studies also found that the phosphatase activity increased with the increase of chlorpyrifos concentration, but gradually decreased more than 20 days later (Madhuri and Rangaswamy, 2002). However, the phosphatase activity was inhibited after repeated application of chlorpyrifos and quetisulfur (Pandey and Singh, 2006). Yao et al. (2006) found that upon the application of acetamiprid, the dehydrogenase activity was enhanced, while the soil urease and catalase had little changes. Kalyani et al. (2010) and Perucci et al. (2000) found that imidazoline and endosulfan could activate the activity of luciferin diacetate hydrolase. Kalyani et al. (2010) and Dutta et al. (2010) found that chlorpyrifos and ethion could also activate the activity of luciferase diacetate hydrolase, while endosulfan could increase the activity of aryl sulfatase. Gundi et al. (2005) found that quintiofos,

monocrotophos and cypermethrin alone or combined could improve the activity of dehydrogenase. According to the study by Fragoeiro and Magan (2008), the application of pesticides such as simazine, trifluralin and dieldrin reduced the activity of soil dehydrogenase. Pandey and Singh (2006) and Singh and Singh (2005) found that the activity of alkaline phosphomonesterase and dehydrogenase in peanut field was first decreased and then increased upon the application of imidacloprid, endosulfan and diazinon, and finally tended to be stable.

(2) The invertase, phosphatase, dehydrogenase, catalase and urease in soil responded differently to the herbicide application. For example, the invertase activity was activated by glyphosate and paraquat, and inhibited by carbaryl (Singh and Singh, 2005), while urease activity was not affected by the 2 pesticides. The activity of phosphatase was inhibited by Furadan, Temik and Talstar herbicides, and that of dehydrogenase was inhibited by Talstar, Kung Fu, Monocrotophos, Omethoate and Acetochlor (Cai et al., 2007). The activity of catalase was inhibited by the rice herbicide Mefenacet. The greater the concentration, the stronger the inhibition. Alachlor can inhibit the activity of the lipase in soil (Dzantor and Felsot, 1991). Atrazine can lower the soil activity of β-glucosidase (Voets et al., 1974).

(3) The enzymes are sensitive to bactericide application include the catalase, invertase, urease, acid phosphatase and alkaline phosphatase in soil. The activity of soil invertase was inhibited by Chlorothalonil, Carbendazim mixture and Cyfluthrin. However, Carbendazim and imidacloprid activated the enzyme activity at low concentration and inhibited it at high concentration. The activity of catalase, urease and acid phosphatase was inhibited by the high concentration of Validamycin, while the low concentration of it had little effect on the enzyme activity (Qian et al., 2007). The activity of acid phosphatase, alkaline phosphatase, urease, catalase and invertase in soil was inhibited by Chlorothalonil (Yu et al., 2006). Higher doses of bactericide can inhibit the activity of dehydrogenase (Adolphe and Michael, 2002; Bello et al., 2008).

1.5.2.2 Indication of soil enzymes on industrial pollutants

Because industrial sewage and sludge are rich in some valuable organic and inorganic nutrients, they can improve the soil fertility after treatment. Therefore, they are often applied for farmland irrigation in agricultural field (Antolin et al., 2005). Studies have found that the application of sewage and sludge can boost the soil biological activity (Saviozzi et al.,

1999), increase the basement respiration and microbial biomass, thus cause corresponding changes in enzyme activity. For example, Kizilkaya et al. (2004) found that in the soil of irrigation sewage and sludge, the activity of urease, phosphomonoesterase, and aryl sulfatase increased. Sewage and sludge in wheat field can boost the activity of BAA-protease, urease, phosphatase and β-glucosidase in soil (Antolin et al., 2005). However, some studies also showed that the enzyme activity decreased upon the application of sewage and sludge. Filip et al. (1999) found that in soil irrigated with sewage, the activity of β-glucosidase, β-acetylaminoglycosidase, protease and phosphatase were inhibited. The different effects of sewage and sludge on soil enzyme activity are closely related to their application volume and composition. For example, some studies found that sewage and sludge, if containing heavy metals, would seriously inhibit the activity of soil dehydrogenase. Kannan and Oblisami (1990) found that irrigation with pulp-contaminated water would increase the activity of soil invertase and dehydrogenase; Li et al. (2019) found that the activity of dehydrogenase, catalase and polyphenol oxidase in soil were significantly positively correlated with the content of petroleum hydrocarbons in soil, while the activity of soil urease was significantly negatively correlated with the content of petroleum hydrocarbons in soil. Li et al. (2019) also found that oiled water had a certain stimulating effect on the catalase, polyphenol oxidase and urease in soil.

Municipal solid wastes are ones produced during production, construction, daily life and other human activities. The harmful substances contained in them can change the soil structure and quality, affect the activity of microorganisms in them, change the enzyme activity accordingly. Existing studies showed that municipal solid waste can affect the activity of β-glucosidase, urease, alkaline phosphatase, dehydrogenase, catalase, cellulase and protease in soil. For example, Garcia Gil et al. (2000) found that municipal solid waste compost increased the activity of dehydrogenase and catalase in farmland soil. Li et al. (2019) found that with the deepening of landfill depth, the activity of urease and cellulase in soil tended to decrease, while the activity of catalase and protease changed little. Felipe et al. (2008) found that municipal solid waste can lead to the increase of β-glucosidase and alkaline phosphatase activity. Carmine et al. (2001) found that municipal solid waste compost could increase the activity of dehydrogenase and nitrate reductase in soil. Claire et al. (1995) found that the activity of peroxidase in soil increased upon contamination of the municipal solid waste compost, but little effect was put on the amidase and urease.

It is well known that heavy metals have long-term toxic effects in the ecosystem and a negative impact on the catalytic process of soil enzymes (Kizilkaya et al., 2004). They disrupt the enzyme activity by inhibiting their catalytically active groups, changing the conformation of proteins, or participating in the formation of disrupting enzyme complexes with other metals (Eivazi and Tabatabai, 1990). Marzadori et al. (1996), Aoyama and Naguno (1996) pointed out that the urease, acid phosphatase and dehydrogenase, all sensitive to heavy metal pollution in soil, can serve as indicators of heavy metal pollution in soil. Soil dehydrogenase activity is more susceptible to the accumulation of Pb and Cd in soil, which is suitable for reflecting the pollution degree of Pb and Cd in soil. Parkhust (1997) found that aryl sulfatase could be used as a sensitive indicator of heavy metal pollution in soil. It has also been found that the activity of L-glutaminase, cellulase and β-glucosidase was inhibited by Hg, Ag, Cr and Cd (Deng and Tabatabai, 1995). Previous study found that the activity of catalase decreased significantly with the increase of Cd and Zn compound concentrations in medium loamy moist cinnamon soil; the polyphenol oxidase activity decreased after the mixed pollution of Cu, Zn, Pb and Cd in the red soil of mining regions; the deep red soil contaminated by Ni, Cu and Cd, they also found that the increase in Ni concentration inhibited the activity of soil invertase, the protease activity decreased significantly with the increase of Cu concentration, and Cd had an obvious inhibitory effect on the protease activity. Frankenberger and Tabatabai (1981, 1991a, 1991b) found that heavy metals such as Hg, Ag, Cr and Cd inhibited the activity of L-asparaginase and thioesterase. Bardgett et al. (1994) also found that these 4 heavy metals inhibited the activity of acid phosphatase and urease. Al-Khafaji and Tabatabai (1979) also found they had an inhibitory effect on aryl sulfatase. Füsun and Esin (2008) found that the activity of aryl sulfatase, alkaline phosphatase and urease was significantly negatively correlated with the total content of Pb, Cr and Mn.

However, some studies also showed that enzyme activity had a positive role in the remediation of heavy metal contaminated soils. For example, activity of catalase increased with the increase of Pb concentration in the medium loamy moist cinnamon soil contaminated by Cd, Zn and Pb (Deng and Tabatabai, 1995). Gao et al. (2010) found that Cd, Cu and Ni all showed a certain activating effect on the phosphatase activity. Mikanova (2006) found that protease and urease activity was weakly inhibited by heavy metals. The pollution of red soil by heavy metals (Cu, Zn, Pb and Cd) had little effect on soil invertase.

1.5.3 Biosensors

Electrochemical enzyme sensors are the most widely used and most numerous type of biosensors functioning based on the monitoring of electroactive species produced or consumed by enzyme-catalyzed reactions. In 1962, Clark and Lyons (1962) firstly proposed the principle and design of enzyme sensor on the basis of oxygen electrode. In 1967, Updike and Hicks (1967) developed the first glucose oxidase sensor based on platinum electrode for quantitative detection of serum glucose content. These studies are considered as milestones in the development of biosensors, especially enzyme sensors. Since then, enzyme sensors, especially electrochemical ones, have attracted great attention and driven extensive research by scientists in various fields, thus embracing a rapid development. This section focuses on its application in environmental pollution monitoring.

Environmental monitoring involves many complex steps such as sampling, sample handling and transportation of the sample to designated laboratories. The big challenge faced by in situ environmental monitoring is the need for an accurate, sensitive, specific, rapid and easy-to-operate analytical instrument to detect pollutants. Biosensors, especially enzyme sensors, are widely used because they can monitor environmental pollutants in real time, in situ, selectively, sensitively and quickly (Badihi-Mossberg et al., 2007). Among them, electrochemical enzyme sensors can be used to monitor organic pollutants, inorganic pollutants and heavy metals in the environment.

1.5.3.1 Organic pollutant

Phenolic compounds in the environment, mainly derived from wastewater of chemical dyeing and papermaking sectors and pesticide degradation products, are compounds with "trigenic effect" that widely exist in the environment. Therefore, it is necessary to develop a simple and effective monitoring method for phenolic compounds. For their advantages of specificity of enzyme-catalyzed reaction, long-term stability of immobilized enzyme, rapidness and sensitivity of electrochemical analysis (Guo and Lakshmikantham, 1988), electrochemical enzyme sensors have attracted increasing attention, and various types are constantly being developed for the monitoring of phenolic substances. Since phenols can act as electron donors for oxidized tyrosinase, laccase and peroxidase, or as electron acceptors, in their oxidized state, for reduced glucose dehydrogenase and cellobiose dehydrogenase, the tyrosinase (Mita et al., 2007), laccase (Tan et al., 2009), peroxidase (Korkut et al., 2008) and pyrroloquinoline quinone based on glucose dehydrogenase or cellobiose dehydrogenase have been used for the monitoring of phenolic compounds (Yaropolov et al., 1995). In the 1990s, some scholars adsorbed both laccase and tyrosinase on graphite electrodes to construct a dual-enzyme system enzyme sensor for continuous flow analysis of phenolic compounds (Yaropolov et al., 1995). Then, a study reported that 8 kinds of crosslinking agents were used to covalently immobilize the tyrosinase on the surface of the aminated carbon electrode, among which the glutaraldehyde crosslinking method showed a high sensitivity, operation stability and storage stability. When used in the determination of catechol and 4-chlorophenol, this method showed a low limit of detection and good repeated use (Wang and Hasebe, 2009). For example, Tan et al. (2009) used a laccase sensor, prepared by immobilizing laccase in the composite film of chitosan multiwalled carbon nanotubes functionalized with glutaraldehyde, to detect catechol. They found that the linear response of the sensor to catechol ranged from $0.1\mu mol \cdot L^{-1}$ to $50\mu mol \cdot L^{-1}$, and the limit of detection was $20nmol \cdot L^{-1}$. Also, it showed good stability and reproducibility.

Pesticides are environmental pollutants that widely exist in water, air, soil, plants and foods. The detection of pesticides, especially organophosphorus pesticides, is a field where electrochemical enzyme sensors are widely used (Dzyadevych et al., 2005). At present, there are two main types of electrochemical enzyme sensors used in the detection of organophosphrous pesticides: inhibitory enzyme sensors and hydrolytic enzyme sensors. Organophosphorus pesticides are widely used as agricultural insecticides, which can lead to cholinergic dysfunction and death by inhibiting the lipase activity, and pose a serious threat to human and animal health (Zhao et al., 2009). Therefore, the detection of trace organophosphorus pesticides is of great significance for the protection of human health. Inhibitory enzyme sensors based on cholinesterase have been widely used for the detection of organophosphor pesticides due to their sensitivity (Solé et al., 2003). The non inhibitory enzyme sensor based on organophosphorus hydrolases is another type for direct monitoring of organophosphorus pesticides. Organophosphorus hydrolases can catalyze the hydrolysis of organophosphorus pesticides, such as paraoxon and parathion, by forming low-toxic products (such as p-nitrophenol and diethyl phosphate). Organophosphorus hydrolases are widely used to produce sensors for monitoring organophosphorus pesticides which have the following advantages: they can be used as substrates rather than inhibitors of enzyme reactions; compared with cholinesterase, organophosphorus hydrolases have strong specificity, good selectivity, less loss of enzyme activity in the reaction and repeated use (Wanekaya et al., 2008). Studies have shown (Russell et al., 1999) that the hydrolysis reaction catalyzed by organic phosphorus hydrolases can lead to the breakage of P-O, P-S, P-F and P-CN bonds in organic phosphorus pesticides. To date, there have been quite extensive studies on noninhibitory enzyme sensors based on organophosphorus hydrolase. For example, Du et al. (2010) deposited gold nanoparticles on the surface of the premodified glass carbon electrode of the multiwall carbon nanotube, then covalently bonded methyl parathion hydrolase to the modified electrode through CdTe quantum dots. Among them, the introduction of multiwalled carbon nanotube and gold nanoparticles increased the surface area and showed a synergistic effect on enzyme catalysis, with CdTe quantum dots serving as a carrier to load a large number of enzymes. Therefore, in the absence of mediator, the enzyme sensor established featured a detection limit of methyl parathion of up to $1.0ng \cdot mL^{-1}$, showing good reproducibility and stability, which can be reused and is suitable for continuous monitoring. Apart from organophosphorus pesticides, the electrochemical enzyme sensor can also be used to detect other pesticides (such as Carbofuran, atrazine, 2, 4-D and thiomacarb). For example, Qu et al. (2010) used the multifilm modified electrode of acetylcholinesterase/polyamid-amine-gold/carbon nanotube prepared with layer-by-layer self-assembly technology to detect the Carbofuran in water samples, found that the linear response range of the enzyme sensor used to Carbofuran was 4.8×10^{-9}—$9.0 \times 10^{-7} mol \cdot L^{-1}$, the detection limit can reach $4.0 \times 10^{-9} mol \cdot L^{-1}$, and the electrode had high sensitivity, good stability and reproducibility. Kim et al. (2008) prepared the tyrosinase-gold nanoparticle-modified glass carbon electrode by covalently combining the self-assembled monolayer film modified by gold nanoparticles with the tyrosinase, and the electrode prepared can be used for the detection of pesticides (such as 2, 4-D, atrazine and Ziram) in continuous flow system. It was found that under the optimized conditions of the continuous flow system, the response of the electrode followed the inhibition mechanism in the concentration range from $0.001ng \cdot mL^{-1}$ to $0.5ng \cdot mL^{-1}$.

1.5.3.2 Inorganic salt

Nitrate is an important indicator used in water quality analysis, which indicates the eutrophication in lakes and coastal waterways (Sohail and Adeloju, 2008). Excessive intake of nitrate can lead to serious health problems. The World Health Organization (WHO) and the US Environmental Protection Agency (USEPA) set the maximum allowable limit of nitrate in drinking water as $44mg \cdot L^{-1}$. Therefore it is necessary to develop a reliable method for nitrate determination in living organisms and environmental media. Traditional nitrate analysis methods can be divided into three categories: spectrophotometry, ion chromatography and electrochemical methods. In electrochemical methods, biosensors are of great significance in the detection of nitrate due to their good selectivity and rapid detection. Over the past decade, there have been an increasing number of studies on the nitrate reductase electrodes (Sohail and Adeloju, 2009). The working principle of this method is as follows: during the enzyme-catalyzed reaction, nitrate is reduced to nitrite while nitrate reductase is oxidized, and electrons are transferred from the electrode surface to the active site of nitrate reductase through medium to reduce nitrate reductase. Then the concentration of nitrate is determined according to the response of the current (Cosnier et al., 2008). Screen printed enzyme sensors based on nitrate reductase can detect nitrate residue at a level lower than stipulated by EU legislation in drinking water, and show good storage stability. Thus it provides a reliable and inexpensive method for the analysis of nitrate ions in drinking water. For example, in the presence of nicotinamide adenine dinucleotide, nitrate reductase is embedded in the polypyrrole film prepared by electrochemical polymerization to prepare an enzyme sensor. This sensor has a good detection limit for nitrate, and its minimum detection concentration is $15\mu mol \cdot L^{-1}$ (Sohail and Adeloju, 2008). In addition, the study on the application of polypyrrole-nitrate reductase-Azuanine A biosensor in the detection of nitrate in water discovered that the sensor had a high sensitivity to nitrate, a detection concentration of as low as $15\mu mol \cdot L^{-1}$ and a response speed of (2—4s) (Sohail and Adeloju, 2009).

Nitrite ion is the intermediate product of the nitrification and denitrification process during the biological treatment of wastewater. In addition, the substance is widely present in food, posing a carcinogenic risk to human beings. The standard value of nitrite was first stipulated in GB 5749—2022 *Standards for Drinking Water Quality*. Therefore the determination of nitrite ion is of great practical significance (Kamyabi and Aghajanloo, 2008). Nitrite reductase is widely used in the detection of nitrite. Silveira et al. (2010) prepared a new electrochemical enzyme sensor, which uses pyrolytic graphite as electrode and requires no intermediary substances, for the detection of nitrite in complex samples. To ensure the analytical capability of the enzyme sensor, the nitrite reductase was successfully fixed in porous silica glass prepared through the sol-gel process. The results showed that the enzyme sensor's sensitivity to nitrite could reach $430mA\cdot mol^{-1}\cdot L\cdot cm^{-2}$, with a detectable concentration as low as $120nmol\cdot L^{-1}$. Prepared using glutaraldehyde crosslinking agent to fix the cytochromatin nitrite reductase in the multiwalled carbon nanotube-polyamides-chitosan nanocomposite membrane, the enzyme sensor responded to nitrite in the two ranges of $0.1-29\mu mol\cdot L^{-1}$ and $29-254\mu mol\cdot L^{-1}$ in (5s) and had a detection limit of $10nmol\cdot L^{-1}$ (Chen et al., 2009). In addition, electrochemical enzyme sensors prepared based on HRP (Zazoua et al., 2009) and catalase (Salimi et al., 2007) can also be used for the detection of nitrite. Research results showed that these enzyme sensors had a favorable detection limit and sensitivity for nitrite.

Heavy metals such as Cu, Cd, Hg and Zn are highly toxic, which can accumulate in living organisms to produce toxic effects. In drinking water, they would threaten the human health. Therefore it is urgent to work out a method for rapid in situ detection of trace levels of heavy metals (Bontidean et al., 2000). Compliant with the above requirements, electrochemical methods can be used for the detection of heavy metals (Ghica and Brett, 2008). Electrochemical enzyme sensors have the greatest potential of detecting heavy metals in water, soil and atmosphere. Electrochemical enzyme sensor depends on the enzyme inhibition or activation methods for heavy metal detection. Enzyme inhibition analysis can quickly determine the heavy metals in environmental samples, as metal ions usually combine with mercaptan groups in enzyme structure to change enzyme conformation, thus affecting the catalytic activity of enzymes. Inhibitory enzyme sensors are more sensitive to the detection of inhibitors than the detection of enzyme substrates, their detection limits are usually far lower than the maximum allowable amount in environmental samples and the value obtained by traditional methods (Bontidean et al., 1998). Inhibitory electrochemical enzyme sensors for heavy metal detection that have been studied include those built based on glucose oxidase (Ghica and Brett, 2008), alkaline phosphatase (Berezhetskyy et al., 2008), urease and invertase. As revealed by the study using the electrochemical enzyme sensor, produced based on the inhibition of alkaline phosphatase, to detect the heavy metal ions Cd^{2+}, Co^{2+}, Zn^{2+}, Ni^{2+} and Pb^{2+} in water, the detection limits of the enzyme sensor for the 5 heavy metal ions were $0.5mg\cdot L^{-1}$, $2mg\cdot L^{-1}$, $2mg\cdot L^{-1}$, $5mg\cdot L^{-1}$ and $40mg\cdot L^{-1}$, respectively. The sensor maintains a stable performance for one month as maximum when stored in a buffer solution at $4℃$. Ghica and Brett produced an electrochemical enzyme sensor for detecting heavy metal cations based on the inhibition of glucose oxidase. The sensors produced using the glucose oxidase to crosslink at the surface of polyneutral red film prepared with the electrochemical polymerization method feature detection limits of $1\mu g \cdot L^{-1}$, $6\mu g \cdot L^{-1}$, $3\mu g \cdot L^{-1}$ and $9\mu g \cdot L^{-1}$ for Cd^{2+}, Cu^{2+}, Pb^{2+} and Zn^{2+}, respectively. Studies also showed that the inhibition of Cd^{2+} by this sensor was reversible and competitive. For Cu^{2+} and Pb^{2+}, it had reversible and competitive inhibition effects, while for Zn^{2+}, it performed a noncompetitive inhibition. In addition, when used to detect Hg^{2+}, Cu^{2+} and other ions in the water, electrochemical enzyme sensors have also shown good sensitivity and stability (Tsai and Doong, 2007).

1.6　Conclusion

This chapter introduces the types and sources of enzymes in the environment (mainly the soil), elaborates the enzyme-catalyzed reaction kinetics, the action of enzymes, and the key factors affecting enzyme activity, summarizes the application of enzyme in the study of environmental sciences, and highlights enzyme's role in the changes of environmental quality, sewage treatment and monitoring of environmental pollution. Besides, basic principles, procedures, and applications of in situ zymography were described and its combined use with other novel techniques were also introduced.

How to rationally utilize and protect the soil, groundwater resources and environment has become an important task for scholars and governments of all countries in the world. Progresses have been made in linking the enzyme activity with the soil productivity/fertility and groundwater quality. However, as one of the research focuses of environmental sciences, the existential state and biochemical kinetic characteristics of enzymes in the groundwater-soil environment shall be highlighted. Also, it will be the direction of future development of enzymology to use enzymatic knowledge to solve practical problems regarding modern environment, agriculture, and ecology.

In the following chapters, the activities, functions and effects of various enzymes in soil and groundwater environment will be introduced from the following dimensions, in order to provide reference and basis for the application of enzymes in soil and groundwater environment.

(1) The response of enzyme kinetics to climate warming;

(2) The effects of application of biomass materials on environmental quality and temporal-spatial distribution characteristics of enzyme activity;

(3) The effects of heavy metals and nano-metal oxides on environmental quality and enzyme activity;

(4) The effects of the application of novel carbon-based remediation materials on the improvement of polluted environment and the enzyme activity; (5) The characteristics and mechanism of the effect of enzyme activity on plant rhizosphere.

Main References

Acosta-Martínez, V., Reicher, Z., Bischoff, M., et al., 1999. The role of tree leaf mulch and nitrogen fertilizer on turfgrass soil quality. Biology and Fertility of Soils, 29(1): 55–61. Available from: https: doi.org/10.1007/s003740050524.

Adolphe, M., Michael, S., 2002. Effects of the phenylamide fungicides, mefenoxam and metalaxyl, on the microbiological properties of a sandy loam and a sandy clay soil. Biology and Fertility of Soils, 35(6): 393–398. Available from: https: doi.org/10.1007/s00374-002-0485-1.

Ahmadi, K., Razavi, B.S., Maharjan, M., et al., 2018. Effects of rhizosphere wettability on microbial biomass, enzyme activities and localization. Rhizosphere, 7: 35–42. Available from: https: doi.org/10.1016/j.rhisph.2018.06.010.

Ajwa, H., Dell, C., Rice, C., 1999. Changes in enzyme activities and microbial biomass of tallgrass prairie soil as related to burning and nitrogen fertilization. Soil Biology and Biochemistry, 31(5): 769–777. Available from: https: doi.org/10.1016/S0038-0717(98)00177-1.

Al-Khafaji, A.A., Tabatabai, M.A., 1979. Effects of trace elements on arylsulfatase activity in soils. Soil Science, 127(3): 129–133.

Antolín, M., Pascual, I., García, C., et al., 2005. Growth, yield and solute content of barley in soils treated with sewage sludge under semiarid Mediterranean conditions. Field Crops Research, 94: 224–237. Available from: https: doi.org/10.1016/j.fcr.2005.01.009.

Aon, M.A., Cabello, M.N., Sarena, D.E., et al., 2001. Spatio-temporal patterns of soil microbial and enzymatic activities in an agricultural soil. Applied Soil Ecology, 18(3): 239–254. Available from: https: doi.org/10.1016/S0929-1393(01)00153-6.

Aoyama, M., Nagumo, T., 2012. Factors affecting microbial biomass and dehydrogenasc activity in apple orchard soils with heavy metal accumulation. Soil Science and Plant Nutrition, 42(4): 821–831. Available from: https: doi.org/10.1080/00380768.1996.10416629.

Ayten, K., Sema, C.C., Oguz, C.T., et al., 2011. Soil enzymes as indication of soil quality. Soil Biology, Soil Enzymology, 22(1): 119–148.

Badiane, N.N.Y., Chotte, J.L., Pate, E., et al., 2001. Use of soil enzyme activities to monitor soil quality in natural and improved fallows in semi-arid tropical regions. Applied Soil Ecology, 18(3): 229–238. Available from: https: doi.org/10.1016/S0929-1393(01)00159-7.

Badihi-Mossberg, M., Buchner, V., Rishpon, J., 2007. Electrochemical biosensors for pollutants in the environment. Electroanalysis, 19: 2015–2028. Available from: https: doi.org/10.1002/elan.200703946.

Baldrian, P., Vetrovsky, T., 2012. Scaling down the analysis of environmental processes: Monitoring enzyme activity in natural substrates on a millimeter resolution scale. Applied and Environmental Microbiology, 78(9): 3473–3475. Available from: https: doi.org/10.1128/AEM.07953-11.

Bandick, A.K., Dick, R.P., 1999. Field management effects on soil enzyme activities. Soil Biology and Biochemistry, 31(11): 1471–1479. Available from: https: doi.org/10.1016/S0038-0717(99)00051-6.

Bardgett, R.D., Speir, T.W., Ross, D.J., et al., 1994. Impact of pasture contamination by copper, chromium, and arsenic timber preservative on soil microbial properties and nematodes. Biology and Fertility of Soils, 18(1): 71–79.

Beck, T., 1984. Methods and application of soil microbiological analysis at the Landensanstalt fur Bodenkultur und Pflanzenbau (LBB) in munich for the determination of some aspects; of soil fertility. In: 5th Symp. On Soil Biology (Bucharest, Romania. February 1981). Bucharest: Romanian National Soc Of Soil Sei. 13–20.

Bello, D., Trasar-Cepeda, C., Leirós, M.C., et al., 2008. Evaluation of various tests for the diagnosis of soil contamination by 2, 4, 5-trichlorophenol (2, 4, 5-TCP). Environmental Pollution, 156(3): 611–617. Available from: https: doi.org/10.1016/j.envpol.2008.06.024.

Berezhetskyy, A.L., Sosovska, O.F., Durrieu, C., et al., 2008. Alkaline phosphatase conductometric biosensor for heavy-metal ions determination. IRBM, 29(2–3): 136–140. Available from: https: doi.org/10.1016/j.rbmret.2007.12.007.

Beulke, S., Malkomes, H.P., 2001. Effects of the herbicides metazachlor and dinoterb on the soil microflora and the degradation and sorption of metazachlor under different environmental conditions. Biology and Fertility of Soils, 33(6): 467–471. Available from: https: doi.org/10.1007/s003740100354.

Bontidean, I., Berggren, C., Johansson, G., et al., 1998. Detection of heavy metal ions at femtomolar levels using protein-based biosensors. Analytical Chemistry, 70(19): 4162–4169. Available from: https: doi.org/10.1021/ac9803636. PMID: 9784752.

Bontidean, I., Lloyd, J., Hobman, J., et al., 2000. Study of bacterial metal resistance protein-based sensitive biosensors for heavy metal monitoring. ACS Symposium Series,762: 102–112.

Burns, R.G., Dick, R.P., 2001. Enzymes in the environment: Ecology, activity and applications[M]. New York: Mar- cel Dekker, Inc.1: 7–22

Cai, X.Y., Sheng, G.Y., Liu, W.P., 2007. Degradation and detoxification of acetochlor in soils treated by organic and thiosulfate amendments. Chemosphere, 66(2): 286–292. Available from: https: doi.org/10.1016/j.chemosphere.2006.05.011.

Caldwell, B.A., 2005. Enzyme activities as a component of soil biodiversity: A review. Pedobiologia International Journal of Soil Biology, 49(6): 637–644. Available from: https: doi.org/10.1016/j. pedobi.2005.06.003.

Cervelli, S., Nannipieri, P., Sequi, P., 1978. Interactions between agrochemicals and soil enzymes. In: Burns R G. Soil Enzymes. New York: Academic Press: 251–293.

Chen, Q.P., Ai, S.Y., Zhu, X.B., et al., 2009. A nitrite biosensor based on the immobilization of Cytochrome c on multi-walled carbon nanotubes–PAMAM–chitosan nanocomposite modified glass carbon electrode. Biosensors and Bioelectronics, 24(10): 2991–2996. Available from: https: doi.org/10.1016/j.bios.2009.03.007. Epub 2009 Mar 17. PMID: 19345570.

Choi, N.S., Kim, B.H., Park, C.S., et al., 2009. Multiple-layer substrate zymography for detection of several enzymes in a single sodium dodecyl sulfate gel. Analytical Biochemistry, 386(1): 121–122. Available from: https: doi.org/10.1016/j.ab.2008.11.007.

Clark, L.C., Lyons, C., 1962. Electrode systems for continuous monitoring in cardiovascular surgery. Annals of the New York Academy of Sciences, 102: 29–45. Available from: https: doi.org/10.1111/j.1749–6632.1962.tb13623.x. PMID: 14021529.

Cosnier, S., Da Silva, S., Shan, D., et al., 2008. Electrochemical nitrate biosensor based on poly (pyrrole–viologen) film–nitrate reductase–clay composite. Bioelectrochemistry, 74(1): 47–51.

Available from: https: doi.org/10.1016/j.bioelechem.2008.04.011.

Crecchio, C., Curci, M., Mininni, R., et al., 2001. Short-term effects of municipal solid waste compost amendments on soil carbon and nitrogen content, some enzyme activities and genetic diversity. Biology and Fertility of Soils, 34(5): 311–318. Available from: https: doi.org/10.1007/s003740100413.

Davison, W., Zhang, H., 1994. *In situ* speciation measurements of trace components in natural waters using thin-film gels. Nature, 367: 546–548. Available from: https: doi.org/10.1038/367546a0.

Defo, M., Njiné, T., Nola, M., et al., 2011. Microcosm study of the long term effect of endosulfan on enzyme and microbial activities on two agricultural soils of Yaounde-Cameroon. African Journal of Agricultural Research, 6: 2039–2050.

Deng, S.P., Tabatabai, M.A., 1995. Cellulase activity of soils: Effect of trace elements. Soil Biology and Biochemistry, 27(7): 977–979. Available from: https: doi.org/10.1016/0038-0717(95)00005-Y.

Dick, R.P., 1994. Soil enzyme activities as Indicators of soil quality. In: Doran J W, Coleman D C , Bezdicek D F. Defining soil quality for a sustainable environment. Madison: Soil Science Society of America Special Publication: 107–124.

Dick, R.P., 1997. Soil enzyme activities as integrative indicators of soil health. In: Pnkrst C, Doube BM, Gupta VVSR. Biological Indicators of Soil Health. Wallingford, Oxon, UK: CAB Internation: 121–156.

Dick, R.P., Rasmussen, P.E., Kerle, E.A., 1988. Influence of long-term residue management on soil enzyme activities in relation to soil chemical properties of a wheat-fallow system. Biology and Fertility of Soils, 6(2): 159–164.

Dick, R.P., Sandor, J.A., Eashc, N.S., 1994. Soil enzyme activities after 1500 years of terrace agriculture in the Coka Valley, Perú. Agriculture Ecosystems and Environment, 50(2): 123–131.

Dick, R.P., Breakwill, D., Turco, R., 1996. Soil enzyme activities and biodiversity measurements as integrating biological indicators. In: Doran J W, Jones A J. Handbook of methods for assessment of soil quality. Madison: SSSA Special Publication 49 Soil Science Society of America Special Publication: 247–272.

Dick, W.A., Tabatabai, M.A., 1992. Significance and potential uses of soil enzymes. In: Meeting F B. Soil Microbial Ecology. New York, Basel, Hong K: Marcel Decker: 95–127.

Du, D., Chen, W.J., Zhang, W.Y., et al., 2010. Covalent coupling of organophosphorus hydrolase loaded quantum dots to carbon nanotube/Au nanocomposite for enhanced detection of methyl parathion. Biosensors and Bioelectronics, 25(6): 1370–1375. Available from: https: doi.org/10.1016/j.bios.2009.10.032.

Duan, C.J., Fang, L.C., Yang, C.L., et al., 2018. Reveal the response of enzyme activities to heavy metals through in situ zymography. Ecotoxicology and Environmental Safety, 156: 106–115. Available from: https: doi.org/10.1016/j.ecoenv.2018.03.015.

Dutta, M., Sardar, D., Pal, R., et al., 2010. Effect of chlorpyrifos on microbial biomass and activities in tropical clay loam soil. Environmental Monitoring and Assessment, 160(1–4): 385–391. Available from: https: doi.org/10.1007/s10661-008-0702-y.

Dzantor, E.K., Felsot, A., 1991. Microbial responses to large concentrations of herbicides in soil. Environmental Toxicology and Chemistry, 10(5): 649–655.

Dzyadevych, S.V., Soldatkin, A.P., Arkhypova, V.N., et al., 2005. Early-warning electrochemical biosensor system for environmental monitoring based on enzyme inhibition. Sensors and Actuators B: Chemical, 105(1): 81–87. Available from: https: doi.org/10.1016/j.snb.2004.02.039.

Eivazi, F., Tabatabai, M.A., 1990. Factors affecting glucosidase and galactosidase activities in soils. Soil Biology and Biochemistry, 22(7): 891–897.

Emsley, J., 1992. Enzyme takes toxins out of the soil. New Scientist, 135: 15.

Felipe, B., Ellen, K., Teresa, H., et al., 2008. Long-term effect of microbial abundance and humus-associated enzyme activities under semiarid conditions. MedSci Entry for Microbial Ecology, 55: 651–661. Available from: https: doi.org/10.1007/s00248-007-9308-0.

Filip, Z., Kanazawa, S., Berthelin, J., 1999. Characterization of effects of a long-term wastewater irrigation on soil quality by microbiological and biochemical parameters. Journal of Plant Nutrition and Soil Science, 162(4): 409–413. Available from: https: doi.org/10.1002/(SICI)1522-2624(199908)162:4< 409::AID-JPLN409>3.0.CO;2-#.

Fragoeiro, S., Magan, N., 2008. Impact of *Trametes versicolor* and *Phanerochaete chrysosporium* on differential breakdown of pesticide mixtures in soil microcosms at two water potentials and associated respiration and enzyme activity. International Biodeterioration & Biodegradation, 62(4): 376–383. Available from: https: doi.org/10.1016/j.ibiod.2008.03.003.

Frankenberger, W.T. , Tabatabai, M.A., 1981. Amidase activity in soils: IV. Effects of trace elements and pesticides 1. Soil Science Society of America Journal, 45(6): 1120–1124.

Frankenberger, W.T., Tabatabai, M.A., 1991. Factors affecting L-asparaginase activity in soils. Biology and Fertility of Soils, 11(1): 1–5.

Frankenberger, W.T., Tabatabai, M.A., 1991. L-Asparaginase activity of soils. Biology and Fertility of Soils, 11(1): 6–12.

Füsun, G., Esin, E., 2008. The effects of heavy metal pollution on enzyme activities and basal soil respiration of roadside soils. Environmental Monitoring and Assessment, 145(1–3):127–133. Available from: https: doi.org/10.1007/s10661-007-0022-7.

Gao, Y,, Zhou, P., Mao, L., et al., 2010. Assessment of effects of heavy metals combined pollution on soil enzyme activities and microbial community structure: Modified ecological dose–response model and PCR-RAPD. Environmental Earth Sciences, 60(3): 603–612. Available from: https: doi.org/10.1007/s12665-009-0200-8.

García-Gil, J.C., Plaza, C., Soler-Rovira, P., et al., 2000. Long-term effects of municipal solid waste compost application on soil enzyme activities and microbial biomass. Soil Biology and Biochemistry, 32(13): 1907–1913. Available from: https: doi.org/10.1016/S0038-0717(00)00165-6.

Ge, T.D., Wei, X.M., Razavi, B.S., et al., 2017. Stability and dynamics of enzyme activity patterns in the rice rhizosphere: Effects of plant growth and temperature. Soil Biology and Biochemistry, 113: 108–115. Available from: https: doi.org/10.1016/j.soilbio.2017.06.005.

Ghica, M.E., Brett, C.M.A., 2008. Glucose oxidase inhibition in poly(neutral red) mediated enzyme biosensors for heavy metal determination. Microchimica Acta, 163(3): 185–193. Available from: https: doi.org/10.1007/s00604-008-0018-1.

Gianfreda, L., Antonietta Rao, M., Piotrowska, A., et al., 2005. Soil enzyme activities as affected by anthropogenic alterations: Intensive agricultural practices and organic pollution. Science of the Total Environment, 341(1–3): 265–279. Available from: https: doi.org/10.1016/j.scitotenv.2004.10.005.

Gianfreda, L., Rao, M.A., 2008. Interactions between xenobiotics and microbial and enzymatic soil activity. Critical Reviews in Environmental Science and Technology, 38(4): 269–310. Available from: https: doi.org/10.1080/10643380701413526.

Giles, C.D., Dupuy, L., Boitt, G., et al., 2018. Root development impacts on the distribution of phosphatase activity: Improvements in quantification using soil zymography. Soil Biology and Biochemistry, 116: 158–166. Available from: https: doi.org/10.1016/j.soilbio.2017.08.011.

Graham, M.H., Haynes, R.J., 2005. Organic matter accumulation and fertilizer-induced acidification interact to affect soil microbial and enzyme activity on a long-term sugarcane management experiment. Biology and

Fertility of Soils, 41(4): 249–256. Available from: https: doi.org/10.1007/s00374-005-0830-2.

Guber, A., Blagodatskaya, E., Juyal, A., et al., 2021. Time-lapse approach to correct deficiencies of 2D soil zymography. Soil Biology and Biochemistry, 157: 108225. Available from: https: doi.org/10.1016/j.soilbio.2021.108225.

Guhr, A., Borken, W., Spohn, M., et al., 2015. Redistribution of soil water by a saprotrophic fungus enhances carbon mineralization. Proceedings of the National Academy of Sciences of the United States of America, 112(47): 14647–14651. Available from: https: doi.org/10.1073/pnas.1514435112.

Gundi, V.A.K.B., Narasimha, G., Reddy, B.R., 2005. Interaction effects of insecticides on microbial populations and dehydrogenase activity in a black clay soil. Journal of Environmental Science and Health Part B, Pesticides, Food Contaminants, and Agricultural Wastes, 40(2): 269–283. Available from: https: doi.org/10.1081/PFC-200045550.

Guo, D.J., Lakshmikantham, V., 1988. Nonlinear Problems in Abstract Cones. Boston: Academic Press.

Heitkötter, J., Marschner, B., 2018. Soil zymography as a powerful tool for exploring hotspots and substrate limitation in undisturbed subsoil. Soil Biology and Biochemistry, 124: 210–217. Available from: https: doi.org/10.1016/j.soilbio.2018.06.021.

Hoang, D., Razavi, B., Kuzyakov, Y., et al., 2016. Earthworm burrows: Kinetics and spatial distribution of enzymes of C-, N- and P- cycles. Soil Biology \& Biochemistry, 99: 94–103. Available from: https: doi.org/10.1016/j.soilbio.2016.04.021.

Hoang, D.T.T., Pausch, J., Razavi, B.S., et al., 2016. Hotspots of microbial activity induced by earthworm burrows, old root channels, and their combination in subsoil. Biology and Fertility of Soils, 52(8): 1105–1119. Available from: https: doi.org/10.1007/s00374-016-1148-y.

Hughes, A.J., Herr, A.E., 2010. Quantitative enzyme activity determination with zeptomole sensitivity by microfluidic gradient-gel zymography. Analytical Chemistry, 82(9): 3803–3811. Available from: https: doi.org/10.1021/ac100201z.

Jastrzebska, E., 2011. The effect of chlorpyrifos and teflubenzuron on the enzymatic activity of soil. Polish Journal of Environmental Studies, 20: 903–910.

Jiang, P., Liu, J., You, S., et al., 2021. Application Prospect and Future Perspective of In-situ Zymography of Soil Enzymes. Chinese Journal of Soil Science, 52, 454–461. (In Chinese)

Kaare, J., Carsten, J., Vigdis, T., et al., 2001. Pesticide effects on bacterial diversity in agricultural soils: a review. Biology and Fertility of Soils, 33(6): 443–453. Available from: https: doi.org/10.1007/s003740100351.

Kalam, A., Tah, J., Mukherjee, A.K., 2004. Pesticide effects on microbial population and soil enzyme activities during vermicomposting of agricultural waste. Journal of Environmental Biology, 25(2): 201–208.

Kalyani, S.S., Sharma, J., Dureja, P., et al., 2010. Influence of endosulfan on microbial biomass and soil enzymatic activities of a tropical alfisol. Bulletin of Environmental Contamination and Toxicology, 84(3): 351–356. Available from: https: doi.org/10.1007/s00128-010-9943-x.

Kamyabi, M.A., Aghajanloo, F., 2008. Electrocatalytic oxidation and determination of nitrite on carbon paste electrode modified with oxovanadium(IV)-4-methyl salophen. Journal of Electroanalytical Chemistry, 614(1):157–165.

Kandeler, E., Luxhøi, J., Tscherko, D., et al., 1999. Xylanase, invertase and protease at the soil–litter interface of a loamy sand. Soil Biology and Biochemistry, 31(8): 1171–1179. Available from: https: doi.org/10.1016/S0038-0717(99)00035-8.

Kannan, K., Oblisami, G., 1990. Influence of paper mill effluent irrigation on soil enzyme activities. Soil

Biology and Biochemistry, 22(7): 923–926.

Khosrozadeh, S., Guber, A., Kravchenko, A., et al., 2022. Soil oxidoreductase zymography: Visualizing spatial distributions of peroxidase and phenol oxidase activities at the root-soil interface. Soil Biology and Biochemistry, 167: 108610. Available from: https: doi.org/10.1016/j.soilbio.2022.108610.

Kim, G.Y., Shim, J., Kang, M.S., et al., 2008. Preparation of a highly sensitive enzyme electrode using gold nanoparticles for measurement of pesticides at the ppt level. Journal of Environmental Monitoring, 10(5): 632–637. Available from: https: doi.org/10.1039/b800553b. Epub 2008 Apr 4. PMID: 18449400.

Kizilkaya, R., Askin, T., Bayrakli, B., et al., 2004. Microbial characteristics of soils contaminated with heavy metals. European Journal of Soil Biology, 40: 95–102. Available from: https: doi.org/10.1016/j.ejsobi.2004.10.002.

Kleiner, D.E, Stetler-Stevenson, W.G., 1994. Quantitative zymography: Detection of picogram quantities of gelatinases. Analytical Biochemistry, 218(2): 325–329. Available from: https: doi.org/10.1006/abio.1994.1186.

Klibanov, A.M., Tu, T.M., Scott, K.P., 1983. Peroxidase-catalyzed removal of phenols from coal-conversion waste waters. Science, 221(4607): 259–261. Available from: https: doi.org/10.1126/science.221.4607.259-a.

Koch, O., Tscherko, D., Kandele,r E., 2007. Temperature sensitivity of microbial respiration, nitrogen mineralization, and potential soil enzyme activities in organic alpine soils. Global Biogeochemical Cycles, 21(4): B4017-1-B4017-11. Available from: https: doi.org/10.1029/2007GB002983.

Korkut, S., Keskinler, B., Erhan, E., 2008. An amperometric biosensor based on multiwalled carbon nanotube-poly(pyrrole)-horseradish peroxidase nanobiocomposite film for determination of phenol derivatives. Talanta, 76(5): 1147–1152. Available from: https: doi.org/10.1016/j.talanta.2008.05.016.

Kurzbaum, E., Kirzhner, F., Armon, R., 2010. A simple method for dehydrogenase activity visualization of intact plant roots grown in soilless culture using tetrazolium violet. Plant Root, 4: 12–16.

Kuzyakov, Y., Xu, X.L., 2013. Competition between roots and microorganisms for nitrogen: Mechanisms and ecological relevance. The New Phytologist, 198(3): 656–669. Available from: https: doi.org/10.1111/nph.12235.

Kuzyakov, Y., Razavi, B.S., 2019. Rhizosphere size and shape: Temporal dynamics and spatial stationarity. Soil Biology and Biochemistry, 135: 343–360. Available from: https: doi.org/10.1016/j.soilbio.2019.05.011.

Li, C., Ding, S.M., Yang, L.Y., et al., 2019. Planar optode: A two-dimensional imaging technique for studying spatial-temporal dynamics of solutes in sediment and soil. Earth-Science Reviews, 197: 102916. Available from: https: doi.org/10.1016/j.earscirev.2019.102916.

Liu, S.B., Razavi, B.S., Su, X., et al., 2017. Spatio-temporal patterns of enzyme activities after manure application reflect mechanisms of niche differentiation between plants and microorganisms. Soil Biology and Biochemistry, 112: 100–109. Available from: https: doi.org/10.1016/j.soilbio.2017.05.006.

Ma, X.M., Liu, Y., Shen, W.J., et al., 2021. Phosphatase activity and acidification in lupine and maize rhizosphere depend on phosphorus availability and root properties: Coupling zymography with planar optodes. Applied Soil Ecology, 167: 104029. Available from: https: doi.org/10.1016/j.apsoil.2021.104029.

Ma, X.M., Razavi, B., Holz, M., et al., 2017. Warming increases hotspot areas of enzyme activity and shortens the duration of hot moments in the root-detritusphere. Soil Biology \& Biochemistry, 107: 226–233. Available from: https: doi.org/10.1016/j.soilbio.2017.01.009.

Ma, X.M., Zarebanadkouki, M., Kuzyakov, Y., et al., 2018. Spatial patterns of enzyme activities in the rhizosphere: Effects of root hairs and root radius. Soil Biology and Biochemistry, 118: 69–78.

Available from: https: doi.org/10.1016/j.soilbio.2017.12.009.

Ma, X.M., Mason-Jones, K., Liu, Y., et al., 2019. Coupling zymography with pH mapping reveals a shift in lupine phosphorus acquisition strategy driven by cluster roots. Soil Biology and Biochemistry, 135: 420–428. Available from: https: doi.org/10.1016/j.soilbio.2019.06.001.

Madhuri, R.J., Rangaswamy, V., 2002. Influence of selected insecticides on phopshatase activity in groundnut (Arachis hypogeae L.) soil. Journal of Environmental Biology, 23: 393–397.

Maloney, S.W., Manem, J., Mallevialle, J., et al., 1986. Transformation of trace organic compounds in drinking water by enzymic oxidative coupling. Environmental Science & Technology, 20(3): 249–253. Available from: https: doi.org/10.1021/es00145a004.

Marinari C., Moscatelli, M.C, et al., 2013. Soil development and microbial functional diversity: Proposal for a;methodological approach. Geoderma, 192(1): 437–445. Available from: https: doi.org/10.1016/j.geoderma.2012.08.023

Marzadori, C., Ciavatta, C., Montecchio, D., et al., 1996. Effects of lead pollution on different soil enzyme activities. Biology and Fertility of Soils, 22(1): 53–58.

Mclaren, A.D.,1975. Soil as a system of humus and clay immobilized enzymes. Chemica Scripta, 8: 97–99.

Mikanova, O., 2006. Effects of heavy metals on some soil biological parameters. Journal of Geochemical Exploration, 88(1–3): 220–223. Available from: https: doi.org/10.1016/j.gexplo.2005.08.043.

Mita, D.G., Attanasio, A., Arduini, F., et al., 2007. Enzymatic determination of BPA by means of tyrosinase immobilized on different carbon carriers. Biosensors and Bioelectronics, 23(1): 60–65. Available from: https: doi.org/10.1016/j.bios.2007.03.010.

Nannipieri, P., Bollag, J., 1991. Use of enzymes to detoxify pesticide-contaminated soils and waters. Journal of Environmental Quality, 20: 510–517.

Nannipieri, P., Ceccanti, B., Grego, S., 1990. Ecological significance of biological activity in soil. In: Bollag JM, Stotzky G. Soil Biochemistry. New York, Basel: Marcel Dekker: 293–355.

Nayak, D.R., Babu, Y.J., Adhya, T.K., 2007. Long-term application of compost influences microbial biomass and enzyme activities in a tropical Aeric Endoaquept planted to rice under flooded condition. Soil Biology and Biochemistry, 39(8): 1897–1906. Available from: https: doi.org/10.1016/j.soilbio.2007.02.003.

Pan, D., Hill, A.P., Kashou, A., et al., 2011. Electrophoretic transfer protein zymography. Analytical Biochemistry, 411(2): 277–283. Available from: https: doi.org/10.1016/j.ab.2011.01.015.

Pandey, S., Singh, D.K., 2006. Soil dehydrogenase, phosphomonoesterase and arginine deaminase activities in an insecticide treated groundnut (*Arachis hypogaea* L.) field. Chemosphere, 63(5): 869–880. Available from: https: doi.org/10.1016/j.chemosphere.2005.07.053.

Parkhust, C.E., 1997. Biological Indicators of Soil Health. United Kingdom: Oxon.

Perucci, P., 1992. Enzyme activity and microbial biomass in a field soil amended with municipal refuse. Biology and Fertility of Soils, 14(1): 54–60.

Perucci, P., Dumontet, S., Bufo, S.A., et al., 2000. Effects of organic amendment and herbicide treatment on soil microbial biomass. Biology and Fertility of Soils, 32(1): 17–23. Available from: https: doi.org/10.1007/s003740000207.

Petersen, D.G, Blazewicz, S.J., Firestone, M., et al., 2012. Abundance of microbial genes associated with nitrogen cycling as indices of biogeochemical process rates across a vegetation gradient in Alaska. Environmental Microbiology, 14(4): 993–1008. Available from: https: doi.org/10.1111/j.1462-2920.2011.02679.x.

Prietzel, J., 2001. Arylsulfatase activities in soils of the Black Forest/Germany—Seasonal variation and effect of $(NH_4)_2SO_4$ fertilization. Soil Biology and Biochemistry, 33(10): 1317–1328. Available from:

https: doi.org/10.1016/S0038-0717(01)00037-2.

Qian, H.F., Hu, B.L., Wang, Z.Y., et al., 2007. Effects of validamycin on some enzymatic activities in soil. Environmental Monitoring and Assessment, 125(1): 1–8. Available from: https: doi.org/10.1007/s10661-006-9211-z.

Qu, Y.H., Sun, Q., Xiao, F., et al., 2010. Layer-by-Layer self-assembled acetylcholinesterase PAMAM-Au on CNTs modified electrode for sensing pesticides. Bioelectrochemistry, 77(2): 139–144. Available from: https: doi.org/10.1016/j.bioelechem.

Rain, M.S., Lakshmi, K.V., Devi, P.S., 2008. Impact of chlorpyrifos on soil enzyme activities in agricultural soil. Asian Journal of Microbiology, Biotechnology and Environmental Science, 10(2): 295–300.

Razavi, B.S., Zarebanadkouki, M., Blagodatskaya, E., et al., 2016. Rhizosphere shape of lentil and maize: Spatial distribution of enzyme activities. Soil Biology and Biochemistry, 96: 229–237. Available from: https: doi.org/10.1016/j.soilbio.2016.02.020.

Razavi, B.S., Hoang, D.T.T., Blagodatskaya, E., et al., 2017. Mapping the footprint of nematodes in the rhizosphere: Cluster root formation and spatial distribution of enzyme activities. Soil Biology and Biochemistry, 115: 213–220. Available from: https: doi.org/10.1016/j.soilbio.2017.08.027.

Razavi, B.S., Zhang, X.C., Bilyera, N., et al., 2019. Soil zymography: Simple and reliable? Review of current knowledge and optimization of the method. Rhizosphere, 11. Available from: https: doi.org/DOI:10.1016/j.rhisph.2019.100161.

Russell, R.J., Pishko, M.V., Simonian, A.L., et al., 1999. Poly(ethylene glycol) hydrogel-encapsulated fluorophore-enzyme conjugates for direct detection of organophosphorus neurotoxins. Analytical Chemistry, 71(21): 4909–4912.

Russo, V.M., Biles, C.L., 2003. Fertilizer rate and β-galactosidase and peroxidase activity in pepper fruit at different stages and years of harvest. Plant Foods for Human Nutrition, 58(3): 231–239. Available from: https: doi.org/10.1023/B:QUAL.0000040312.26474.29.

Salimi, A., Noorbakhsh, A., Ghadermarzi, M., 2007. Amperometric detection of nitrite, iodate and periodate at glassy carbon electrode modified with catalase and multi-wall carbon nanotubes. Sensors and Actuators B: Chemical, 123(1): 530–537. Available from: https: doi.org/10.1016/j.snb.2006.09.054.

Sanaullah, M., Razavi, B.S., Blagodatskaya, E., et al., 2016. Spatial distribution and catalytic mechanisms of β-glucosidase activity at the root-soil interface. Biology and Fertility of Soils, 52(4): 505–514. Available from: https: doi.org/10.1007/s00374-016-1094-8.

Sannino, F., Gianfreda, L., 2001. Pesticide in fluence on soil enzymatic activities. Chemosphere, 45: 417–425. Available from: https: doi.org/10.1016/S0045-6535(01)00045-5.

Saviozzi, A., Biasci, A., Riffaldi, R., et al., 1999. Long-term effects of farmyard manure and sewage sludge on some soil biochemical characteristics. Biology and Fertility of Soils, 30(1): 100–106. Available from: https: doi.org/10.1007/s003740050594.

Serra-Wittling, C., Houot, S., Barriuso, E., 1995. Soil enzymatic response to addition of municipal solid-waste compost. Biology and Fertility of Soils, 20(4): 226–236. Available from: https: doi.org/10.1007/BF00336082.

Silveira, C.M., Gomes, S.P., Araújo, A.N., et al., 2010. An efficient non-mediated amperometric biosensor for nitrite determination. Biosensors and Bioelectronics, 25(9): 2026–2032. Available from: https: doi.org/10.1016/j.bios.2010.01.031.

Singh, J., Singh, D.K., 2005. Dehydrogenase and phosphomonoesterase activities in groundnut (*Arachis hypogaea* L.) field after diazinon, imidacloprid and lindane treatments. Chemosphere, 60(1): 32–42.

Available from: https: doi.org/10.1016/j.chemosphere.2004.11.096.

Sohail, M., Adeloju, S.B., 2008. Electroimmobilization of nitrate reductase and nicotinamide adenine dinucleotide into polypyrrole films for potentiometric detection of nitrate. Sensors and Actuators B: Chemical, 133(1): 333–339. Available from: https: doi.org/10.1016/j.snb.2008.02.032.

Sohail, M., Adeloju, S.B., 2009. Fabrication of Redox-Mediator Supported Potentiometric Nitrate Biosensor with Nitrate Reductase. Electroanalysis, 21, 1411–1418. Available from: https: doi.org/ 10.1002/elan.200804542.

Sohal, H.S., Srivastava, A.K., 1994. Environment and biotechnology. New Delhi: Ashish pubishing house.

Solé, S., Merkoçi, A., Alegret, S., 2003. Determination of toxic substances based on enzyme inhibition. part I. electrochemical biosensors for the determination of pesticides using batch procedures. Critical Reviews in Analytical Chemistry, 33(2): 89–126. Available from: https: doi.org/10.1080/727072334.

Speir, T.W., Ross, D.J., Feltham, C.W., et al., 1992. Assessment of the feasibility of using CCA (copper, chromium and arsenic)-treated and boric acid-treated sawdust as soil amendments. Plant and Soil, 142(2): 249–258.

Spohn, M., Carminati, A., Kuzyakov, Y., 2013. Soil zymography–A novel *in situ* method for mapping distribution of enzyme activity in soil. Soil Biology and Biochemistry, 58: 275–280. Available from: https: doi.org/10.1016/j.soilbio.2012.12.004.

Spohn, M., Kuzyakov, Y., 2013. Distribution of microbial- and root-derived phosphatase activities in the rhizosphere depending on P availability and C allocation–Coupling soil zymography with 14C imaging. Soil Biology and Biochemistry, 67: 106–113. Available from: https: doi.org/ DOI: 10.1016/j.soilbio.2013.08.015.

Spohn, M., Kuzyakov, Y., 2014. Spatial and temporal dynamics of hotspots of enzyme activity in soil as affected by living and dead roots: a soil zymography analysis. Plant and Soil, 379(1): 67–77. Available from: https: doi.org/10.1007/s11104-014-2041-9.

Stefanic, G., Eliade, G., Chirnogeanu, I., 1984. Researches concerning a biological index of soil fertility. In: 5th Symp. on Soil Biology (Bucharest, Romania February). Bucharest: 1981. Romanian National Society of Soil Science, Bucharest. 35–45.

Tan, Y.M., Deng, W.F., Ge, B., et al., 2009. Biofuel cell and phenolic biosensor based on acid-resistant laccase–glutaraldehyde functionalized chitosan–multiwalled carbon nanotubes nanocomposite film. Biosensors and Bioelectronics, 24(7): 2225–2231. Available from: https: doi.org/10.1016/j.bios.2008.11.026.

Taylor, J., Wilson, B., Mills, M., et al., 2002. Comparison of microbial numbers and enzymatic activities in surface soils and subsoils using various techniques. Soil Biology and Biochemistry, 34(3): 387–401. Available from: https: doi.org/10.1016/S0038-0717(01)00199-7.

Tsai, H.C., Doong, R.A., 2007. Preparation and characterization of urease-encapsulated biosensors in poly(vinyl alcohol)-modified silica Sol-gel materials. Biosensors & Bioelectronics, 23(1): 66–73. Available from: https: doi.org/10.1016/j.bios.2007.03.017.

Updike, S.J., Hicks, G.P., 1967. The enzyme electrode. Nature, 214(5092):986-8. Available from: https: doi.org/ 10.1038/214986a0.

Vandooren, J., Geurts, N., Martens, E., et al., 2013. Zymography methods for visualizing hydrolytic enzymes. Nature Methods, 10(3): 211–220. Available from: https: doi.org/10.1038/nmeth.2371.

Verstraete, W., Voets, J.P., 1977. Soil microbial and biochemical characteristics in relation to soil management and fertility. Soil Biology and Biochemistry, 9(4): 253–258.

Visser, S., Parkinson, D., 1992. Soil biological criteria as indicators of soil quality: Soil microorganisms.

American Journal of Alternative Agriculture, 7(1–2): 33–37.

Voets, J.P., Meerschman, P., Verstraete, W., 1974. Soil microbiological and biochemical effects of long-term atrazine applications. Soil Biology and Biochemistry, 6(3): 149–152.

Voothuluru, P., Braun, D.M., Boyer, J.S., 2018. An *in vivo* imaging assay detects spatial variability in glucose release from plant roots. Plant Physiology, 178(3): 1002–1010. Available from: https: doi.org/10.1104/pp.18.00614.

Wanekaya, A.K., Chen, W., Mulchandani, A., 2008. Recent biosensing developments in environmental security. Journal of Environmental Monitoring: JEM, 10(6): 703–712. Available from: https: doi.org/10.1039/b806830p.

Wang, Y., Hasebe, Y., 2009. Carbon felt-based biocatalytic enzymatic flow-through detectors: Chemical modification of tyrosinase onto amino-functionalized carbon felt using various coupling reagents. Talanta, 79(4): 1135–1141. Available from: https: doi.org/10.1016/j.talanta.2009.02.028.

Wei, X.M., Ge, T.D., Zhu, Z.K., et al., 2019a. Expansion of rice enzymatic rhizosphere: Temporal dynamics in response to phosphorus and cellulose application. Plant and Soil, 445(1): 169–181. Available from: https: doi.org/10.1007/s11104-018-03902-0.

Wei, X.M., Razavi, B.S., Hu, Y.J., et al., 2019b. C/P stoichiometry of dying rice root defines the spatial distribution and dynamics of enzyme activities in root-detritusphere. Biology and Fertility of Soils, 55(3): 251–263. Available from: https: doi.org/10.1007/s00374-019-01345-y.

Wilkesman, J., Kurz, L., 2009. Protease analysis by zymography: A review on techniques and patents. Recent Patents on Biotechnology, 3(3): 175–184.

Wood, T.G., 1991. Field investigations on the decomposition of leaves of *Eucalyptus deiegatensis* in relation to environmental factors. Pedobiologia, 14: 343–371

Yakovchenko, V., Sikora, L.J., Kaufman, D.D., 1996. A biologically based indicator of soil quality. Biology and Fertility of Soils, 21(4): 245–251. Available from: https: doi.org/10.1007/BF00334899.

Yao, X.H., Min, H., Lü, Z.H., et al., 2006. Influence of acetamiprid on soil enzymatic activities and respiration. European Journal of Soil Biology, 42(2): 120–126. Available from: https: doi.org/10.1016/j.ejsobi.2005.12.001.

Yaropolov, A.I., Kharybin, A.N., Emnéus, J., et al., 1995. Flow-injection analysis of phenols at a graphite electrode modified with co-immobilised laccase and tyrosinase. Analytica Chimica Acta, 308(1–3): 137–144. Available from: https: doi.org/10.1016/0003-2670(94)00404-A.

Yu, Y.L., Shan, M., Fang, H., et al., 2006. Responses of soil microorganisms and enzymes to repeated applications of chlorothalonil. Journal of Agricultural and Food Chemistry, 54(26): 10070–10075. Available from: https: doi.org/10.1021/jf0624190.

Zazoua, A., Hnaien, M., Cosnier, S., et al., 2009. A new HRP/catalase biosensor based on microconductometric transduction for nitrite determination. Materials Science and Engineering: C, 29(6): 1919–1922. Available from: https: doi.org/10.1016/j.msec.2009.03.008.

Zhang, Y., Li, C., Sun, Q., et al., 2021. Phosphorus acquisition strategy of *Vallisneria natans* in sediment based on *in situ* imaging techniques. Environmental Research, 202: 111635. Available from: https: doi.org/10.1016/j.envres.2021.111635.

Zhao, W., Ge, P.Y., Xu, J.J., et al., 2009. Selective detection of hypertoxic organophosphates pesticides via PDMS composite based acetylcholinesterase-inhibition biosensor. Environmental Science and Technology, 43(17): 6724–6729. Available from: https: doi.org/10.1021/es900841n.

CHAPTER 2

Global climate change and enzyme activities

2.1 Global climate change and its mutual effects with the environment

2.1.1 Overview of global climate change

Global warming triggered by human activities has become an indisputable fact. Global climate change marks one of the most complex challenges facing mankind in the 21st century. In the past two decades, climate change has topped the world's ten most urgent environmental problems, leaving palpable impacts on our everyday life—increase of extreme climate events, destruction of ecological environment, and variation of animal and plant behavior, among others. Climate change will exert profound and enormous impacts on the global ecosystem and socioeconomic system, most of which are negative or unfavorable. Climate change is becoming an increasingly important issue for governments under the unremitting call of the scientists and international organizations. To avoid the tendency towards radical environmentalism that independent panels might bring, the World Meteorological Organization and the United Nations Environment Programme established the Intergovernmental Panel on Climate Change (IPCC) in 1988, which included experts in the intergovernmental framework to conduct climate change assessments and provided authoritative information on climate change. So far, the IPCC has published six assessment reports on climate change. According to IPCC's sixth assessment report, climate change will intensify in all regions in the coming decades. The report postulates that if the global temperature rises by 1.5℃, heat waves will increase, warm seasons will be extended and cold seasons shrunk. When global temperature rises by 2℃, extreme heat will impose adverse effects on agricultural production and human health (IPCC, 2021). However, it is not just about temperature as climate change is bringing different combinations of changes to different regions, all of which intensify with further warming. The intensification of water cycle changes in rainfall characteristics and extreme sea-level events are a few examples.

Many studies have shown that the driving forces of climate change mainly include natural forcing factors and human forcing factors (Liu et al., 2017; Li et al., 2019). The former mainly includes volcanic activities and Earth orbit parameters, while the latter covers two aspects: First, due to the burning of fossil fuels and other activities since the industrial era, the concentration of greenhouse gases in the atmosphere has increased, resulting in the formation of aerosols that further affect the Earth's atmospheric radiation budget balance. The fifth phase of the Coupled Model Intercomparison Project 5 (CMIP5) of IPCC showed that the increase of aerosol in East Asia resulted in a decrease of about $1.05\,^{\circ}\!C$ in surface temperature during 1985−2005, and different aerosol components had greatly distinct impacts on climate in different regions. Second, the change of land use mode alters the characteristics of underlying surface, which leads to the change of energy, momentum and water transmission between ground and air, then the regional climate change. Studies have shown that land use change is an important driving factor for the increase of daily maximum temperature during summer. In summary, research and observational data in recent decades indicate that human activities are the principal driving forces of climate change.

2.1.2　Impacts of global climate change

2.1.2.1　Ecosystem

Climate change is the main driving force of the evolution of the terrestrial ecosystem. Having imposed many impacts on the distribution of vegetation zone, vegetation stability, forest productivity desertification, rocky desertification, permafrost degradation and biodiversity of the terrestrial ecosystem, climate change has become a major threat to maintaining the stability and adaptive ability of the terrestrial ecosystem (Figure 2.1). The terrestrial ecosystem is highly sensitive to climate changes such as temperature change and precipitation fluctuation (Nolan et al., 2018). In particular, climate warming leads to increasing evapotranspiration of terrestrial ecosystem, aggravating soil water deficit, and laying a significant negative effect on vegetation productivity in arid areas (Fu et al., 2013). Desertification, rocky desertification, salinization and permafrost degradation may further aggravate the decline of vegetation productivity and weaken the stability and sustainability of terrestrial ecosystem's services. From the perspective of species, climate change affects the structure of species community and changes the geographical distribution and migration path of species, and the rapid change of habitat leads to the disappearance of some species from their original distribution area due to habitat degradation or loss. The increase of harmful organisms and biological invasions intensifies the instability of terrestrial ecosystem. Some studies have also shown that climate changes faster than the transfer of species' spatial distribution and the change of species' community composition, resulting in the lag of species' response to climate change. In addition, carbon sequestration and oxygen release are important ways for forests to cope with climate change and reduce carbon emissions. Impacts of extreme climate events and the increase of forest pests and diseases have led to the decline of forest carbon sinks, posing complex challenges to maintaining and increasing ecosystem carbon sinks in the future. Taking soil and groundwater as examples, this book summarizes the impact of climate change on the soil and groundwater environment.

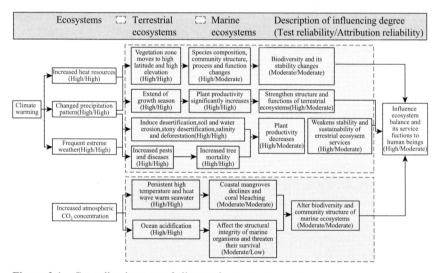

Figure 2.1 Cascading impacts of climate change on ecosystems.

2.1.2.2 Soil

Given that chemical elements in plants mostly come from soil, soil nutrients are very important for the growth and development of plants, having direct impact on their physiological, ecological and production capacity. A mass of studies have shown that under the climate change, biogeochemical reactions in soil system will undergo a series of changes, which further affect the soil-plant nutrient cycling process (Nam et al., 2013).

According to the global climate carbon cycle model, it is predicted that the increase of CO_2 concentration in atmosphere will increase the organic carbon content in soil since the increase of CO_2 concentration will improve the photosynthetic efficiency and water use efficiency of plants, thus promoting plants growth. Plants then bring more carbon to the soil through root exudates and litter. However, the rise of carbon volume may stimulate the activity of soil microorganisms, thus facilitating the decomposition of organic carbon (Van Kessel et al., 2000). Therefore, no consensus has been reached on the effect of the increasing CO_2 concentration on the organic carbon reserve in soil. The change of CO_2 concentration in soil had a certain effect on soil nitrogen. Some scholars argue that when CO_2 concentration increases, the root exudates will increase, thus providing more nutrients for microorganisms and accelerating their growth and reproduction to eventually accelerate the turnover of soil nitrogen. Meanwhile, the increase of CO_2 concentration can reduce the stomatal conductance of plants, which is conducive to maintaining soil moisture and promoting enzyme activity, thus accelerating the mineralization of organic nitrogen (Moorhead and Linkins, 1997). Some studies have also found that the increase of CO_2 concentration will reduce the content of ammonium nitrogen in soil, which maybe because under high CO_2 concentration, soil microorganisms obtain more carbon sources to improve the fixation of inorganic nitrogen, thereby reducing the content of available nitrogen in soil. Therefore, with the increase of CO_2 concentration, the mineralization and solidification of soil nitrogen turn out to be multifactor coupled processes. In addition, the increase of CO_2 concentration in atmosphere will reduce the content of available phosphorus in soil, which may be because the increase of CO_2 concentration boosts the activity of phosphatase. This in turn increases the content of available phosphorus, thus promoting the uptake of phosphorus by plants (Saiya-Cork et al., 2002).

Rising temperatures give rise to an enormous loss of soil carbon. This is because warming will increase the microbial activity and thus accelerate the decomposition of organic carbon in soil (McKane et al., 1997). However, some studies also found that in regions with an annual average temperature of $10-20\,^\circ\!C$, the rising temperature will improve the organic carbon reserve in soil. Therefore, the impact of temperature rise on soil organic carbon (SOC) may vary greatly from region to region. Temperature is the most important abiotic factor for nitrogen mineralization in soil. Studies showed that with the rise of soil temperature, nitrogen mineralization was enhanced (Robinson et al., 2004) and available nitrogen content increased, which further promoted the growth of surface vegetation biomass and the nitrogen assimilation. The vegetation increased would consume more CO_2 as it absorbs nitrogen. Therefore the accelerated nitrogen cycling in soil may also ease the impact of elevated CO_2 concentrations. However, some studies have found that with the rise in temperature, the nitrate nitrogen content in soil decreases, which maybe because the plant absorption covers up the net mineralization of increased nitrogen. In general, the temperature rise may enhance the availability of soil nitrogen.

In addition, the global climate change would cause the change of precipitation patterns which directly dictate the moisture content in soil to affect the soil aeration, thus inducing changes of the organic carbon mineralization. Many studies have shown that the organic carbon reserve in soil is on the rise with the increase of precipitation. Some studies have also found that in regions of black soil, when the moisture content in soil stays above 70% of the maximum field capacity, the decomposition of organic matters is accelerated, and the content of organic matters is rapidly declining. However, when the precipitation decreases and water content in soil is too low, the activity of soil microorganisms and enzymes would decrease, thus inhibiting the carbon mineralization in soil. Therefore the previous studies on the response of soil carbon turnover to moisture conditions have shown inconsistent results, which request further studies. In addition, when the precipitation increases significantly, the nitrogen mineralization in soil accelerates significantly, the nitrogen uptake by plants and microorganisms increases, and the available nitrogen content in soil decreases remarkably (Øygarden et al., 2014). Meanwhile, the increase of precipitation will accelerate the leaching loss of nitrate nitrogen and lower the availability of nitrogen. However, other studies have also found that there is no significant relationship between precipitation and nitrogen mineralization rate in soil, perhaps because the plant uptake of nitrogen and above-ground net primary productivity are unaffected (Jongen et al., 2013). However, with the rise of average annual precipitation and temperature, the total phosphorus content in soil showed a significant declining trend. For example, in tropical monsoon regions, when the annual precipitation rises by nearly six times, the total phosphorus content in soil drops by about 2/3 (Miller et al., 2001). In tropical and subtropical regions, the high temperature and rainy climate accelerates the soil weathering and phosphorus leaching, thereby lowering the total phosphorus content in soil. However, some studies have also found that temperature rise had no significant effect on the net mineralization rate of soil phosphorus (Rinnan et al., 2007).

The stoichiometric characteristics of soil nutrients can characterize the balance of soil nutrients. For example, the soil C:N can characterize the soil quality. The higher the value, the slower the mineralization of soil organic matters. The soil C:P represents the phosphate mineralization capacity in soil; the soil N:P characterizes the nitrogen and phosphorus availability in soil. The effects of global climate change on the stoichiometric characteristics of soil nutrients are complicated. Studies have shown that climate warming and drying may increase the soil C:P and N:P. However, some studies have also shown that C:N and C:P in soil significantly declined against the temperature rise. However, the increase of precipitation will improve the stoichiometric ratio of soil nutrients, perhaps because the rainy environment is conducive to the accumulation of carbon and nitrogen in soil. Moreover, against the global climate change, the driving factors of stoichiometry characteristics of soil nutrients are relatively complex, and plants, microorganisms and soil structure may all have an impact on them. Therefore, the response of soil nutrients' stoichiometry characteristics to climate change and the driving mechanism behind it need to be further studied. This is of great significance for maintaining the soil nutrient balance.

2.1.2.3 *Water resource*

Nowadays, the world is facing the challenge of a serious shortage of water resources, and deteriorating problems of water security management. In the context of global warming, the spatial and temporal distribution characteristics of water resources have significantly changed, and their impact has gradually begun to be concerned by research institutions and scholars around the world. Since 1985, the International Hydrographic Organization has repeatedly analyzed and studied the impact of climate change on water resources, systematically explored water resources' response mechanisms to climate change, and secured certain achievements. Climate change affects the storage, circulation and spatial-temporal distribution of water resources, exposing the water resource system to higher vulnerability and risks. On the one hand, climate warming causes the glacier to shrink continuously, and the water resources stored in the glacier reservoir are released in large quantities in the short term, which may intensify the risk of glacial lake outburst.

Meanwhile, the runoff volume of the glacier replenishment river increases significantly in the short term, and the long-term loss of the glacier water resource reserves is serious. Finally, glacier runoff is further decreased to dictate the sustainable utilization of downstream water resources. On the other hand, climate warming intensifies the water cycle and evapotranspiration, making the spatial and temporal distribution of water resources more uneven, and creating a sudden, multiple and concurrent trend of extreme drought and flood events. At the catchment scale, the impacts of climate change on river runoff show significant spatial heterogeneity. According to the prediction of the *China's Climate and Ecological Environment Evolution: 2021*, due to the impact of climate changes, the vulnerability of water resources in China will generally climb by 2030, the area of moderately vulnerable and above regions will be significantly expanded, and the area of extremely vulnerable regions will also be further expanded. In particular, the vulnerability in the northwest and northeast regions will rise significantly. From the global point of view, the precipitation has increased in the past 50 years, but the precipitation pattern changes differently in various regions. In the 20th century, the precipitation in the middle and high latitudes of the northern hemisphere grew by 0.5%−1.0% per decade; in the tropics it grew by 0.2%−0.3% per decade; and in the subtropical regions, it dropped by 0.3% per decade. No comparable systemic changes were found in the vast regions of the southern hemisphere. However, it was dominated by surface water resources (Amorim et al., 2020; Anjum et al., 2019).

2.1.2.4 *Groundwater*

One immediate consequence of global warming is the major change in the water cycle. With changes in rainfall and evaporation, water resources will be redistributed around the globe. Climate change affects not only the replenishment and circulation of groundwater, but also the water-rock interaction through the changes in chemical composition and surface temperature of atmospheric rainfall, thus changing the quality of groundwater. Torbert et al. (1996) emphasized that the impact of increased atmospheric CO_2 concentration on groundwater quality in the agro-ecosystem was a blank field in the study of global climate change. The rise of CO_2 concentration in atmosphere will lead to the change of plant growth conditions and C/N ratio, which then affect the C and N cycling in soil, and eventually the concentration of nitrate leaching into groundwater. Allen et al. (2004) studied the sensitivity of the Grand Forks aquifer in southern British Columbia, Canada to the climate change.

Groundwater is an important fresh water resource. Against the severe shortage of fresh water resources, it is of great significance to study the impact of climate change on groundwater resources and the sustainable utilization of water resources. Chen et al. (2002) pointed out that any change in climate parameters may act on the recharge conditions of groundwater systems, thus affecting the sustainable utilization of groundwater. The trend of climbing average annual temperature and decreasing precipitation in the Canadian grasslands in the past 40–50 years has triggered worries about the sustainable utilization of groundwater in this region. To this end, they presented an empirical model linking climate parameters to groundwater levels, based on simplified flow and water equilibrium models. The model has been successfully applied in the upper limestone aquifer in southern Manitoba, Canada. Loáiciga et al. (2000) simulated the changes of spring flow in Edwards BFZ karst aquifer in Texas, one of the largest groundwater systems in the United States, under the conditions of $2 \times CO_2$ climate and 25% increase in groundwater exploitation in three scenarios: extreme lack of water, near average supply, and beyond average supply. It was concluded that under the climate of $2 \times CO_2$, the water resources in this region were seriously threatened unless the exploitation activities were carefully controlled. Brouyeres et al. (2004) established a comprehensive hydrological model including groundwater flow in Geer Basin, Belgium to evaluate the impact of climate change on groundwater.

With the development of science and technology, scholars at home and abroad have made in-depth research on the law of water resource evolution under climate change. The United States Geological Survey (USGS) found that groundwater systems were remarkably responsive to climate change, and even small changes in precipitation could cause large fluctuations in the water levels. Chen et al. (2004) analyzed the correlation between precipitation, temperature and groundwater level in Manitoba and found a strong correlation among the three; in the shallow buried area of groundwater, the temperature made a more obvious impact on groundwater level. Bekele and Knapp (2010) used the SWAT hydrological model to study the dynamic changes of groundwater under different climate scenarios in the Fox Basin, and found that the groundwater recharge sources changed significantly under climate change. Timothy (2011) studied the influence of global warming on the exploitation of groundwater resources, and predicted the dynamic change trend of groundwater level in extreme drought years. Bates et al. (2008) and Mizyed (2009) found that changes in precipitation and evaporation had an impact not only on groundwater level, but also on groundwater quality.

In the 1980s, China began to study the effects of climate change on water resources. Compared with surface water, the hydrology process of groundwater is more complex and changeable, bringing a more obvious delayed response to climate change. Therefore, the impact of climate change on the groundwater system becomes a new hot spot for the researches today. As of today, there are two main methods to study the dynamic change law of groundwater under climate change: one is to establish the correlation between groundwater and various climate elements by long-sequence trend analysis of the observation data (e.g., groundwater level, precipitation, evaporation and temperature) in the study region, and use the correlation to analyze the change law of groundwater under different climate scenarios; the other one is to build a numerical simulation model of groundwater in the study region to predict precipitation, evaporation, temperature and other data under different climate scenarios, and predict the future evolution of groundwater by modifying the source and sink terms in the numerical model.

The mentioned researches suggest that there is a close relationship between the change of groundwater level and the climate change, and climatic factors such as precipitation, evaporation and temperature are the main drivers of the change of groundwater level. However, different regions have different degrees of sensitivity to the climate change, and the dynamic changes of groundwater demonstrate different trends.

2.1.3 Feedback of soil and groundwater environment to global climate change

2.1.3.1 Soil

The pedosphere is the one at the junction of Earth's lithosphere, atmosphere, hydrosphere and biosphere. It is not only a repository of natural resources that human beings live on and a habitat for human beings and organisms, but also a hub of material and energy exchange between organisms in the ecosystem and the environment. The important role of the pedosphere in global climate change, especially in the global carbon cycle, can be summarized into two aspects. First, soil is a huge store and converter of matters. Every year about 5.5×10^{11}t plant organisms are formed and decomposed in soil on Earth, 90% of which are transferred into the atmosphere as gas, and 10% into intermediate products preserved in the pedosphere. Carbon is a basic element for photosynthesis of organisms. Plants convert CO_2 into organic carbon through photosynthesis, and then input it into soil in the form of natural litter and root secretion. After decomposition, it is finally returned to the atmosphere as CO_2, CH_4 and other gases. The organic matters in agro-ecosystem input into soil in the natural litter form varied to different crop species, accounting for about 18% of the net biomass on average. The organic carbon input into the soil by root precipitation also takes a considerable part, and wheat and other crops can account for 5%−24% of the net carbon sequestration. The organic carbon input into the soil by root precipitation is easily decomposed by microorganisms, and most of which is released in the form of CO_2. It is obvious that the pedosphere is an important reservoir of carbon in the form of soil organic matter. The average carbon storage in soil within a 1m depth is 24.6kg·m^{-3} (Bohn, 1976). The global storage of SOC is 2.5×10^{12}, which accounts for 3/4 of the total carbon content in the whole biosphere. Therefore once the pedosphere is destroyed, its huge reserve of organic carbon will enter the atmosphere or other earth subsystems to aggravate the ascending of CO_2. Second, soil respiration releases large amounts of organic carbon into the atmosphere as CO_2. The amount of CO_2 released by soil respiration is quite considerable, say up to $640\text{g/(m}^2\text{·a)}$ in farmland ecosystem and $400-640\text{g/(m}^2\text{·a)}$ in grassland ecosystem. It is estimated that $(0.8-4.6) \times 10^{15}$g of carbon is released from soil into the atmosphere each year. Therefore, small changes in soil respiration will trigger significant changes in CO_2 concentration in atmosphere, thus affecting the global warming and other climatic factors (e.g., precipitation and radiation) that are associated with higher CO_2 concentrations.

2.1.3.2 *Groundwater*

As an active participant in the global material cycle and energy exchange, groundwater affects and is affected by environmental changes. For example, the study of Zhang et al. showed that over-exploitation of groundwater also stood for additional release of CO_2. Based on the average total carbon content in water, it is estimated that for every $1 \times 10^8 \text{m}^3$ of groundwater mined, its CO_2 consumption by evaporation is about the same as that emitted by burning 9000t of coal. In addition, the dissolution and precipitation of carbonate caused by the interaction between water and rock in karst area has a material influence on the global carbon cycle. In the water cycle, the groundwater system functions to turn the "pulse-like" rainfall into continuous water flow underground for a longer stay, and the water-rock interaction gives it the material and energy needed to maintain the near surface life. Thus groundwater is a crucial factor to maintain the balance of ecological environment. Currently, the impact of the global descending of groundwater level on the environment may be an issue that cannot be ignored. For this reason, some researchers propose the concept of "ecological water level", that is, the interval of groundwater level required to maintain the stability of regional ecosystem. If the groundwater level is higher than the upper limit of the interval, the surface soil will be salinized; if it is lower than the lower limit, it will cause the river cutoff, vegetation death, lake drying, farmland abandonment and soil desertification.

2.1.4 Summary and outlook

In recent decades, many major research projects have been implemented in the world, such as the Man and the Biosphere Programme (MAB) and the International Geosphere-Biosphere Programme (IGBP). Global Change and Terrestrial Ecosystem (GCTC) and IPCC directly or indirectly relate to the research on global climate change. These studies have promoted the development of relevant sciences and modern technologies. The current researches on global climate change have driven the development, interpenetration and crossover of a mass of basic disciplines (such as ecology, biological geosciences and atmospheric chemistry). The implementation of global climate change researches will certainly promote the progress of industries such as agriculture, environmental protection, energy conservation and water conservation. Although many disciplines have different opinions on the time scale of global climate change, the reasons for its occurrence, and how to understand the global warming and various scientific issues, global warming is an indisputable fact, with space for dispute only in the extent of warming, causes or regional distribution, especially the forecast of climate change in the future. Therefore, for studies on the impact of global climate change in the future on soil and groundwater environment, it is suggested to highlight the key and difficult issues listed below.

(1) The increase of CO_2 concentration, the rise of temperature and the change of rainfall are the three most important ecological factors of the effects of global climate change on the agricultural production and agro-ecosystem. By far, in most experiments concerning the impact of climate change on agro-ecosystem, single-factor levels, such as doubling CO_2 concentration and temperature rise by 1.0℃ or 2.0℃, are used for researches (Qin and Chen, 2005). Based on the gradual improvement of the existing experimental methods and technical means, it is suggested that in the future studies, the increase of CO_2 concentration, temperature rise, rainfall change and other factors be comprehensively considered, and the process of global climate change be demonstrated, with focus laid on the impact of the comprehensive effects on the growth and development of crops.

(2) The impact of climate change on groundwater is a research subject worthy of attention, but the current researches are mostly limited to the change of groundwater quantity under different climate change conditions, with the research methods and prediction accuracy to be further improved. As an active participant in the global material cycle and energy exchange, groundwater affects and is affected by environment. Currently, the impact of the global descending of groundwater level on the environment may be an issue that cannot be ignored.

(3) The quantitative research on the relationship between climate change and soil, nutrients and microbial characteristics should be paced up. This will certainly provide a more accurate scientific basis for predicting the investment costs in agricultural production in the future. However, in the long-time scale of climate change, there are great difficulties in the sampling of soil samples. How to innovate experimental methods and means for quantitatively studying the changes of N, P, K and organic matters in soil remains a difficult research topic. In addition, higher attention should be paid to the following aspects in future studies: ① From the perspective of molecular biology, the variation trend of the dominant rhizosphere microbial population caused by the increase of CO_2 concentration; ② studies on the interaction between root exudates and rhizosphere microorganisms with the increased CO_2 concentration; ③ studies on the response of plants to long-term CO_2 concentration in atmosphere from the perspective of microorganisms; ④ The relationship between the direct and indirect effects of increasing CO_2 concentration on microorganisms.

2.2 Temperature sensitivity of enzyme activities

2.2.1 Introduction

Microorganisms in the natural environment cope with changing conditions that demand a wide range of metabolic adaptations (Neidhardt et al., 1990). Among the most challenging environments are high latitude and cold ecosystem, which are threatened by global warming (Davidson and Janssens, 2006). Warming has a fundamental impact on microbial activity, metabolism and enzyme activities (Allison et al., 2010; Van Gestel et al., 2013; Zimmermann and Bird, 2012). Extracellular enzymes are essential to microbial metabolism and soil functioning, as they depolymerize organic compounds, thus generate soluble oligomers and monomers that can be transported into the cells (Blagodatskaya et al., 2016; Wallenstein et al., 2010).

Enzyme activity is a saturating function of substrate concentration and is described by the Michaelis-Menten relationship (Michaelis and Menten, 1913). Both parameters of the Michaelis-Menten equation, the maximal catalytic reaction rate at a given temperature (V_{max}) and the half-saturation constant (K_m) are temperature-sensitive (Davidson et al., 2006; Davidson and Janssens, 2006), they usually increase with temperature (Stone et al., 2012). Various enzymes have different temperature sensitivities, and changes in soil temperature may also alter the relative rates of decomposition of different components of organic matter (Koch et al., 2007; Wallenstein et al., 2010; Stone et al., 2012; Razavi et al., 2015). Therefore the temperature sensitivity of enzymes responsible for organic matter decomposition is the most crucial parameter for predicting the effects of global warming on the nutrient and carbon cycles (Davidson et al., 2006; Davidson and Janssens, 2006).

The temperature sensitivity of V_{max} is directly related to the activation energy for enzyme reaction (Davidson and Janssens, 2006). Activation energies are parameters that mechanistically link enzyme kinetics and temperature response through the Arrhenius equation (Wallenstein et al., 2010). Based on the Arrhenius law, when activation energy is low, the exponential term will tend to 1 and consequently the reaction will become temperature independent (Marx et al., 2007). In the other words, the lower the activation energy, the lower the temperature sensitivity of reaction rate. Enzymes catalyze biochemical reactions by lowering their activation energy (Gerlt and Gassman, 1993). Thus, a super-efficient enzyme will bring the activation energy to zero (Marx et al., 2007). This is important because, in the context of cold-adapted microorganisms, one way to maintain decomposition processes at low temperatures would be to develop enzymes that are temperature-independent (Marx et al., 2007).

Microbial physiology is evolutionarily selected for the most efficient enzyme systems (Allison et al., 2010; Hochachka and Somero, 2002). Moreover, the activities of extracellular enzymes could be adapted to different temperature regimes (German et al., 2012) with the goal of maintaining critical enzymatic functions. There is evidence for biogeographical patterns in enzyme temperature sensitivity (Huston et al., 2000; Feller, 2003; German et al., 2012). Many studies have observed that cold-adapted microorganisms can produce cold-adapted enzymes that catalyze reactions at lower temperatures with lower activation energy and higher binding affinity (i.e., low K_m) (Fields, 2001; Bradford, 2013) than their mesophilic counterparts (Gerday et al., 1997). Importantly, microbial adaptation and acclimation strategies have physiological costs (Schimel et al., 2007) and can reduce enzyme catalytic efficiency—determined as V_{max}/K_m (Stone et al., 2012; Tischer et al., 2015).

The parameters of enzyme kinetics—specifically K_m, which determines the binding affinity of the enzyme to substrate—are indicative of enzyme flexibility (the capacity for quick conformation change) (Somero, 1975). The increased flexibility would cause the cold-adapted enzyme to spend more time maintaining conformations that are not optimal for substrate binding (Siddiqui and Cavicchioli, 2006). This can be measured as a gradual increase of K_m with temperature (Fields, 2001). Key to effective enzymatic function is tradeoffs between functional capacity and enzyme flexibility, which covary with habitat temperature (Somero, 1995; Fields, 2001; Tokuriki et al., 2012). Conformational flexibility and enzyme function are closely related, and organisms have evolved to produce enzymes with thermal optima at their habitat temperature. For example, more flexible enzyme systems are expected under cold conditions, while greatly reduced enzyme flexibility (i.e., low temperature sensitivity of K_m) is predicted in warmer climates (Johns and Somero, 2004; Dong and Somero, 2009; Bradford, 2013).

Furthermore, as enzyme systems are altered by climate warming, different sets of isoenzymes (i.e., enzymes with the same function but different conformations and structures) are expected to be expressed at cold and warm temperatures (Somero, 1978; Bradford, 2013; Razavi et al., 2016). Isoenzymes with higher temperature optima can be produced by the same microbial species adapted to warming (Hochachka and Somero,

2002). Alternatively, isoenzymes can be expressed as a result of changes in microbial community structure caused by warming (Baldwin and Hochachka, 1970; Vanhala et al., 2011). In both cases, temperature sensitivity of catalytic reactions is dependent on enzyme isoforms. Nonetheless, all these mechanisms suggest that microbes prefer to produce enzymes that maintain optimal activity under native soil conditions.

Despite intensive discussion on the mechanisms of enzyme temperature sensitivity, it remains unclear how the functional characteristics of enzymes in cold-adapted soil will be altered by temperature increases. This is extremely important because it provides evidence of the response of cold-adapted soil microbes and the fate of huge amounts of SOC stored in these ecosystems by the acceleration of enzymatic decomposition in a warmer world. In addition, there is a lack of studies on the catalytic efficiency of soil enzymes in cold ecosystems as affected by warming.

This study was designed to test the effects of intensive warming on the catalytic properties of soil enzymes in a cold-adapted environment. We hypothesized that maximal reaction rate will be insensitive to intensive warming at high temperature range (H1); the substrate affinity (K_m) will remain constant at elevated temperature due to the expression of enzymes with less flexibility (H2). To test our hypothesis, we collected soil from the Qinghai-Xizang Plateau and incubated the samples for one month over a temperature range of 0–40℃ (with 5℃ steps) and determined the kinetics, temperature sensitivities and activation energy of six enzymes involved in decomposition of soil organics: cellobiohydrolase and β-glucosidase, which are commonly measured as enzymes responsible for consecutive stages of cellulose degradation (German et al., 2011); xylanase, which degrades xylooligosaccharides into xylose and is thus responsible for breaking down hemicelluloses (Chen et al., 2012); acid phosphatase, which catalyzes the hydrolysis of organic phosphate compounds to phosphate esters (Eivazi and Tabatabai, 1977; Malcolm, 1983). Activities of tyrosine aminopeptidase and leucine aminopeptidase were analyzed to assess the hydrolysis of peptide bonds (Koch et al., 2007; Chen et al., 2012).

2.2.2 Temperature sensitivity of soil enzymes

The V_{max} values increased with temperature for all enzymes (Figure 2.2). Changes in V_{max}-Q_{10} were not gradual over the whole range of temperatures tested, and were clearly pronounced between 0℃ and 15℃ (Figure 2.3). The magnitude of the temperature response varied between enzymes, ranging from 1.3 to 3.8, which corresponds to E_a values of 19kcal·mol^{-1} to 53kcal·mol^{-1} (Figure 2.4). For all enzymes, E_a was higher in the low temperature range (0−20℃) and strongly decreased from 25℃ to 40℃ (Figure 2.4). The fitting of V_{max} to the Arrhenius model demonstrated higher E_a values for cellobiohydrolase and xylanase compared to proteases, phosphatase and β-glucosidase.

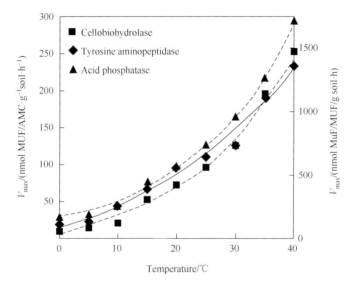

Figure 2.2 Enzyme activity as a function of temperature demonstrates a gradual increase for cellobiohydrolase, tyrosine aminopeptidase and phosphatase within the range of nine temperatures. *Each enzyme was assayed at a range of substrate concentrations (8 concentrations) at each of nine temperatures. Values are means of four replications (±SE). (Activity of the other three enzymes is presented in Table 2.1). Credit: Reprinted with permission from Razavi et al., 2017.

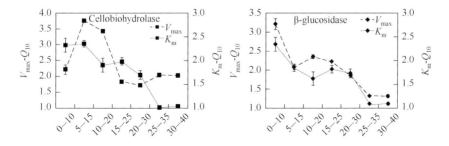

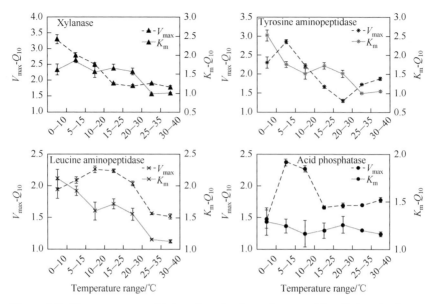

Figure 2.3 Temperature sensitivity of maximal reaction rate (V_{max}-Q_{10}) and substrate affinity (K_m-Q_{10}) of six enzymes as a function of temperature with 5°C increments. *Values are means of four replications ($\pm$SE). Credit: Reprinted with permission from Razavi et al., 2017.

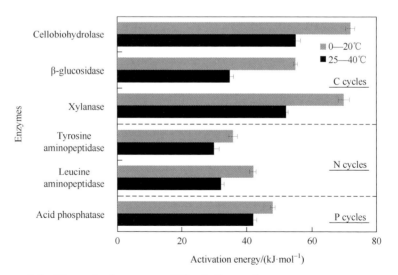

Figure 2.4 The activation energy (E_a) of all tested enzymes at two temperature ranges: low (0−20℃) and high (25−40℃). Credit: Reprinted with permission from Razavi et al., 2017.

2.2.3 Response of substrate affinity to temperature

The changes in K_m-Q_{10} were not gradual over the range of temperatures tested, and were maximal between 0℃ and 15℃ (Figure 2.3, Table 2.1). The Q_{10} values for K_m varied in the narrower range of 1.0 to 2.5, that is 1.5 times lower compared to V_{max}-Q_{10}. The K_m-Q_{10} demonstrated two enzyme-specific patterns: ①Decrease of K_m-Q_{10} for the whole temperature ranges; this pattern corresponded to enzymes of the C and N cycles; ②The pattern observed for phosphatase K_m-Q_{10} was nearly constant for the whole temperature range.

Table 2.1 K_m and catalytic efficiency (V_{max}/K_m) of β-glucosidase, Xylanase and Leucine aminopeptidase.

Enzymes	Temperature/℃	V_{max}/(nmol MUF/AMC·g^{-1}soil·h^{-1})	K_m/(μmol·g^{-1}soil)	V_{max}/K_m
β-glucosidase	0	30±1.3	6±0.3	4.9±0.3
	5	80±1.8	12±0.4	6.8±0.3
	10	98±1.7	14±0.4	6.8±0.2
	15	166±1.8	22±0.3	7.6±0.1
	20	230±2.7	23±0.3	9.8±0.2
	25	369±2.6	40±0.3	9.3±0.1
	30	429±7.7	41±0.7	10.6±0.3
	35	487±6.3	43±0.3	11.3±0.2
	40	564±14	44±0.8	12.7±0.4
Xylanase	0	5±0.1	5.3±0.9	1.0±0.2
	5	11±0.3	6.5±1.4	1.7±0.4
	10	17±0.3	8.6±1.4	2.0±0.3
	15	31±0.3	12±1.1	2.6±0.2
	20	42±0.6	13±1.9	3.2±0.4
	25	59±0.3	20±0.9	2.9±0.1
	30	77±0.4	21±1.3	3.7±0.2
	35	113±0.4	20±1.5	5.6±0.4
	40	138±0.5	21±1.7	6.6±0.5
Leucine aminopeptidase	0	33±0.4	5±1.3	6.3±1.5
	5	50±0.6	7.8±1.0	6.4±0.8
	10	64±0.5	11±0.9	5.8±0.5
	15	105±0.4	15±0.6	7.0±0.3
	20	145±0.6	18±0.7	8.3±0.3
	25	204±0.7	25±1.5	8.0±0.5
	30	295±1	27±1.1	10.8±0.4
	35	365±1	29±0.7	12.5±0.3
	40	448±2	31±1.1	14.7±0.5

The temperature effect on K_m revealed a distinct threshold with a significant decrease in the affinity of all enzymes to substrate at temperatures above 25 ℃ (Figure 2.5). Cellobiohydrolase, β-glucosidase and xylanase demonstrated stepwise increases of K_m values at low to

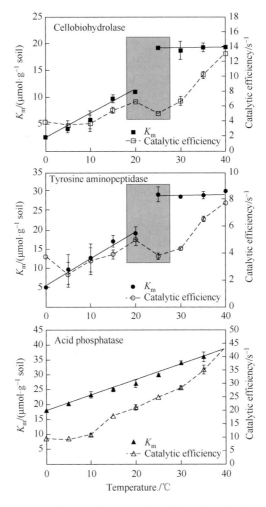

Figure 2.5 K_m and catalytic efficiency (V_{max}/K_m) of cellobiohydrolase, tyrosine aminopeptidase and phosphatase. *Shading indicates temperature ranges with extreme K_m increases accompanied by decreases in catalytic efficiency. (K_m and catalytic efficiency of the other three enzymes are presented in Table 2.1). Credit: Reprinted with permission from Razavi et al., 2017.

moderate temperatures (0–20℃). The K_m values of these enzymes strongly increased (by around 40%) between 20℃ and 25℃ (Figure 2.5). After such an extreme increase, the K_m values did not change significantly up to 40℃ (Figure 2.5). The changes of phosphatase's K_m followed a pattern different to that of the enzymes involved in carbohydrate decomposition and proteases. Phosphatase demonstrated slightly increased K_m values across the whole temperature range (0–40℃), (Figure 2.5).

Thus, the K_m of all C and N cycle enzymes changed significantly within psychrophilic and mesophilic temperatures, while substrate affinity was relatively constant within the elevated range (25–40℃).

Most soil studies and models tacitly accept the gradual (according to Q_{10}) increase of reaction rates (and consequently process intensities) with temperature. Both V_{max} and K_m increased with temperature for all tested enzymes, although the increase was not linear and indicated different temperature sensitivities of V_{max} and K_m (Figure 2.2 and Figure 2.5). The Q_{10} values of reaction rates varied from 1.9 to 3.8 within the low temperature range and decreased to 1.3 at higher temperature. Similarly, the activation energy of all tested enzymes was higher at low and moderate temperatures (0–20℃) compared to elevated levels (25–40℃). This general reduction of temperature sensitivity confirms theoretical predictions (Davidson and Janssens, 2006) and experimental observations on reduced reaction rate Q_{10} values at elevated temperature (Tjoelker et al., 2001; Razavi et al., 2015). In line with previous studies, activation energy and temperature sensitivity of enzymes responsible for complex C-compound degradation (i.e., Xylanase and cellobiohydrolase) were higher compared to β-glucosidase (Craine et al., 2010; Conant et al., 2011). However, contrary to our hypothesis (H1), reaction rates of enzymes that degrade low quality polymers respond to temperature changes (i.e., V_{max}–Q_{10}=2) even in warm temperature ranges. This might explain why recalcitrant C compounds decomposed faster under warming (Knorr et al., 2005).

Following the strong increase at 20℃, the K_m remained nearly constant from 25℃ to 40℃, while the maximal enzyme activity (V_{max}) gradually increased with temperature. Alternatively, constant K_m can be explained by an expression of multiple isoenzymes each with a different temperature optimum (Somero, 1995; Bradford, 2013). Such isoenzyme expression leads to an optimal balance between the static character of the enzyme (responsible for high efficiency at constant optimal temperature) and functional capacity, given similar flexibilities of isoenzymes under their respective optimal working conditions (Zavodsky et al., 1998; Conant et al., 2011; Razavi et al., 2016).

Sudden and strong changes in K_m at 25℃ indicated a switch from cold- and moderate-to warm-adapted enzyme systems with decreased substrate affinity. In fact, such an increase was responsible for the reduced temperature sensitivity and catalytic efficiency of overall enzyme function. A constant K_m value from 25℃ to 40℃ was accompanied by a gradual increase of catalytic efficiency with temperature. This pattern supports our hypothesis (H2). This means that the production of enzymes with higher stability and efficiency is a preferred microbial strategy in the studied cold-adapted soil (Stone et al., 2012; Razavi et al., 2016). From another point of view, the adaptive modifications to carry out protein, cellulose and hemicellulose catalysis at mesophilic and thermophilic temperatures are unnecessary in this soil.

We found a gradual increase of K_m from 0−40℃ (phosphatase) and for all other tested enzymes from 0−20℃. This could be a consequence of increased enzyme flexibility, that is, the capacity for quick conformation changes ensuring a fast rate of catalytic reaction by changing temperature. We also assume that the gradual increase of K_m with increasing temperature reflected stepwise expression of isoenzymes. Proteases and cellulolytic enzymes demonstrated constant K_m from 25℃ to 40℃, which was in line with the previous findings of Fields and Somero (1998) and the theoretical prediction of Bradford (2013) regarding the stability of enzyme systems at high temperatures. A strong increase in K_m by 40%−50% at high temperatures (25℃ versus 20℃) reflected a two-fold reduction of the enzyme-substrate affinities. However, such temperature thresholds seemed to be higher in temperate climates (30℃), (Razavi et al., 2016) compared to highland areas like Xizang (25℃).

It should be noted that the K_m measured in this study could be considered as "apparent" K_m (German et al., 2012; Stone et al., 2012) because of the confounding effects of many temperature-sensitive processes in soil (e.g., desorption and adsorption reactions which affect V_{max} and K_m values) (Nannipieri and Gianfreda, 1998; Davidson and Janssens, 2006).

2.2.4 Catalytic efficiency of enzymes as affected by temperature

The catalytic efficiency of the enzymes (V_{max}/K_m) increased from cold to moderate temperatures (0–20℃). Further extreme increases in K_m at the 25℃ threshold were always accompanied with a sharp decrease in the catalytic efficiency of enzymes of the C and N cycles (Figure 2.5), and leveled off above 25℃. In contrast, the catalytic efficiency of phosphatase increased gradually from 0℃ to 40℃ (Figure 2.5).

Catalytic efficiency demonstrated a general trend of gradual increasie at both cold and warm temperatures (Figure 2.5 and Figure 2.6). The only remarkable exception occurred at 25℃, where a strong increase in K_m was accompanied by a significant decrease in catalytic efficiency (Figure 2.6). Thus, decoupled responses of V_{max} and K_m to temperature would result in irregular increases of catalytic efficiency . Such decoupled responses reflect, not the separate optimization of K_m and V_{max}, but rather evolutionary "compromises" through which an optimal balance between these parameters was achieved. Quite simply, if catalytic properties were to be maintained under a particular thermal regime, the "goal" that must be met would be the expression of isoenzymes with similar K_m values (Somero, 1978). Thus, V_{max} and K_m values were adjusted according to

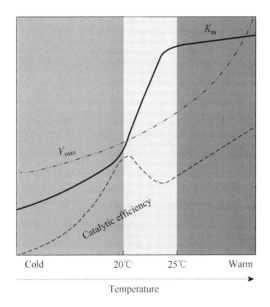

Figure 2.6 Generalized thermal responses of enzyme catalytic properties to a temperature increase. *The scheme explains that catalytic efficiency gradually increases with temperature at both cold and warm temperatures except at 25℃, where a strong increase in K_m occurs. Credit: Reprinted with permission from Razavi et al., 2017.

temperature to ensure the improvement of catalytic efficiency within a wide range of temperatures. Maintaining the high binding affinity to substrate (constant K_m) ensured efficient enzyme conformation within the unaccustomed temperate range. Remarkably, high substrate availability in Xizang soil facilitates such a compromise and the achievement of the "goal".

Previous studies on pure cultures demonstrated a decline in catalytic efficiency between 20-30℃ (Siddiqui and Cavicchioli, 2006). However, the pattern observed here demonstrated that the complex composition of microbial communities in soil maintained decomposition of SOM within an extremely wide range of temperatures under a climate with a large annual temperature variation of −24.8℃ to 24.1℃.

2.2.5 Summary

Overall, the changes in enzyme kinetics in response to elevated soil temperature revealed here indicated altered enzyme systems and a possible shift in community composition (Bárcenas-Moreno et al., 2009). Further evidence of sharp changes in K_m could be an indicator of isoenzyme expression (Baldwin and Hochachka, 1970) due to a major shift in species dominance above 25℃ (Khalili et al., 2011; Bradford, 2013). These temperatures were extremely unusual for the original microbial community under the natural climate, with annual temperature of 2.4℃ (Bárcenas-Moreno et al., 2009).

Finally, enzymes that degrade low quality polymers are temperature-sensitive within the whole range of temperatures (0—40℃); soil microorganisms are able to maintain stable or flexible enzyme systems with low or high substrate affinity within wide temperature ranges to ensure efficient enzymatic functions under diurnally and annually varying temperatures. This ensures the easier adaptation of microbially driven decomposition to changing climate. Static substrate affinity at warmer temperatures proved that acclimation involves expressing more stable enzymes at the warmer temperature, potentially with the same K_m but not necessarily. We conclude that considering the temperature thresholds of strong changes in enzyme-based processes is crucial to modeling the consequences of warming for C, N and P cycles and predicting the fate of soil carbon stocks in a warmer world.

Main References

Allen, D.M., Mackie, D.C., Wei, M., 2004. Groundwater and climate change: A sensitivity analysis for the Grand Forks aquifer, southern British Columbia, Canada. Hydrogeology Journal, 12(3): 270–290.

Available from: https: doi.org/10.1007/s10040-003-0261-9.

Allison, S.D., Wallenstein, M.D., Bradford, M.A., 2010. Soil-carbon response to warming dependent on microbial physiology. Nature Geoscience, 3: 336–340. Available from: https: doi.org/10.1038/ngeo846.

Amorim, P.B.D., Souza, K.I.S.D., Chaffe, P.L.B., 2020. A web-based tool for synthesis assessments of the impacts of climate change on water resources. Environ Model Softw, 133: 104848. Available from: https: doi.org/10.1016/j.envsoft.2020.104848.

Anjum, M.N., Ding, Y.J., Shangguan, D.H., 2019. Simulation of the projected climate change impacts on the river flow regimes under CMIP5 RCP scenarios in the westerlies dominated belt, northern Pakistan. Atmospheric Research, 227(5): 233–248. Available from: https: doi.org/10.1016/j.atmosres.2019.05.017.

Baldwin, J., Hochachka, P.W., 1970. Functional significance of isoenzymes in thermal acclimatization Acetylcholinesterase from trout brain. The Biochemical Journal, 116(5): 883–887.

Bárcenas-Moreno, G., Gómez-Brandón, M., Rousk, J., et al., 2009. Adaptation of soil microbial communities to temperature: Comparison of fungi and bacteria in a laboratory experiment. Global Change Biology, 15(12): 2950–2957. Available from: https: doi.org/ 10.1111/j.1365-2486.2009.01882.x.

Bates, B., Kundzewicz, Z., Wu, S., et al., 2008. Climate change and water technical paper VI of the intergovernmental panel on Climate Change. Intergovernmental panel on climate change secretariat, Geneva, 24: 333–335.

Blagodatskaya, E., Blagodatsky, S., Khomyakov, N., et al., 2016. Temperature sensitivity and enzymatic mechanisms of soil organic matter decomposition along an altitudinal gradient on Mount Kilimanjaro. Scientific Reports, 6: 22240. Available from: https: doi.org/ 10.1038/srep22240.

Bohn, H. L., 1976. Estimate of organic carbon in world soils. Soil Science Society of America Journal, 40(3): 468–470.

Bradford, M.A., 2013. Thermal adaptation of decomposer communities in warming soils. Frontiers in Microbiology, 4: 333. Available from: https: doi.org/ 10.3389/fmicb.2013.00333.

Brouyère, S., Carabin, G., Dassargues, A., 2004. Climate change impacts on groundwater resources: Modelled deficits in a chalky aquifer, Geer Basin, Belgium. Hydrogeology Journal, 12(2): 123–134. Available from: https: doi.org/10.1007/s10040-003-0293-1.

Buyer, J.S., Drinkwater, L.E., 1997. Comparison of substrate utilization assay and fatty acid analysis of soil microbial communities. Journal of Microbiological Methods, 30(1): 3–11. Available from: https: doi.org/10.1016/S0167-7012(97)00038-9.

Chen, Z.H., Grasby, S.E., Osadetz, K.G., 2002. Predicting average annual groundwater levels from climatic variables: An empirical model. Journal of Hydrology, 260(1–4): 102–117. Available from: https: doi.org/10.1016/S0022-1694(01)00606-0.

Chen, Z.H., Grasby, S.E., Osadetz, K.G., 2004. Relation between climate variability and groundwater levels in the upper carbonate aquifer, southern Manitoba, Canada. Journal of Hydrology, 290(1–2): 43–62. Available from: https: doi.org/10.1016/j.jhydrol.2003.11.029.

Chen, R.R, Blagodatskaya, E., Senbayram, M., et al., 2012. Decomposition of biogas residues in soil and their effects on microbial growth kinetics and enzyme activities. Biomass and Bioenergy, 45: 221–229. Available from: https: doi.org/ 10.1016/j.biombioe.2012.06.014.

Conant, R.T., Ryan, M.G., Ågren, G.I., et al., 2011. Temperature and soil organic matter decomposition rates: Synthesis of current knowledge and a way forward. Global Change Biology 17, 3392–3404. Available from: https: doi.org/10.1111/j.1365-2486.2011.02496.x.

Craine, J.M., Fierer, N., McLauchlan, K.K., 2010. Widespread coupling between the rate and temperature sensitivity of organic matter decay. Nature Geoscience, 3: 854–857. Available from: https: doi.org/ 10.1038/ngeo1009.

Davidson, E.A., Janssens, I.A., 2006. Temperature sensitivity of soil carbon decomposition and feedbacks to climate change. Nature, 440(7081): 165–173. Available from: https: doi.org/ 10.1038/nature04514.

Davidson, E.A., Janssens, I.A., Luo, Y.Q., 2006. On the variability of respiration in terrestrial ecosystems: Moving beyond Q_{10}. Global Change Biology, 12(2): 154–164. Available from: https: doi.org/ 10.1111/j.1365-2486.2005.01065.x.

Dong, Y.W., Somero, G.N., 2009. Temperature adaptation of cytosolic malate dehydrogenases of limpets (genus Lottia): Differences in stability and function due to minor changes in sequence correlate with biogeographic and vertical distributions. The Journal of Experimental Biology Moving beyond Q_{10}. Global Change Biology, 12(2): 169–177. Available from: https: doi.org/ 10.1242/jeb.024505.

Eivazi, F., Tabatabai, M.A., 1977. Phosphatases in soils. Soil Biology and Biochemistry, 9: 167–172. Available from: https: doi.org/10.1016/0038-0717(77)90070-0.

Bekele, E.G., Knapp, H.V., 2010. Watershed modeling to assessing impacts of potential climate change on water supply availability. Water Resources Management, 24(13): 3299–3320. Available from: https: doi.org/10.1007/s11269-010-9607-y.

Feller, G., 2003. Molecular adaptations to cold in psychrophilic enzymes. Cellular and Molecular Life Sciences: CMLS, 60(4): 648–662. Available from: https: doi.org/ 10.1007/s00018-003-2155-3.

Fields, P.A., 2001. Review: Protein function at thermal extremes: Balancing stability and flexibility. Comparative Biochemistry and Physiology Part A: Molecular & Integrative Physiology, 129(2–3): 417–431. Available from: https: doi.org/ 10.1016/S1095-6433(00)00359-7.

Fields, P.A., Somero, G.N., 1998. Hot spots in cold adaptation: Localized increases in conformational flexibility in lactate dehydrogenase A4 orthologs of Antarctic notothenioid fishes. Proceedings of the National Academy of Sciences of the United States of America, 95(19): 11476–11481. Available from: https: doi.org/10.1073/pnas.95.19.11476.

Fu, G., Zhang, X.Z., Zhang, Y.J., et al., 2013. Experimental warming does not enhance gross primary production and above-ground biomass in the alpine meadow of Tibet. Journal of Applied Remote Sensing, 7(1): 073505.

Gerday, C., Aittaleb, M., Arpigny, J.L., et al., 1997. Psychrophilic enzymes: A thermodynamic challenge. Biochimica et Biophysica Acta, 1342(2): 119–131. Available from: https: doi.org/10.1016/S0167-4838(97)00093-9

Gerlt, J.A., Gassman, P.G., 1993. An explanation for rapid enzyme-catalyzed proton abstraction from carbon acids: importance of late transition states in concerted mechanisms. Journal of the American Chemical Society, 115(24):11552–11568.

German, D.P., Weintraub, M.N., Grandy, A.S., et al., 2011. Optimization of hydrolytic and oxidative enzyme methods for ecosystem studies. Soil Biology and Biochemistry, 43(7): 1387–1397. Available from: https: doi.org/10.1016/j.soilbio.2011.03.017.

German, D.P., Marcelo, K.R.B., Stone, M.M., et al., 2012. The Michaelis–Menten kinetics of soil extracellular enzymes in response to temperature: A cross-latitudinal study. Global Change Biology, 18(4): 1468–1479. Available from: https: doi.org/10.1111/j.1365-2486.2011.02615.x.

Hochachka, P.W., Somero, G.N., 2002. Biochemical Adaptation. New York: Oxford University Press.

Huston, A.L., Krieger-Brockett, B.B., Deming, J.W., 2000. Remarkably low temperature optima for extracellular enzyme activity from Arctic bacteria and sea ice. Environmental Microbiology, 2(4):

383–388. Available from: https: doi.org/ 10.1046/j.1462-2920.2000.00118.x.

IPCC, 2021. IPCC sixth assessment report: Climate change 2021: the physical science basis. Available from: https://www.ipcc.ch/report/ar6/wg1/#SPM.

Johns, G.C., Somero, G.N., 2004. Evolutionary convergence in adaptation of proteins to temperature: A4-lactate dehydrogenases of Pacific damselfishes (Chromis spp.) . Molecular Biology and Evolution, 21(2): 314–320. Available from: https: doi.org/10.1093/molbev/msh021.

Jongen, M., Lecomte, X., Unger, S. et al., 2013. Precipitation variability does not affect soil respiration and nitrogen dynamics in the understorey of a Mediterranean oak woodland. Plant and Soil, 372(1): 235–251. Available from: https: doi.org/10.1007/s11104-013-1728-7.

Khalili, B., Nourbakhsh, F., Nili, N., et al., 2011. Diversity of soil cellulase isoenzymes is associated with soil cellulase kinetic and thermodynamic parameters. Soil Biology and Biochemistry, 43(8): 1639–1648. Available from: https: doi.org/10.1016/j.soilbio.2011.03.019.

Knorr, W., Prentice, I.C., House, J.I., et al., 2005. Long-term sensitivity of soil carbon turnover to warming. Nature, 433(7023): 298–301. Available from: https: doi.org/10.1038/nature03226.

Koch, O., Tscherko, D., Kandeler, E., 2007. Temperature sensitivity of microbial respiration, nitrogen mineralization, and potential soil enzyme activities in organic alpine soils. Global Biogeochemical Cycles, 21(4): B4017-1-B4017-11. Available from: https: doi.org/ 10.1029/2007GB002983.

Li K., Jacob D. J., Liao H., et al., 2019. Anthropogenic drivers of 2013-2017 trends in summer surface ozone in China. Proceedings of the National Academy of Sciences of the United States of America, 116(2): 422–427. Available from: https: doi.org/10.1073/pnas.1812168116

Lipson, D.A., Schadt, C.W., Schmidt, S.K., 2002. Changes in soil microbial community structure and function in an alpine dry meadow following spring snow melt. Microbial Ecology, 43(3): 307–314. Available from: https: doi.org/ 10.1007/s00248-001-1057-x.

Liu C., Hu H.B., Zhang Y., et al., 2017. The direct effects of aerosols and decadal variation of global sea surface temperature on the East Asian summer precipitation in cam3.0. Journal of Tropical Meteorology, 23(2): 217–228. Available from: https: doi.org/10.16555/j.1006-8775.2017.02.010.

Loáiciga, H.A., Maidment, D.R., Valdes, J.B., 2000. Climate-change impacts in a regional karst aquifer, Texas, USA. Journal of Hydrology, 227(1–4): 173–194. Available from: https: doi.org/10.1016/S0022-1694(99)00179-1

Malcolm, R.E., 1983. Assessment of phosphatase activity in soils. Soil Biology and Biochemistry, 15(4): 403–408. Available from: https: doi.org/ 10.1016/0038-0717(83)90003-2.

Marx, J.C., Collins, T., D'Amico, S., et al., 2007. Cold-adapted enzymes from marine Antarctic microorganisms. Marine Biotechnology, 9(3): 293–304. Available from: https: doi.org/10.1007/ s10126-006-6103-8

McKane, R.B., Rastetter, E.B., Shaver, G.R., et al., 1997. Climatic effects on tundra carbon storage inferred from experimental data and a model. Ecology, 78(4): 1170. Available from: https: doi.org/10.1890/ 0012-9658(1997)078[1170:CEOTCS]2.0.CO;2

Michaelis, L., Menten, M.L., 1913. Die kinetik der invertinwirkung. Biochemisches Zeitschrift, 49: 334–336.

Miller, A.J., Schuur, E.A.G, Chadwick, O.A., 2001. Redox control of phosphorus pools in Hawaiian montane forest soils. Geoderma, 102(3):219–237. Available from: https: doi.org/10.1016/S0016-7061(01)00016-7.

Mizyed, N, 2009. Impacts of climate change on water resources availability and agricultural water demand in the west bank. Water Resources Management, 23(10): 2015–2029. Available from: https: doi.org/ 10.1007/s11269-008-9367-0.

Moorhead, D.L., Linkins, A, 1997. Elevated CO_2 alters below ground exoenzyme activities in tussock tundra. Plant and Soil, 189(2): 321–329. Available from: https: doi.org/10.1023/A:1004246720186.

Nam, H.S., Kwak, J.H., Lim, S.S. et al., 2013. Fertilizer N uptake of paddy rice in two soils with different fertility under experimental warming with elevated CO_2. Plant Soil, 369(1): 563–575. Available from: https: doi.org/10.1007/s11104-013-1598-z.

Nannipieri, P., Gianfreda, L., 1998. Kinetics of enzyme reactions in soil environments. In: Huang, P.M., Senesi, N., Buffle, J. (eds.), Structure and Surface Reactions of Soil Particles. John Wiley & Sons: 449–479.

Neidhardt, F.C., Ingraham, J.L., Schaechter, M., 1990. Physiology of the bacterial cell: A molecular approach. Sunderland: Sinauer Associates.

Nolan, C., Overpeck, J.T., Allen, J.R.M., et al., 2018. Past and future global transformation of terrestrial ecosystems under climate change. Science, 361(6405): 920–923.

Øygarden, L., Deelstra, J., Lagzdins, A., et al., 2014. Climate change and the potential effects on runoff and nitrogen losses in the Nordic–Baltic Region. Agriculture, Ecosystems & Environment, 198: 114–126. Available from: https: doi.org/10.1016/j.agee.2014.06.025.

Razavi, B.S., Blagodatskaya, E., Kuzyakov, Y., 2015. Nonlinear temperature sensitivity of enzyme kinetics explains canceling effect: A case study on loamy haplic Luvisol. Frontiers in Microbiology, 6: 1126. Available from: https: doi.org/ 10.3389/fmicb.2015.01126.

Razavi, B.S., Blagodatskaya, E., Kuzyakov, Y., 2016. Temperature selects for static soil enzyme systems to maintain high catalytic efficiency. Soil Biology and Biochemistry, 97: 15–22. Available from: https: doi.org/ 10.1016/j.soilbio.2016.02.018.

Razavi, B.S., Liu, S.B., Kuzyakov, Y., 2017. Hot experience for cold-adapted microorganisms: Temperature sensitivity of soil enzymes. Soil Biology and Biochemistry 105, 236–243. Available from: https: doi.org/10.1016/j.soilbio.2016.11.026

Rinnan, R., Michelsen, A., Bååth, E., et al., 2007. Mineralization and carbon turnover in subarctic heath soil as affected by warming and additional litter. Soil Biology and Biochemistry, 39(12): 3014–3023. Available from: https: doi.org/10.1016/j.soilbio.2007.05.035.

Robinson, C.H., Saunders, P.W., Madan, N.J., et al., 2004. Does nitrogen deposition affect soil microfungal diversity and soil N and P dynamics in a high Arctic ecosystem? Global Change Biology, 10(7): 1065–1079. Available from: https: doi.org/10.1111/j.1529-8817.2003.00793.x

Saiya-Cork, K.R., Sinsabaugh, R.L., Zak D.R., 2002. The effects of long term nitrogen deposition on extracellular enzyme activity in an *Acer saccharum* forest soil. Soil Biology and Biochemistry, 34(9): 1309–1315. Available from: https: doi.org/10.1016/S0038-0717(02)00074-3.

Schimel, J., Balser, T.C., Wallenstein, M., 2007. Microbial stress-response physiology and its implications for ecosystem function. Ecology, 88(6): 1386–1394. Available from: https: doi.org/ 10.1890/06-0219.

Siddiqui, K.S., Cavicchioli, R., 2006. Cold-adapted enzymes. The Annual Review of Biochemistry,75: 403–433. Available from: https: doi.org/10.1146/ annurev.biochem.75.103004.142723.

Somero, G.N., 1975. The roles of isozymes in adaptation to varying temperatures. In: Isozymes. Amsterdam: Elsevier:221–234.

Somero, G.N., 1978. Temperature adaptation of enzymes: Biological optimization through structure function compromises. Annual Review of Ecology and Systematics, 9: 1–29. Available from: https:

doi.org/10.1146/annurev.es.09.110178.000245.

Somero, G.N., 1995. Proteins and temperature Annual Review Physiology, 57: 43–68. Available from: https: doi.org/ 10.1146/annurev.ph.57.030195.000355.

Stone, M.M., Weiss, M.S., Goodale, C.L., et al, 2012. Temperature sensitivity of soil enzyme kinetics under N-fertilization in two temperate forests. Global Change Biology, 18(3): 1173–1184. Available from: https: doi.org/ 10.1111/j.1365-2486.2011.02545.x.

Timothy, R., 2011. Beneath the surface of global change: Impacts of climate change on groundwater. Journal of Hydrology, 405(3–4): 532–560. Available from: https: doi.org/10.1016/j.jhydrol.2011.05.002

Tischer, A., Blagodatskaya, E., Hamer, U., 2015. Microbial community structure and resource availability drive the catalytic efficiency of soil enzymes under land-use change conditions. Soil Biology and Biochemistry, 89: 226–237. Available from: https: doi.org/10.1016/j.soilbio.2015.07.011.

Tjoelker, M.G., Oleksyn, J., Reich, P.B., 2001. Modelling respiration of vegetation: evidence for a general temperature-dependent Q10. Global Change Biology 7, 223–230. Available from: https: doi.org/ 10.1046/j.1365-2486.2001.00397.x.

Tokuriki, N., Jackson, C.J., Afriat-Jurnou, L., et al., 2012. Diminishing returns and tradeoffs constrain the laboratory optimization of an enzyme. Nature Communications, 3: 1257. Available from: https: doi.org/ 10.1038/ncomms2246.

Torbert, H.A., Prior, S.A., Rogers, H.H., et al., 1996. Elevated atmospheric carbon dioxide in agroecosystems affects groundwater quality. Journal of Environmental Quality, 25(4): 720–726. Available from: https: doi.org/10.2134/jeq1996.00472425002500040011x.

Van Gestel, N.C., Reischke, S., Bååth, E., 2013. Temperature sensitivity of bacterial growth in a hot desert soil with large temperature fluctuations. Soil Biology and Biochemistry, 65: 180–185. Available from: https: doi.org/10.1016/j.soilbio.2013.05.016.

Van Kessel, C., Horwath, W.R., Hartwig, U., et al., 2000. Net soil carbon input under ambient and elevated CO_2 concentrations: Isotopic evidence after 4 years. Global Change Biology, 6(4): 435–444. Available from: https: doi.org/10.1046/j.1365-2486.2000.00318.x.

Vanhala, P., Karhu, K., Tuomi, M., et al., 2011. Transplantation of organic surface horizons of boreal soils into warmer regions alters microbiology but not the temperature sensitivity of decomposition: Warming, soil biology and decomposition. Global Change Biology, 17(1): 538–550. Available from: https: doi.org/10.1111/j.1365-2486.2009.02154.x.

Wallenstein, M., Allison, S.D., Ernakovich, J., et al., 2010. Controls on the temperature sensitivity of soil enzymes: a key driver of in situ enzyme activity rates. In: Shukla, G., Varma, A. (Eds.), Soil Enzymology. Berlin Heidelberg: Springer:245–258.

Wu, Z.Y., Wang, H., Shan, J.J., 2022. Climate change risks faced by China's ecological security and countermeasures. Journal of Central South University of Forestry & Technology (Social Sciences), 16(4): 25–33. (In Chinese)

Závodszky, P., Kardos, J., Svingor, A., et al., 1998. Adjustment of conformational flexibility is a key event in the thermal adaptation of proteins. Proceedings of the National Academy of Sciences of the United States of America, 95(13): 7406–7411. Available from: https: doi.org/ 10.1073/pnas.95.13.7406

Zimmermann, M., Bird, M.I., 2012. Temperature sensitivity of tropical forest soil respiration increase along an altitudinal gradient with ongoing decomposition. Geoderma, 187: 8–15. Available from: https: doi.org/10.1016/j.geoderma.2012.04.015.

CHAPTER 3

Response of enzyme activities to manure applications

3.1 Impact of manure on soil biochemical properties: a global synthesis

3.1.1 Introduction

Soils are a key reservoir of global biodiversity and the basis of global food production (FAO, 2015). The increasing human population combined with shrinking agricultural land further highlights the necessity to maintain soil health and quality (Thangarajan et al., 2013). Arable soils are being rapidly degraded due to erosion, structure degradation, nutrient depletion, loss of soil organic carbon (SOC) and other threats (Fan et al., 2012). In the Anthropocene, manure that is mainly excreted by animals or derived from plant residues is an environmentally-friendly soil amendment to remediate soil degradation and thereby increase the stability of crop production and agro-ecosystem functioning. Manure application not only increases aggregate stability and soil porosity (Haynes and Naidu, 1998; Karami et al., 2012) and decreases bulk density (Edmeades, 2003), but also improves the soil biochemical properties (Jiang et al., 2018; Luo et al., 2018; Saha et al., 2008a).

The application of manure is able to increase SOC and nutrient content [e.g., nitrogen (N), phosphorus (P)] (Liang et al., 2014; Nicolas et al., 2012; Ros et al., 2006; Zhang et al., 2015a, 2015b). For instance, field application of cattle manure for 46 years increased the SOC and total nitrogen (TN) content by 34% and 31%, respectively (Giacometti et al., 2014). The application of swine manure at a rate of 41 $Mg \cdot ha^{-1}$ for 15 years increased the total P content from $123 mg \cdot kg^{-1}$ to $458 mg \cdot kg^{-1}$ (Zhang et al., 2015a, 2015b). The application of poultry manure (Tejada, 2009) or composted cattle manure (Zaller and Köpke, 2004) increased the microbial C by more than 200% or 27%, respectively. Despite the positive effects of manure on soil microbial indices, great variability in the response of soil biochemical properties after manure application exists, especially for enzyme activities (Foster et al., 2016). For instance, positive effects of green manure (i.e., *Trifolium pratense*, L. and *Brassica napus*, L.) or composted cattle manure on β-glucosidase, urease, phosphatase, dehydrogenase and sulfatase were observed (Tejada et al., 2008; Tripathy et al., 2008), while a negative response of chitinase to green and composted swine manure was found in a 6-month field experiment (Liu et al., 2017a, 2017b). In contrast to the positive/negative responses, Giacometti et al. (2014) found no response of acid phosphatase activities to a 46-year application of green/cattle manure. In particular, the observed contradictory effects of manure application on enzymes that are involved in the C cycle indicate that different C sources may be depleted by fertilization, which affects the activity of enzymes involved in the C cycle. While each of these results makes sense individually, they do not fit together into a consistent theory across a broad range of soils and management systems. Abundant unexplained responses seen in some systems but not in others suggest a major knowledge gap and the necessity of multi-meta-analysis (Gurevitch et al., 2018).

Despite a wide range of studies on the influence of manure application on soil biochemical parameters (Luo et al., 2018; Kallenbach and Grandy, 2011; Maillard et al., 2014; Thangarajan et al., 2013; Webb et al., 2010; Kallenbach and Grandy, 2011), there are only rare attempts to provide a more comprehensive, mechanistic understanding based on soil characteristics (Jiang et al., 2018). For instance, by analyzing the impacts of manure application on soil microbial C and N in croplands, Ren et al. (2018) found a positive correlation between annual C and N input from manure and microbial C and N, respectively. A recent study using a large database more accurately identified the effects of manure on SOC, TN and biological parameters as compared to mineral fertilizers (Luo et al., 2018). However, the conclusion by Luo et al. (2018), which showed that manure application results in the increase of soil enzyme activities, is vulnerable to uncertainty in the calculation of average extracellular enzyme activities. The averaging of extracellular enzyme activities that are used as proxies of a specific substrate or nutrient acquisition is based on an assumption that the sum of the major C- or nutrient acquiring enzyme activities is a better indicator of the total C- or nutrient acquisition than are individual enzymes (Bell et al., 2014). However, this approach masks the diverse responses of individual enzymes that are involved in the same nutrient cycle. For instance, in a northwestern Himalayan ecosystem, continuous application of composted cattle manure at rates of 3700kg·ha^{-1} and 5500kg·ha^{-1} increased the alkaline phosphatase but caused a decline in the acid phosphatase activity (Saha et al., 2008a). In addition, recent meta-analyses (Luo et al., 2018; Maillard et al., 2014) did not differentiate manure-only application from the application of manure combined with mineral fertilizers. Considering both strategies together led to inaccurate quantification of the manure's effect on soil organic matter and biochemical properties (Luan et al., 2019). The effects of manure chemical composition (i.e., the C and N content, the C/N ratio) on microbe-mediated soil functioning were not evaluated but were proposed for future investigation (Luo et al., 2018). These characteristics (e.g., types, forms of production and application) are critical in maintaining soil fertility, but their roles in controlling soil properties have yet to be summarized (Ali et al., 2019; Maillard et al., 2014). Thus, a more comprehensive evaluation of manure application's effects on soil biochemical properties was required from both agronomic and environmental perspectives. This helps us understand the consequences of agricultural management strategies for food production.

Here, we generalized studies worldwide that report the effects of manure application on SOC, TN, microbial C, microbial N and activities of seven enzymes (i.e., C-cycling: β-1, 4-glucosidase; energy-acquiring: dehydrogenase; N-cycling: urease, N-acetyl-β-D-glucosaminidase; P-cycling: acid phosphatase and alkaline phosphatase; S-cycling: sulfatase). The impacts of key explanatory variables were evaluated, that is, climatic factors [mean annual temperature (MAT), mean annual precipitation (MAP)], soil properties [initial soil pH, type (e.g., Alfisols, Entisols and so on), texture], management (alone or combined with mineral fertilizers, field or lab, duration) and manure characteristics [type (e.g., cattle, swine and so on), composted or non-composted, dry or wet application]. The relationship between basic manure chemical composition (i.e., C, N, P and K content and their stoichiometry) and their impact on soil properties was a specific focus. Our main objectives were: ①comprehensively evaluate and quantify the effects of manure on soil biochemical properties; ②verify the reliability of using the average extracellular enzyme activities as proxies of a specific substrate or nutrient acquisition; ③clarify the influence of the above-mentioned explanatory variables on manure impacts. We hypothesized that: ①the activity of enzymes involved in acquisition of the same element may respond differently to manure application because of their different functions in complex C and nutrient cycles; ②the combined application of mineral fertilizers with manure lowers the impact of manure on soil biochemical properties; ③manure characteristics significantly affect soil properties.

3.1.2 Nutrient composition of manure and its effects on soil properties

Green manure has the highest organic carbon(OC) content [(407 ± 24)g·kg^{-1}; all data presented on a dry-weight basis] compared to that of the other manure types (Figure 3.1). The TN content in cattle manure is the lowest (15 ± 2)g·kg^{-1} among all manure types. In contrast, the TP content is the lowest in green manure (2.6 ± 0.5)g·kg^{-1}. The unique OC, TN and TP compositions in green manure result in the highest ratios: C/N=42 ± 8.1, C/P=221 ± 32, N/P=12 ± 3.8 (Figure 3.1). The OC content and nutrient compositions of composted manure were nominally lower than those of non-composted manure (Figure 3.1). Composted manure has nominally higher TN (18 ± 2.2)g·kg^{-1} and TP (11 ± 3.4)g·kg^{-1} than non-composted manure ($P>0.05$). This leads to nominally lower C/N (20 ± 2.6) and C/P (60 ± 22) ratios in composted manure ($P>0.05$).

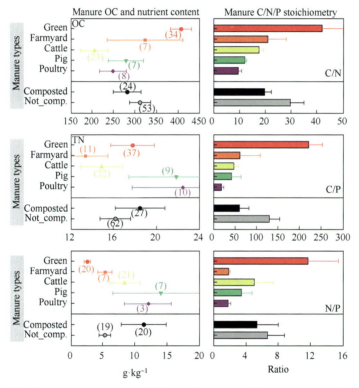

Figure 3.1 Average organic carbon (OC), total nitrogen (TN) and total phosphorus (TP) content (left) and C/N/P stoichiometric ratios (right) of various manure types. Green manure, farmyard manure, cattle manure, pig manure and poultry manure were illustrated according to their relatively larger database. "Composted" represents composted manure and "Not-composted" represents non-composted manure. Whiskers show standard error ($\pm$SE). Credit: Reprinted with permission from Liu et al., 2020.

Manure application increased SOC and TN content by 27%±4.2% and 33%±8.7% (Figure 3.2) compared to soil without manure. Microbial C and N increased by 88%±6.3% and 84%±11%, respectively. No significant shifts of the soil C/N ratio were found (Figure 3.3), but the microbial C/N ratio decreased from 9.49±0.34 to 7.64±0.44 after manure application ($P <$ 0.0001, Figure 3.3). For experiments applying swine manure, the microbial C/N ratio decreased from 8.9±0.5 in the control group to 5.6±0.5 in the manure-treated group.

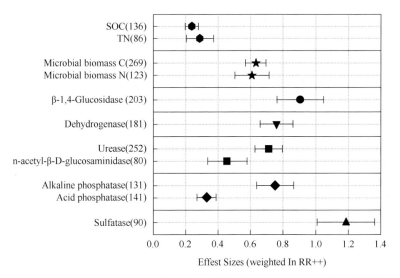

Figure 3.2　Weighted effect sizes of manure application on SOC, TN, microbial biomass C and N content as well as enzyme activities. The number in the parentheses besides each property is the sample size. The thinner black lines were drawn to group C-, energy-, N-, P- and S-acquiring enzymes. SOC and TN represent soil organic carbon and total nitrogen content. Whiskers represent 95% confidence intervals. Credit: Reprinted with permission from Liu et al., 2020.

The ΔpH was negatively correlated with the initial soil pH ($P<0.01$, Figure 3.4). The x-intercept value is approximately 7.47, which indicates that the soil pH will increase when the initial soil pH is lower than 7.47 and decrease when the initial soil pH is higher.

Manure application to clay loam soil induced nominally larger effect sizes of SOC and TN (Figure 3.5). Microbial C and N content had larger effect sizes when manure was applied to sandy loam and sandy clay loam (Figure 3.6). Applying manure in the lab led to larger increases of SOC,

TN, and microbial C and N content than those in the field (Figure 3.5 and Figure 3.6).

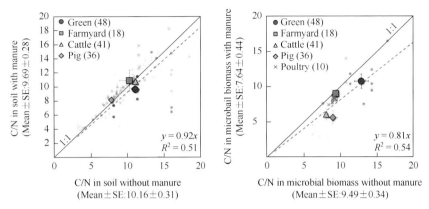

Figure 3.3 Relationship of soil C/N (left) and microbial biomass C/N (right) between soil without and with manure. Manure types are differentiated with colors. The large solid symbols are the means of the small semitransparent ones. Whiskers represent standard errors of means. Black continuous lines are 1:1 lines, and dashed lines reflects the linear regression. The linear regression lines were forced through the origin to reflect whether the ratio of (C/N in soil with manure) to (C/N in soil without manure) is close to 1.0. Credit: Reprinted with permission from Liu et al., 2020.

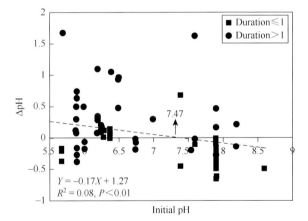

Figure 3.4 Relationships between ΔpH and initial soil pH. Experiments with duration shorter or longer than one year are presented with different colors. Positive ΔpH values show increase of soil pH (decreased acidity) after manure application. Credit: Reprinted with permission from Liu et al., 2020.

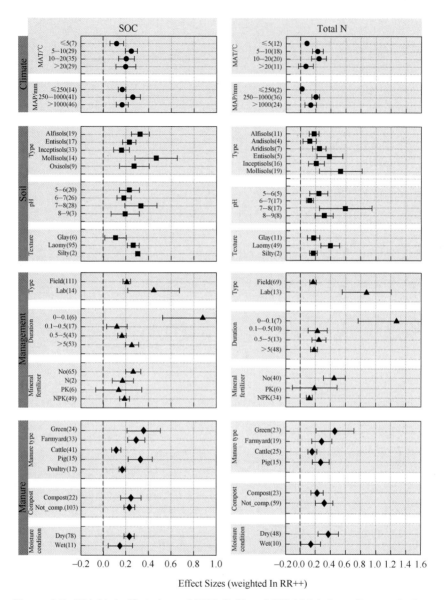

Figure 3.5 Weighted effect sizes of SOC (left) and TN (right) depending on the key explanatory variables (i.e., climatic factors, soil properties, management, and manure characteristics). The number in the parentheses beside each property is the sample size. SOC and TN represent soil organic carbon and total nitrogen content. Whiskers represent 95% confidence intervals. Credit: Reprinted with permission from Liu et al., 2020.

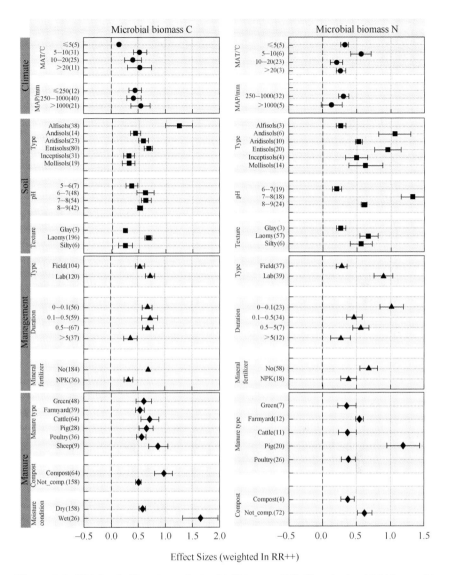

Figure 3.6 Weighted effect sizes of microbial biomass C (left) and microbial biomass N (right) depending on the key explanatory variables (i.e., climatic factors, soil properties, management, and manure characteristics). The number in the parentheses beside each property is the sample size. Whiskers represent 95% confidence intervals. Credit: Reprinted with permission from Liu et al., 2020.

Soil OC and TN content increased by 27%±4.2% and 33%±8.7% (Figure 3.5), respectively, with manure application. These increases as shown in Figure 3.7 are greater than the impact induced by mineral fertilizers (0.1%−8.4% for SOC and 15% for TN; Geisseler et al., 2017; Geisseler and Scow, 2014; Jian et al., 2016; Xiao et al., 2018). This greater effect of manure can be partly due to the C (206−407)g·kg^{-1} and N (13−22)g·kg^{-1} contained in the manure. In addition, manure application can also lower the soil bulk density and increase the porosity and water holding capacity (Eden, 2017). This, together with the loading of OC and nutrients from manure, affects the microbiome and plant growth, which in turn will increase the C input via rhizodeposition and litter input (Liu et al., 2017a, 2017b). Even though the data for other soil nutrients (P and K) were not reviewed, manuring increased soil nutrients status (Figure 3.1). In addition, microbial C and N also strongly increased by 88%±6.3% and 84%±11%, respectively, mainly because of the increased resource availability and physical soil properties for microbial growth following manure application (Chen et al., 2009; de Graaff et al., 2019). The larger increase of microbial C or N compared to the SOC or TN increase implies a higher ratio of microbial C or N to SOC or TN in manured soils, respectively, which could reflect increased soil organic matter quality (Friedel et al., 2006). This highlights

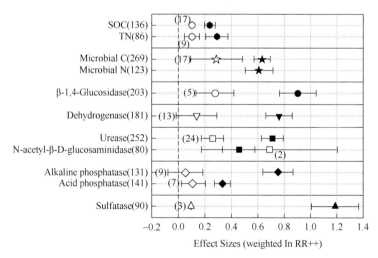

Figure 3.7 Weighted effect sizes of the effects of manure (solid points) and mineral fertilizers (hollow points) on SOC, TN, microbial biomass C and N content as well as enzyme activities. The number in the parentheses beside each property is the sample size of manure application. The number in the parentheses beside each hollow point is the sample size of mineral fertilization. Whiskers represent 95% confidence intervals. Credit: Reprinted with permission from Liu et al., 2020.

the remarkable consequences of manure application for the long-term improvement of soil fertility and the acceleration of C and N cycles.

Soil C/N ratios remained stable with manure application, which indicated the stability of the coupled C and N cycles. An exceptional decline of the soil C/N ratio was common for experiments (as shown in Figure 3.3, as indicated by the red color) shorter than 21 days (Kizilkaya, 2008; Kizilkaya and Hepsen, 2007). Mancinelli et al. (2013) reported a decreasing soil C/N ratio after applying green manure for 0.5–4.5 months and suggested that the incorporation of legume biomass into soil was the main reason, due to its low C/N ratio (Mancinelli et al., 2013). However, a two-year study in leguminous green manure-based cropping systems showed no strong change in the soil C/N ratio (Astier et al., 2006). Based on our large database, green manure application has no strong impact on the soil C/N ratio (Figure 3.3), even though it has the largest C/N ratio (i.e., 42 ± 8.1) oycompared to that of other manure types. The quality (e.g., the C/N ratio) of manure applied to soil controls soil N_2O emissions with a negative relation with C/N ratio (Huang et al., 2004). The application of manure with a lower C/N ratio might accelerate N mineralization rates and cause greater N_2O emissions, thus indicating the likelihood of greater and faster N losses. When the C/N ratio is larger (e.g., >25), the N mineralization activity declines due to increased microbial N immobilization. In both the cases, the soil C and N cycles remain coupled, and the application of manure with different C/N ratios induced no shift in the soil C/N ratio. This stable soil C/N ratio can also be explained by the fine and stable mineral-associated organic matter, which represents the major portion of total soil organic matter and responds slowly to management practices (Cotrufo et al., 2019; Gentsch et al., 2015; Samson et al., 2020).

Unlike the soil C/N ratio, microbial C/N decreased after manure application, especially for swine manure (as indicated by the green color in Figure 3.3). Such a response of microbial C and N indicates greater N immobilization. In particular, the decrease in microbial C/N ratio suggests a shift in microbial community towards bacterial dominance because of their lower C/N ratio as compared to that of fungal biomass (Six et al., 2006). Long-term and continuous increases in bacterial biomass following manure application, instead of fungal biomass, have been well documented (Marschner et al., 2003; Peacock et al., 2001; Rousk and Bååth, 2007). In addition, the C/N ratio of various manure types was a determinant of the microbial C/N ratio or fungal to bacterial growth, with lower C/N ratios of manure (e.g., swine manure) being more beneficial to bacterial growth than to fungal growth (Grosso et al., 2016; Zornoza et al., 2016).

3.1.3 Effects of manure application on enzyme activities

The application of manure increased the activities of β-glucosidase (β-glu), dehydrogenase (Deh), acid and alkaline phosphatase (Acp and Akp), N-acetyl-β-D-glucosaminidase (NAG), urease (Ure) and sulfatase by 147%±15%, 114%±10%, 39%±6%, 112%±12%, 58%±13%, 104%±8.7% and 228%±19%, respectively (Figure 3.2). All the enzyme activities showed either an increase or no change in response to the manure application for all conditions (i.e., climatic factors, soil properties, management and manure characteristics) (Figure 3.8—Figure 3.11). The largest effect sizes for almost all the enzyme activities were found within the MAT and MAP ranges of 10—20℃ and 250—1000mm compared with those of other ranges. The exception was Acp and Akp, which showed lower effect sizes in the MAP range of 250—1000mm compared to that in the MAP range of >1000mm (Figure 3.10). Among all soil types, manure application to Mollisols induces the largest responses of SOC (59%) and TN (77%). Almost all enzyme activities showed the greatest increase when manure was applied to Entisols (Figure 3.8—Figure 3.11). Most enzyme activities had the largest effect sizes when the initial soil pH was in the range of 6—8. In particular, the effect size of Akp decreased when the initial soil pH increased ($P<0.001$, Figure 3.10). The application of manure to sandy loam or sandy clay loam soil induced relatively larger effect sizes of almost all enzyme activities compared to those of other soils (Figure 3.8—Figure 3.11). One exception is for Acp activity, which had the largest effect size when manure was applied in clay loam soil (Figure 3.10 left).

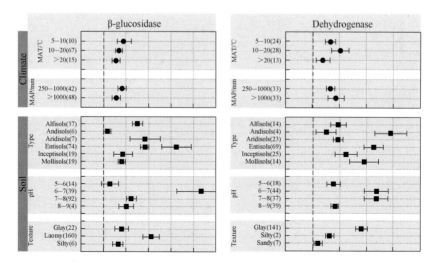

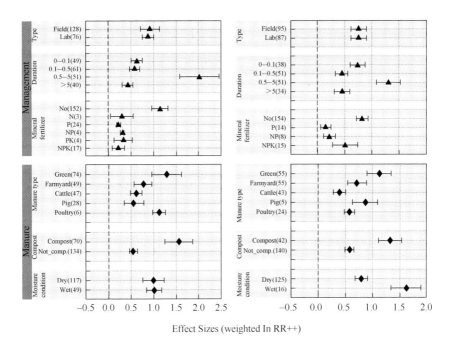

Figure 3.8 Weighted effect sizes of the activity of β-glucosidase (left) and dehydrogenase (right) depending on the key explanatory variables (i.e., climatic factors, soil properties, management, and manure characteristics). The number in the parentheses beside each property is the sample size. Whiskers represent 95% confidence intervals. Reprinted with permission from Liu et al., 2020.

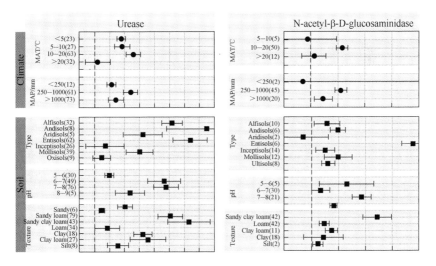

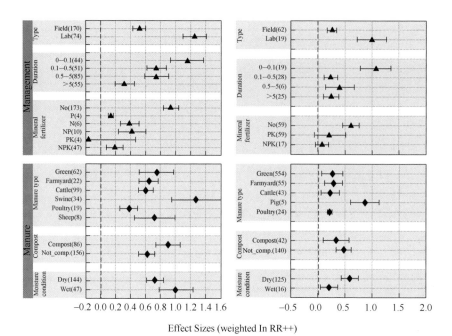

Effect Sizes (weighted In RR++)

Figure 3.9 Weighted effect sizes of the activities of urease (left) and chitinase (right) depending on the key explanatory variables (i.e., climate, soil, management and manure properties). The number in the parentheses besides each property is the sample size. Whiskers represent 95% confidence intervals. Reprinted with permission from Liu et al., 2020.

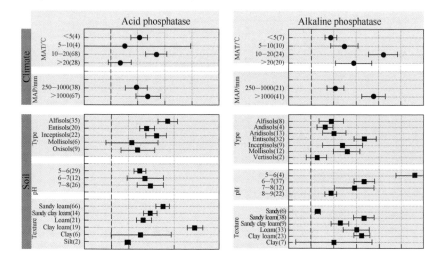

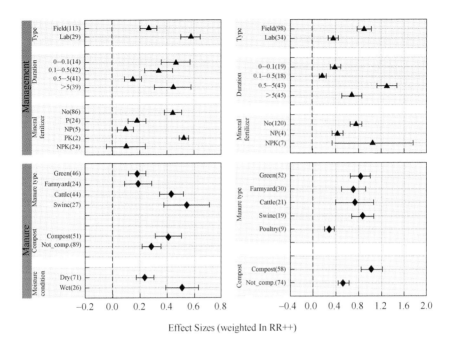

Figure 3.10 Weighted effect sizes of the activities of acid phosphatase (left) and alkaline phosphatase (right) depending on the key explanatory variables (i.e., climate, soil, management, and manure properties). The number in the parentheses besides each property is the sample size. Whiskers represent 95% confidence intervals. Credit: Reprinted with permission from Liu et al., 2020.

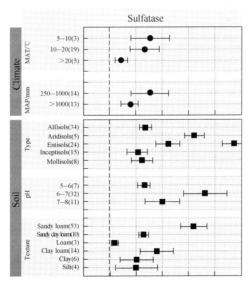

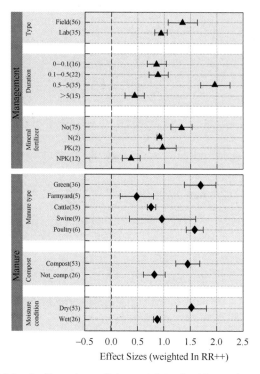

Figure 3.11 Weighted effect sizes of the activity of sulfatase depending on the key explanatory variables (i.e., climate, soil, management, and manure properties). The number in the parentheses besides each property is the sample size. Whiskers represent 95% confidence intervals. Credit: Reprinted with permission from Liu et al., 2020.

Field and lab-controlled conditions had similar effects on B-glu (~148%) and Deh (~110%) (Figure 3.8). The field experiment induced larger increases of Akp (148%±5.6%) and Sul (286%±17%) than the lab experiment, but applying manure in the lab led to larger increases of Ure, Nag and Acp than those in the field (Figure 3.9 and Figure 3.10). When the application of manure was combined with mineral fertilizers (i.e., only N, only P, N+P, P+K and N+P+K), the manure effects on all the enzyme activities were lower than those with the application of manure alone. The composted manure application induced larger effect sizes of most enzyme activities. Swine manure had the largest effect sizes on Ure (258%±39%) and Nag (138%±31%) (i.e., n-cycling-related parameters, Figure 3.9).

The Akp/Acp ratio is an indicator for soil pH adjustment. A significant linear relation was found for Akp/Acp ratio between soil without manure and soil with manure (Figure 3.12). The mean Akp/Acp ratios in soil without manure and soil with manure are 0.80 ± 0.17 and 0.86 ± 0.14, respectively. Manure application rate $(Mg\cdot ha^{-1}\cdot year^{-1})$ was correlated with the $\Delta Akp/Acp$ ratio [here, $\Delta Akp/Acp$ ratio = (Akp/Acp ratio in soil with manure) $-$ (Akp/Acp ratio in soil without manure)] and a quadratic relationship was found ($P<0.0001$; Figure 3.12).

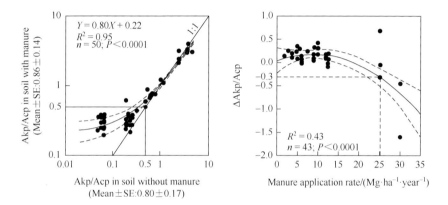

Figure 3.12 Relationship of Akp/Acp ratio in soil without and with manure (left) and relationship between manure application rate/(Mg·ha−1·year−1) and $\Delta Akp/Acp$ ratio (right). "Akp" means alkaline phosphatase activity; "Acp" means acid phosphatase activity. The values in the parentheses represent the mean and standard errors of Akp/Acp ratio in soil with and without manure. "n" means observation numbers. The red dashed lines are the 95% confidence bands. The value "0.5" on the left figure was proposed as the Akp/Acp target ratio with optimal pH for crop production (Dick et al., 2000). The value "−0.3" in the right figure was the difference between the proposed target ratio (i.e., 0.5) and mean Akp/Acp ratio in soil without manure (i.e., 0.8; mean of "X" in left figure) (i.e., 0.5−0.8=−0.3). $\Delta Akp/Acp$ ratio = (Akp/Acp ratio in soil with manure) − (Akp/Acp ratio in soil without manure). Reprinted with permission from Liu et al., 2020.

The effect sizes of nearly all enzyme activities showed that they were positively correlated ($P<0.05$, Table 3.1). Especially, β-glu, Ure and Deh were significantly correlated with other enzymes (Table 3.1). Microbial C was the most pronounced property that positively correlated with the enzyme activity changes ($P<0.05$) (except Sul) as compared to SOC, TN and microbial N.

Table 3.1 Spearman's rank correlation coefficients between response ratios of extracellular enzyme activities, SOC, TN, microbial C and N then continuous explanatory variables (MAP, MAT, pH and duration of trials).

	Soil properties				Enzyme activities							Experimental variables			
	SOC	TN	MBC	MBN	B-glu	Nag	Acp	Akp	Ure	Sul	Deh	MAT	MAP	Duration	Initial pH
SOC		0.78*** (71)	0.31* (69)					0.46* (27)			0.49*** (46)		-0.25* (110)		
TN			0.40* (37)	0.87*** (12)					0.33* (44)		0.43* (25)				
MBC				0.75*** (55)		0.64*** (31)		0.49* (22)	0.50*** (129)		0.61*** (108)	0.26*** (159)	-0.28* (71)	-0.21** (219)	
MBN					0.4*** (104)	0.74*** (41)	0.76*** (52)		0.75*** (34)		0.82*** (79)				
β-glu						0.50*** (51)	0.65*** (78)	0.70*** (64)	0.64*** (154)	0.90*** (81)	0.69* (8)		-0.61*** (76)	-0.61*** (76)	0.49*** (61)
NAG							0.87*** (14)	0.48*** (51)	0.76*** (38)	0.60* (15)	0.57*** (38)		-0.37** (68)	-0.43*** (78)	0.51*** (57)
Acp									0.54*** (81)	0.41** (51)					

continued

	Soil properties				Enzyme activities							Experimental variables			
	SOC	TN	MBC	MBN	B-glu	Nag	Acp	Akp	Ure	Sul	Deh	MAT	MAP	Duration	Initial pH
Akp									0.66*** (74)	0.95*** (37)	0.85*** (79)		0.42*** (87)		−0.55*** (76)
Ure										0.81*** (79)	0.80*** (107)	−0.19* (165)	−0.37*** (166)	−0.37*** (244)	0.44*** (161)
Sul											0.87*** (48)				
Deh															
MAT													0.39*** (369)	−0.30*** (357)	0.36*** (240)
MAP														0.13** (370)	−0.48*** (253)
Duration															−0.37*** (353)
Initial pH															

Note: *$P<0.05$, **$P<0.01$, ***$P<0.001$

107

Among the continuous explanatory variables, MAP was negatively correlated with most of the factors involved in N-cycling-related properties (i.e., TN, microbial N, Ure and Nag) ($P<0.01$), and Akp & Deh increased with the increase in MAP ($P<0.001$). The duration of manure application was only negatively correlated with the β-glu, Ure, Sul and Deh ($P<0.05$).

Manure application increased enzyme activities and this increase was enzyme-specific, even for enzymes that are involved in the cycling of the same element, which confirms our first hypothesis. For instance, Acp increased by approximately 40% with manure application, while the increase of Akp reached 110%. The same phenomenon occurred for Nag (+59%) and Ure (+106%). Acp is produced by both plants and microbes, but Akp originates from soil bacteria, fungi and fauna (Dinkelaker and Marschner, 1992; Tarafdar and Claassen, 1988). This contributed to the greater increase of Akp activity in manure-treated soil as compared to the control group (no manure application, see Table 3.2). Consequently, the summation or averaging of major C (or nutrient) acquiring enzyme activities (as used in Luo et al., 2018 and Jian et al., 2016) as an indicator of the total C (or nutrient) acquisition is problematic. Such summation or averaging approaches disregard the specific functions of individual enzymes. For instance, despite Nag and Ure both being involved in the N cycle, they catalyze different reactions. Nag is mainly responsible for chitin hydrolysis, which is a constituent of the cell walls or structural tissues of fungi, with the end product being acetyl-glucosamine (Rodriguez-Kabana et al., 1983). In contrast, Ure participates in the hydrolysis of urea to NH_3 (Fisher et al., 2017). The response ratios of almost all enzymes are positively correlated, which illustrates the convergence of ratios of specific C, N and P acquisition activities as 1:1:1 (Sinsabaugh et al., 2008). Therefore, instead of calculating the sum or average of the enzyme activities, we propose that the response of one or two enzyme activities (e.g., β-glu, Ure and Deh) to manure application can partly reflect the response of others ($P<0.001$). For accurate quantification, however, it is necessary to identify the responses individually or develop a more proper index to reflect the response.

Table 3.2 Comparisons used during the data analysis.

Comparison	Control	Treatment
1	Unfertilized nil-treatment	Manure-only
2	Mineral fertilized treatment	Manure application with mineral fertilizers

Soil sulfatase activity showed the greatest increase (i.e., 228%±19%) with manure application. To our knowledge, it is the first time that the response of the soil sulfatase activity to manure application has been generalized. Sulfatase catalyzes the hydrolysis of sulfate esters of complex macromolecules, and plays an important role in the transformation of organic sulfates to inorganic form (i.e., SO_4^{2-}) (Acosta-Martínez and Tabatabai, 2000). The release of SO_4^{2-} is especially crucial for sulfur supply in agricultural soils that do not receive pollutant sulfur deposition from the atmosphere (Turner et al., 2016). This greatest response of sulfatase activity may be induced by input of substrates (e.g., ester sulfate) and stimulated by microbial growth under manure application (Bandick and Dick, 1999; Giacometti et al., 2014; Piotrowska and Wilczewski, 2012).

The response ratio of most enzymes was not correlated with that of SOC but was positively related to the microbial C (Table 3.1). Consequently, the effects of manure on enzyme activities were more closely related to the enzyme producers, that is, microbes (Le Bayon and Binet, 2006). Manure was applied at the beginning, but SOC was generally measured at the end of each growing season for the studies with longer duration (>1 growing season). At the sampling time for the SOC analyses, the labile organic matter from manure had mostly decomposed. The remaining soil organic matter is thus mainly comprised of more recalcitrant materials which are not the primary substrate for hydrolytic enzymes. This explains the absence of significant relationships between the responses of the SOC content and enzyme activities after manure application.

3.1.4 Soil pH and its influence on the manuring effect

The soil pH is one of the most important properties regulating enzyme activities and synthesis (Dick et al., 2000) by controlling the microbial community composition (Rousk et al., 2009; Zhalnina et al., 2014) and mediating nutrient availability (Sinsabaugh et al., 2008). The application of manure can increase the pH buffering capacity (Shi et al., 2019), thus indicating that its application will adjust the soil pH to the neutral range (i.e., acidification of alkaline soil or alkalization of acidic soil) (Figure 3.4). This is crucial because most plants grow best near neutral pH (i.e., 6.5−7.5), and the response of microbial activities to manure application was the strongest in this range (Figure 3.6, Figure 3.8−Figure 3.11). This buffering capacity is one of the substantial advantages of applying manure as compared to mineral fertilizers, which commonly lower the soil pH significantly because of the nutrient uptake and release of protons (Richter et al., 1994). Even though the application of manure may also acidify soil by ammonia volatilization or nitrification, addition of several cations (e.g., Ca^{2+}, Mg^{2+}) contributes to buffer the soil solution (Zhang et al., 2015a, 2015b). In addition, bacterial growth decreases at lower pH (e.g., ~4.5) much faster than fungal growth (Grosso et al., 2016; Rousk et al., 2009). The application of manure, on the contrary, mitigates the decrease of bacterial growth but slows the growth of fungi by increasing the soil pH to the neutral range. This further supports our finding that the application of manure narrows the microbial C/N ratio, and our related argument that manured soils are more conducive to bacterial than fungal growth.

This neutralizing effect of manure also explains the negative relationship between the response of Akp and the initial soil pH (i.e., the soil pH before manure was applied, ranging from 4.5 to 8.6 in this study) ($P<0.001$). The activity of Akp increases with the increasing soil pH and has an optimum of ~9.0 (Ekenler and Tabatabai, 2003; Koncki et al., 2005). When the initial soil pH is high and alkaline, the Akp activity is already high, and thus the manure application will induce only a slight increment in the activity. Accordingly, manuring can cause a large increment of Akp activity by increasing the pH of acidic soils. In contrast, the increment of Acp increases with the increasing initial soil pH (Figure 3.10). The contrasted responses of Acp and Akp activities also explain the effect of manure on adjusting the Akp/Acp ratio: In acidic soil, the application of manure will induce a greater increment of Akp than Acp and thus increase the Akp/Acp ratio; in alkaline soil, the increment of Acp will be greater than Akp, thus inducing a decrease in the Akp/Acp ratio (Figure 3.12).

The Akp/Acp ratio of 0.50 was proposed as the proper Akp/Acp target ratio for crop production (Dick et al., 2000). The application of manure may increase the Akp/Acp ratio up to 0.86 ± 0.14 (Figure 3.12), which is 0.36 units higher than the suggested target ratio (i.e., 0.5). This indicates that the effect of manure application on crop production may be limited. Large manure input improves soil productivity (Edmeades, 2003), so we hypothesize that manure application rate may influence the Akp/Acp ratio. The quadratic relationship between manure application rate and ΔAkp/Acp ratio suggests that, to bring the Akp/Acp ratio in soil without manure from 0.80 to the target ratio of 0.5, the manure application rate should correspond to $25Mg\cdot ha^{-1}\cdot year^{-1}$ (Figure 3.12). Consequently, a manure application rate of $\sim25Mg\cdot ha^{-1}\cdot year^{-1}$ would bring the Akp/Acp ratio to the target level for optimal crop production. However, it is still inappropriate to set this rate as the global best rate, even though the data were extracted from studies with different climate, manure types and soil types (e.g., Li et al., 2012; Liang et al., 2015; Saha et al., 2008b; Tamilselvi et al., 2015). To reach a global best rate, much larger database is required.

3.1.5 Effects of soil, climate, management and manure-related factors on the soil biochemical properties

3.1.5.1 *Climate factors*

For most soil enzyme activities, the response to manure was the greatest in the MAT range of 10–20℃, thus indicating that the optimal temperature for soil enzyme function is at this range. By comparison, MAP did not greatly influence the soil properties and enzyme activities, except for the phosphatases, which showed the strongest response to manure application at sites with MAP>1000mm. This specific response can be related to strong P limitation because of strong rate of weathering and leaching (Richardson et al., 2004). High P demand of soil prompts the hydrolysis of organic P to available phosphates by increasing phosphatase activities. The negative correlation between soil initial pH and MAP ($P<0.001$) suggests that the soil acidity increases with precipitation. As discussed earlier, manure addition may help maintaining soil pH closer to neutrality. This, in turn, could help regulate P availability in P-deficient soils.

3.1.5.2 *Soil properties*

Mollisols are among the most fertile soils and are rich in organic matter (Eswaran et al., 2012). Therefore, extra supplies from manure are mostly preserved in Mollisols following manure application and induced the largest SOC and TN increase. However, Mollisols are under severe threat, as they are intensively cultivated for agricultural production (Hatfield et al., 2017; Liu et al., 2012a, 2012b). In the present study, manure application appeared to have some potentials to maintain or remediate fertility of degraded Mollisols.

The strongest increase of enzyme activities after manure application was observed in Entisols. Manure application to Entisols strongly stimulated microbial activities because Entisols are young soils (Kolb et al., 2009). Entisols are poorly developed and are mostly dominated by sand and low OC content, which results in a low cation exchange capacity and weak nutrient retention (Franco-Andreu et al., 2017; Lehmann and Schroth, 2003; Mancinelli et al., 2013). Therefore, stimulation of microbial activities by manure in Entisols increases the risk of nutrient losses from leaching if manure is applied in the absence of plant nutrient uptake.

The greatest increase of microbial C and N as well as almost all soil enzyme activities, as influenced by manure application, occurs in sandy loam or sandy clay loam soils. Both soils are conducive to microbial and plant growth due to its better drainage than that of clay or silty soil and greater nutrient-retaining capacity than sandy soil. An organic amendment has been found to have greater improvement on the physical properties of coarser-textured soil compared to that of finer-textured soil (e.g., hydraulic conductivity, water retention capacity, bulk density, aggregation and aggregate stability) (Aggelides and Londra, 2000; Candemir and Gülser, 2011). This suggests that manure application to sandy loam or sandy clay loam soil plays the greatest role in soil fertility improvements and nutrient cycles. One exceptional greatest increase was Acp activity in clay loam soil when manure was applied. As was shown in our database, most of the sampling points (i.e., 19, Figure 3.10) to calculate the effect size of Acp activity in clay loam soil are from Liang et al.(2015). They attributed the strong increase of Acp activity to the strong secretion of Acp by roots. However, this does not explain the lower response in other soils and thus further investigations are required.

3.1.5.3 *Management characteristics*

Larger increases of SOC, TN, and microbial C and N following manure application were observed in the lab-controlled conditions than in the field, thus indicating that lab studies may overestimate manure effects. Laboratory experiments include pot trials, controlled dark incubation and greenhouse trials. Experimental durations are shorter than those of field experiments, lasting from days to one year (Franco-Andreu et al., 2017; Nicolas et al., 2012; Tripathy et al., 2008). In the short term, a relatively slow decomposition of organic matter and abundant labile organic compounds in manure may contribute to the extreme increase. This argument is also supported by the greater increase of SOC, TN and microbial C when the experimental duration is within 0.1 year. Higher temperature is generally common for experiments under controlled conditions compared to field experiments (Max et al., 2012; Xu et al., 2019). This may also contribute to the overestimation of manure effects on microbial C and N content.

The combined application of manure with mineral fertilizers lowers the manure effects on soil biochemical properties, which is in accordance with the second hypothesis. The major benefit of enzyme production is the release of organic monomers or nutrients that microbes and plant roots can take up. Evolutionary economic mechanisms of enzyme production suggest that microbial communities function similarly to economic units, maximizing their productivity by allocating resources to extracellular C-, N-, and P-releasing enzymes, depending on the substrate quality and nutrient limitations (Allison et al., 2011). Thus, excessive mineral fertilization suppresses the need for enzyme expression. Compared with the effects of mineral fertilizers (e.g., N, P and N+P) on soil biochemical properties, as shown by previous studies (Jian et al., 2016; Luo et al., 2018), the effect of manure + mineral fertilizers was much larger. This means that even though mineral fertilizer attenuates the influence of manure on soil biochemical properties, the combined effect of manure + mineral fertilizers is still stronger than the application of mineral fertilizers alone. Nevertheless, the greater increase of soil biochemical properties in the manure-alone application did not represent the optimal conditions for plant growth. Long-term use of manure was found to have a similar influence on crop production compared to mineral fertilizers when applied at equivalent N rates (Celestina et al., 2019; Chen et al., 2018; Edmeades, 2003). This could be a result of the high competition for nutrients between microbes and plants following manure addition may slow plant growth, although it improves the soil fertility (Kuzyakov and Xu, 2013; Liu et al., 2017a, 2017b).

3.1.5.4 *Manure characteristics*

Composted manure had nominally lower C content but nominally larger N and P content than non-composted manure, thus composting increased the nutrient content and decreases the C/N and C/P ratios (Figure 3.1). Composting reduces pathogens and parasites, weakens seed viability and suppresses soil-borne plant diseases (Larney and Hao, 2007; Mehta et al., 2013). Composting also contributes to humification of organic residues which are more beneficial for plant growth (Cavagnaro, 2015; Mehta et al., 2013; Quilty and Cattle, 2011). These special characteristics of composts cause a greater increase of microbial C and of all enzyme activities as compared with noncomposted manure, revealing stronger organic matter decomposition and nutrient cycling following composted manure application.

Regarding manure types, swine manure application resulted in the greatest increase of N-cycling-related parameters [i.e., microbial N (230%), Ure (258%) and Nag (138%) activities]. This implies that swine manure strongly stimulates N mineralization. This may be explained by the relatively low C/N ratio of swine manure ($\sim 12 \pm 0.9$), which induces a decrease in the soil microbial C/N ratio and increases the N demand of microbes (Grosso et al., 2016; Nicolardot et al., 2001). These results, in accordance with the third hypothesis, emphasize the importance of manure characteristics (e.g., manure types, composted or noncomposted) for manure effects on soil biochemical properties.

3.1.6 Summary

Given that enormous amounts of manure are produced globally, it is necessary to promote nutrient recycling by using manure resources instead of synthetic or nonrenewable mineral fertilizers (Powers et al., 2019). In comparison with previous studies on mineral fertilizers (Jian et al., 2016; Luo et al., 2018), our review demonstrates that manure application is more beneficial for soil fertility. However, manure application does not guarantee better crop production, as suggested by the high Akp/Acp ratio found after manuring (i.e., ~0.86 > the Akp/Acp target ratio of 0.5; Figure 3.12). The quadratic relationship (Figure 3.12) suggested that a manure application rate of $25 \text{Mg·ha}^{-1} \cdot \text{year}^{-1}$ is optimal pertaining to crop growth and soil fertility. Dick et al. (2000) suggested that manuring with additional lime treatment can also help in reducing the Akp/Acp ratio (Dick et al., 2000).

The addition of composted manure, and manure in the absence of fertilizers, had a greater effect on most biochemical properties than noncomposted manure or manure in the presence of fertilizers (Figure 3.13). The application of manure to Entisols also shows a strong response in comparison to other soils. The application of swine manure, especially, may increase the risk of N loss due to its greater effect on N mineralization. Therefore, incorrect timing or strategy of manure application may increase N loss and pollution via NO_3^- leaching or gaseous emissions due to rapid organic matter mineralization in the absence of plant N uptake (Xia et al., 2017). The addition of P to soil by manuring may also threaten water quality because of the role of P in eutrophication of water resources (Tabbara, 2003). Some contaminants (e.g., phthalic acid esters, heavy metals, and so on) toxic to microbes and enzymes may also be introduced into soil following manure application (He et al., 2015; Tejada et al., 2011). The presence of antibiotic residues, human pathogens, and so on may also pose potential health risks to public health (Venglovsky et al., 2009). Consequently, the identification of these potential threats to soil and plant growth before manure application is crucial and should be taken into consideration when designing agricultural management strategies. Furthermore, other important parameters (e.g., soil depth, tillage systems; Shirani et al., 2002; Zaller and Köpke, 2004) may also have significant influence on the manure effect and therefore should be included in further investigation.

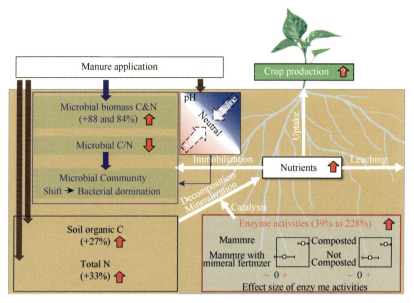

Figure 3.13 Impact of manure application on soil biochemical properties. The red upward and downward arrows represent the increase and decrease of soil biochemical properties. The percentage in the parentheses shows the increment of soil biochemical properties. The embedded figure about pH illustrates the neutralizing effect of manure application on soil, which may be one reason for the microbial community shift. Manure application increases soil C sequestration and accelerates nutrient cycles, but it does not guarantee improvement in crop production because of strong nutrient immobilization. Credit: Reprinted with permission from Liu et al., 2020.

3.2 Spatiotemporal patterns of enzyme activities after manure application reflect mechanisms of niche differentiation between plants and microorganisms

3.2.1 Introduction

The application of livestock manure has been widely accepted as a sustainable management practice in agriculture, providing environmentally and agronomically sound outcomes (Risse et al., 2006; Brandjes et al., 1996; Scotti et al., 2015). Manure incorporation into soil forms a detritusphere abundant in organic carbon (OC) and nutrients (Moore et al., 2004). It is beneficial for improvement of soil quality and crop production (Butler et al., 2013; Calleja-Cervantes et al., 2015; Zaller and Köpke, 2004).

The application strategy is an important aspect of manure management (Webb et al., 2010; Thomsen, 2005). It affects soil-plant-microbial interactions by determining the locations of nutrients or altering soil properties (moisture, O_2 diffusion, bulk density) (Acosta-Martínez and Waldrip, 2014; Zhu et al., 2015). As a consequence, responses of plants and microorganisms vary depending on the manure application strategy. For instance, mixing of manure into soil increased soil microbial biomass (Lovell and Jarvis, 1996; Malik et al., 2013), but no response of soil microbial biomass was observed when manure pats were placed on the soil surface (Lovell and Jarvis, 1996; Cai et al., 2014). Although remarkable increases in plant production have been reported after either incorporating manure into soil (Malik et al., 2013) or broadcasting manure on the soil surface (Aarons et al., 2009; Matilla, 2006), a direct comparison of plant production under various manure application strategies is still lacking.

Enzymes, excreted by both plants and microbes, are early indicators of soil quality and the main mediators of organic matter decomposition (Nannipieri et al., 2007; Sinsabaugh et al., 2008). Assays of enzyme activities have been widely used to investigate the influence of manure application on soil nutrient cycling and microbial activities. Most studies observed significantly increased enzyme activities in soils amended with livestock manures (Liang et al., 2014; Calleja-Cervantes et al., 2015; Bell et al., 2006). However, the study of spatial and temporal responses of enzyme activities requires advanced visualization technology (Acosta-Martínez and Waldrip, 2014).

On the Qinghai-Xizang Plateau, yaks are one of the main species of livestock, and around 40% of their manure is used as fertilizer for cropland and pastures (FAO, 2003; Wang, 2009). However, the impact of yak manure application strategies on the growth of Xizang barley (a staple crop) and on soil enzyme activities remains unknown. Such knowledge could lead to better manure application strategies. We used soil from the Qinghai-Xizang Plateau for better consideration of local nutrient conditions and soil properties, and in the context of prevalent ecosystem degradation (Babel et al., 2014; Hafner et al., 2012).

Here we used direct soil zymography (Razavi et al., 2016) to investigate the impact of different yak manure application strategies on the growth of Xizang barley (*Hordeum vulgare* L.) and on the temporal and spatial patterns of enzyme activities in Xizang soil. We compared manure application strategies using three treatments. ①Localized manure: manure application as a layer in the upper soil; ②Homogenized manure: mixing manure throughout the soil; ③No manure: a control without manure application. Our objectives were to investigate the effects of manure application strategy on plant shoot and root biomass and on the spatial and temporal patterns of soil enzyme activities. Direct soil zymography was used to visualize and quantify the spatial and temporal distribution of enzyme activity for the three enzymes: β-glucosidase, phosphomonoesterase and N-acetyl-glucosaminidase. β-glucosidase is responsible for catalyzing the hydrolysis of terminal 1,4-linked β-D-glucose residues from β-D- glucosides (German et al., 2011), and it is involved in the carbon (C) cycle. Phosphomonoesterase, which catalyzes the hydrolysis of organic phosphorus (P) compounds to inorganic P (Eivazi and Tabatabai, 1977; Malcolm, 1983), is involved in the P cycle. N-acetyl-glucosaminidase (chitinase), which accomplishes the decomposition of chitin to yield low molecular weight chitooligomers (Hamid et al., 2013), is responsible for C- and nitrogen (N) - acquisition. The considerable addition of labile organic compounds and nutrients in manure are expected to greatly influence plant and microorganism activities, and soil enzyme activities. We hypothesized: weaker enzyme activities at the root-soil interface as compared with a strong increase of enzyme activities in the manure-induced detritusphere (H1); stronger stimulation of plant growth by the homogenized manure application strategy (H2).

3.2.2 Temporal response of enzyme activities to manure application strategy

Enzyme activities increased from day 5 to day 25 after manure application, but decreased after 25 days (Figure 3.14—Figure 3.17). Phosphomonoesterase, β-glucosidase and chitinase activities were 47%—104% higher on day 25 than on day 5. However, their activities had decreased 10%—27% by day 45 relative to the activities on day 25. In the control group, these enzyme activities had increased 40%—72% by day 25 compared with their activities on day 5, but showed no significant changes between days 25 and 45 (-12% to $+9\%$, $P>0.05$).

All enzyme activities increased with manure application relative to the control (Figure 3.18). Homogenized manure generally induced larger increases than localized manure. Homogenized manure increased phosphormonoesterase, β-glucosidase and chitinase activity by 6%—41% in comparison with the control (Figure 3.18). In contrast, localized manure induced an increase of phosphomonoesterase and chitinase activities by 7%—29% as compared with the control. Localized manure also increased β-glucosidase activity by 16% and 37% on days 5 and 25, but its activity was 8% lower than the control on day 45 ($P>0.05$).

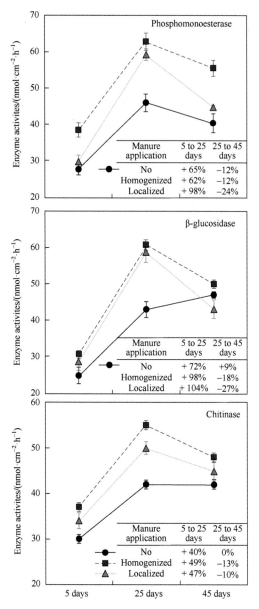

Figure 3.14 Response of phosphomonoesterase (top), β-glucosidase (middle) and chitinase (bottom) activities to manure application strategies over time. The embedded tables show relative changes of enzyme activities between 5 and 25 days, and 25 and 45 days. Error bars represent standard deviations ($\pm$SD). Credit: Reprinted with permission from Liu et al., 2017.

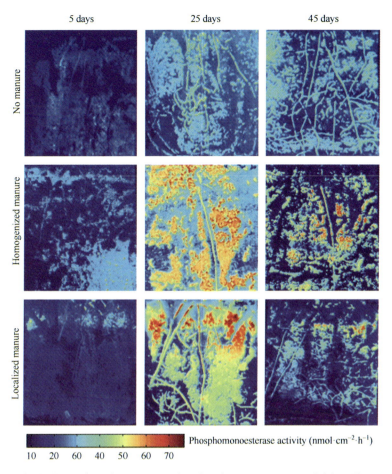

Phosphomonoesterase activity (nmol·cm^{-2}·h^{-1})

10 20 60 40 50 60 70

Figure 3.15 Examples of zymograms for phosphomonoesterase activities. Three rows represent response of activities to three manure application strategies: No manure, Homogenized manure and Localized manure. Figures from left to right are the measurements at days 5, 25 and 45. The color bar corresponds to phosphomonoesterase activity (nmol·cm^{-2}·h^{-1}). Credit: Reprinted with permission from Liu et al., 2017.

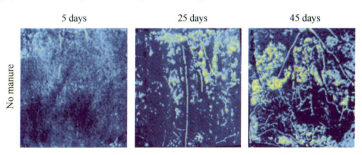

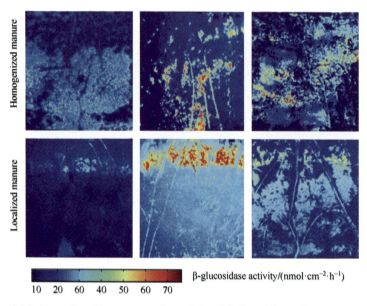

β-glucosidase activity/(nmol·cm^{-2}·h^{-1})

10 20 60 40 50 60 70

Figure 3.16 Examples of zymograms for activity of β-glucosidase. Three rows represent response of activities to three manure application strategies: No manure, Homogenized manure and Localized manure. Figures from left to right are the measurements at days 5, 25 and 45. The color bar corresponds to activity of β-glucosidase(nmol·cm^{-2}·h^{-1}). Credit: Reprinted with permission from Liu et al., 2017.

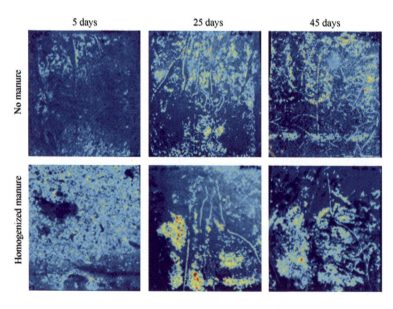

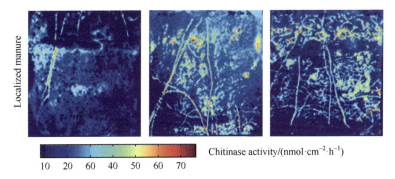

Chitinase activity/(nmol·cm^{-2}·h^{-1})

10   20   60   40   50   60   70

Figure 3.17 Examples of zymograms for activity of chitinase. Three rows represent response of activities to three manure application strategies: No manure, Homogenized manure and Localized manure. Figures from left to right are the measurements at days 5, 25 and 45. We speculated that the higher activity of chitinase at the rhizoplane in soil with localized manure at day 25 may be induced by the interactions between roots and mycorrhizal fungi. The color bar corresponds to activity of chitinase(nmol·cm^{-2}·h^{-1}). Credit: Reprinted with permission from Liu et al., 2017.

The capability of manure to regulate soil biological processes was controlled by the manure application strategy. Homogenized manure induced higher activities of C-, N-, and P-acquisition enzymes than localized manure (Figure 3.18). Three mechanisms drove these differences in response to manure application strategies. The first two mechanisms were manure-induced changes to soil physico-chemical properties (Haynes and Naidu, 1998; Dungait et al., 2009). First, by mixing into soil, manure directly loosened the soil, decreased the bulk density and increased the soil porosity (Celik et al., 2004). Second, labile organic compounds and nutrients in the manure were also sufficiently mixed with soil following homogenized manure application. The third mechanism was the loading of indigenous enzymes and microbes from manure into the soil (Dinesh et al., 1998; Criquet et al., 2007; Tiquia, 2002). These mechanisms provided a favorable environment for soil microbial proliferation and activity in the rhizoboxes with homogenized manure. In contrast, localized manure affected biological processes through gradual leaching of soluble organic substances and mineral nutrients into the soil (Dickinson et al., 1981). Transport of indigenous enzymes from the manure layer into the soil was negligible due to strong adsorption by soil particles (Poll et al., 2006). Consequently, the combined effects of abiotic (e.g., loose soil structure) and biotic factors (e.g., OC, nutrients, enzymes and microbes) induced higher enzyme activities in the rhizoboxes with homogenized manure.

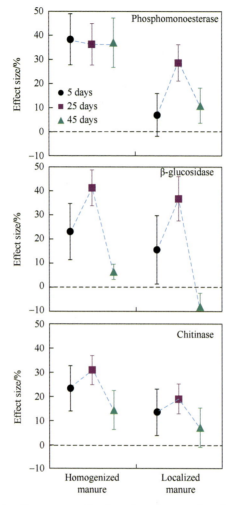

Figure 3.18 Effects of manure application strategies on phosphomonoesterase (top), β-glucosidase (middle) and chitinase (bottom) activities in the whole soil. The effect size shows the change of enzyme activities in soil with homogenized or localized manure addition compared to the control. Error bars represent standard deviations ($\pm$ SD). Credit: Reprinted with permission from Liu et al., 2017. Copyright 2017 Elsevier.

Activities of all tested enzymes demonstrated a consistent pattern over time for both manure application strategies: ①All enzyme activities increased in the first 25 days. Most enzyme activities in the homogenized and localized manure applications were higher than in the control group (Figure 3.14). ②Enzyme activities decreased from day 25 to day 45 in the homogenized and localized manure applications. In contrast, all enzyme activities in the control group remained stable during this period. This indicated that the heightened enzyme activities in the homogenized and localized manure applications were mainly caused by the manure-induced detritusphere. Indeed, manure added quite substantial amounts of labile organic substances to the soil, thereby increasing microbial activity and thus nutrient demand and enzyme expression. Over time, these substances were completely decomposed, resulting in lower microbial activity and thus reductions in enzyme activity. Similarly, studies based on destructive methods demonstrated such short-term acceleration of microbial processes induced by sewage sludge (Criquet et al., 2007; Pascual et al., 2002).

Furthermore, the stable enzyme activities from day 25 to day 45 in the control group (with only the rhizosphere effect of barley, Figure 3.19) demonstrated that the duration of hot moments in the rhizosphere was at least 20 days. This was much longer than the lifetime (only a few days) of hotspots initiated by single releases of root exudate, as evidenced by time-resolved ^{14}C

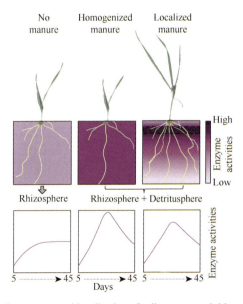

Figure 3.19 General responses and localization of soil enzyme activities to manure application strategies over time. A clear detritusphere extension of enzyme activities was observed below the localized manure. The manure-induced detritusphere stimulated larger increases of enzyme activities than the "No manure" treatment (i.e. only the rhizosphere effect of barley), although the increase lasted less than 45 days. Homogenized manure elevated enzyme activities more than localized manure, while localized manure induced higher shoot and root biomass than homogenized manure. Credit: Reprinted with permission from Liu et al., 2017.

imaging after $^{14}CO_2$ pulse labeling of *Lolium perenne* (Pausch and Kuzyakov, 2011). Therefore, we conclude that continuous inputs of labile organics due to root growth prolonged the duration of hot moments in the rhizosphere.

3.2.3 Spatial response of enzyme activities to manure application strategies

A clear downward extension of enzyme activities from the manure layer into the underlying soil was observed with localized manure application (Figure 3.20). This extension was enzyme-specific: for example, phosphomonoesterase activity extended from 3.1cm on day 5 to 9.2cm on day 25 and finally exceeded 10cm depth on day 45. In comparison, the extension of activity of β-glucosidase was less (3.1cm on day 5, 4.7cm on day 25 and 7.0cm on day 45). Such extension was not seen for chitinase (data not shown). All distances were measured from the top of the rhizobox and included the depth of the manure layer.

The spatial distribution of enzyme activities was noticeably affected by manure application strategy. In localized manure application, enzyme activities in the top manure layer were higher than those in the control group and in the soil below this top manure layer (Figure 3.15–Figure 3.17), which was in consistent with our first hypothesis (H1) about the strong increase of enzyme activities in the detritusphere as compared with the root-soil interface below the manure layer. This means that manure itself had higher enzyme and microbial activities, in support of the mechanism that manure-derived enzymes or microorganisms contributed to the increased enzyme and microbial activities (Calleja-Cervantes et al., 2015; Dinesh et al., 1998). Though indigenous enzymes of manure were all concentrated in the localized manure layer, the nutrients in this layer could be leached downward. Leaching of available nutrients and available organics from the manure layer stimulated microorganisms, and consequently enzyme activities, in the soil below the layer (Kang et al., 2009). Therefore, enzyme activities extended downwards over time because of the redistribution of nutrients and organics (Figure 3.20). Such extension indicated a gradual influence of manure on soil biochemical processes and this influence was also enzyme-specific. For instance, phosphomonoesterase exhibited deeper and faster downward extension (Figure 3.20). In contrast, this extension was narrower for β-glucosidase. The significantly higher ratio of E_R to E_H for phosphomonoesterase (~1.13) in the control group also demonstrated that phosphomonoesterase activity on the rhizoplane was 13% higher than in the soil hotspots, suggesting that the soil was P-deficient (Ren et al., 2016) and thus the plant secreted more phosphomonoesterase to obtain inorganic P for its growth (Hunter et al., 2014). In comparison, β-glucosidase, which is mostly involved in the degradation of cellulose, showed a narrower extension. The wider extension of phosphomonoesterase compared to β-glucosidase activities has also been observed in the rhizosphere (Razavi et al., 2016). In the present study, the detritusphere extension of enzyme activities was much wider than the rhizosphere extension observed in other studies (e.g., Razavi et al., 2016; Tarafdar and Jungk, 1987; Sauer et al., 2006). This is explained by the direction of water fluxes: to the roots in the rhizosphere, but from the manure layer downwards. Consequently, extension of enzyme activities was much faster and wider in the detritusphere than in the rhizosphere, due to vertical diffusion and leaching processes.

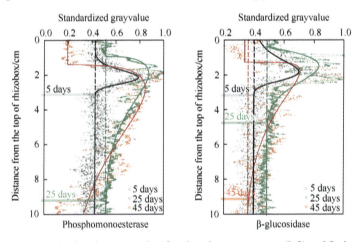

Figure 3.20 The detritusphere extension for phosphomonoesterase (left) and β-glucosidase activities (right) from the initial manure layer at the top (presented as the shaded area between 1.0cm and 2.5cm) over time. The depth from the manure application to the constant level of the regression curve was considered as the detritusphere extension of enzyme activity. This distance at days 5 and 25 was marked by semitransparent strips (black for 5 days and green for 25 days). Due to the limited rhizobox size, the roots started to grow laterally once they reached the bottom, after around 10—15 days of growth, inducing very high root densities at the bottom (ca. 2—3cm). To avoid artefacts from high root densities, we used only the upper 10cm of the membrane. According to the regression, the depth at day 45 already exceeded the membrane boundary (>10cm) and thus was not presented. Five-parameter Weibull regression was used to fit enzyme activities with the distance from the top of the rhizobox. Reprinted with permission from Liu et al., 2017.

3.2.4　Response of plants to manure application strategies

The ratio of E_R to E_H (enzyme activities on the rhizoplane to that in soil hotspots) was below 1.0 following the application of manure (Figure 3.21, $P<0.05$), indicating that average enzyme activities on the rhizoplane were lower than the activities in manure-induced soil hotspots. This ratio did not change over time for homogenized manure application. Phosphomonoesterase and β-glucosidase activities on the rhizoplane were both around 10% lower than that in the soil hotspots, while activity of chitinase was 15% lower. In contrast to the homogenized manure, when localized manure was applied, the E_R to E_H ratio decreased from day 25 to day 45. For instance, the ratio of phosphomonoesterase activities decreased from 0.89 to 0.74, while that of β-glucosidase decreased from 0.79 to 0.68. For chitinase, this ratio had the highest change (from 0.98 to 0.75). For the control, this ratio was always around 1.0, except for that of phosphomonoesterase at day 25 (~1.13).

Manure application strategy had significant impact on shoot and root biomass of barley (Figure 3.22). Localized manure produced the highest shoot and root biomass (3.1 and 6.7 times, respectively, higher than for homogenized manure, $P<0.05$). Localized manure significantly decreased the shoot/root ratio from 2.7 to 1.1 ($P<0.05$), indicating that manure application strategies modified the trade-off between shoot and root biomass.

Following manure application, enzyme activities in the manure-induced detritusphere were higher than on root surfaces. This — in accordance with our first hypothesis (H1)—indicated that the detritusphere became more attractive to microbes than the rhizosphere, because high microbial activities tended to be in the hotspots of bulk soil instead of being balanced between the rhizoplane and soil hotspots, as in the control group ($E_R \approx E_H$). Furthermore, though enzyme activities on the rhizoplane were lower, the reason for this may differ between homogenized and localized manure applications. For the homogenized manure application, tough competition for inorganic and organic nutrients between microbes and roots was initiated as soon as manure-derived microbes and labile substrates were introduced by mixing manure into the soil (Kuzyakov and Blagodatskaya, 2015; Malik et al., 2013; Xu et al., 2006, 2011). This may temporarily reduce plant nutrient availability, depress root growth and explain the lower enzyme activities on the root surface for homogenized manure application. In contrast, with localized manure application, the pre-existing and newly mineralized nutrients were easily leached downward and competition between microbes and plant roots within the localized manure layer was

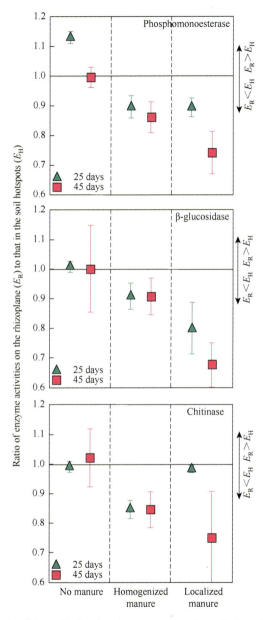

Figure 3.21 Ratio of E_R to E_H for phosphomonoesterase (top), β-glucosidase (middle) and chitinase (bottom). E_R and E_H are the average enzyme activities on the rhizoplane and in the soil hotspots, respectively. The values above 1.0 reflect higher enzyme activities around the roots than in hotspots in root-free soil areas. Error bars represent standard deviations ($\pm$SD). Credit: Reprinted with permission from Liu et al., 2017.

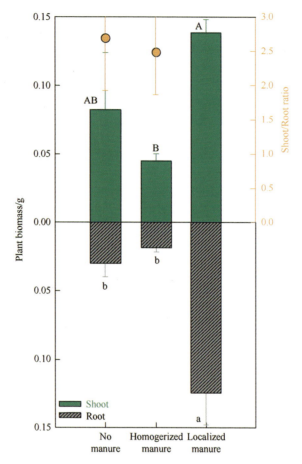

Figure 3.22 Plant biomass and shoot/root ratio under three manure application strategies: No manure, Homogenized manure and Localized manure. The capital and lower-case letters show significant differences between application strategies ($P<0.05$). Error bars represent standard deviations ($\pm$SD). Credit: Reprinted with permission from Liu et al., 2017.

weaker than that of the homogenized manure application. This spatial niche differentiation for the manure microbial community and roots decreased their competition for nutrients and simultaneously increased nutrient uptake, and so, the plant biomass. Both situations were also reflected in the shoot and root biomass at days 45: shoot and root biomass with localized manure application were respectively 3.1 and 6.7 times higher than those of homogenized manure application, and so our second hypothesis (H2) was rejected. Compared to the control, the relatively low plant biomass in the homogenized manure application also indicated that strong competition between microbes and roots existed when manure was homogenized with soil. This significant difference demonstrated that localized manure was more advantageous for barley growth than homogenized manure.

This is especially important on the Qinghai-Xizang Plateau, because soils have been very seriously degraded in the last 30−50 years due to intensive human activities (e.g., overgrazing) and climate change (Chen et al., 2013). This has induced large soil organic carbon and nutrient losses and thus considerably decreased soil fertility. Localized manure application has been found to increase soil ammonium and nitrate concentrations in soils of the Qinghai-Xizang Plateau (Cai et al., 2014; He et al., 2009). However, manure application at the soil surface leads to ammonia volatilization, involving significant nitrogen losses and negative effects on the environment. This is especially important on the Qinghai-Xizang Plateau, because the solar radiation is much higher compared with other regions around the world (Liu et al., 2012a, 2012b), which increases the temperature of manure and accelerates the ammonia volatilization. Alternatively, homogenized manure application may reduce nitrogen losses by avoiding the impact of solar radiation. Therefore, to thoroughly investigate the impact of both manure application strategies, the effects on nitrogen emissions and leaching should be considered.

3.2.5 Summary

For the first time, we elucidated and visualized the impacts of different manure application strategies on enzyme activities in soil in situ, spatially and temporally. The manure-induced detritusphere increased enzyme activities more than the rhizosphere effect of barley alone. Manure-induced hotspots also showed higher enzyme activities than the rhizoplane. Together, these findings demonstrate that microbial activities in the detritusphere are much more stimulated than on the root-soil interface (i.e., rhizosphere and rhizoplane). The detritusphere's vertical extension of phosphomonoesterase activity from the localized manure application was much faster than that of activity of β-glucosidase. Overall, homogenized manure increased enzyme activities more than localized manure. However, localized manure induced 3.1 and 6.7 times higher shoot and root biomass, respectively. We conclude that localized manure application decreases competition for nutrients between microorganisms and roots and simultaneously increases plant performance.

Main References

Aarons, S.R., O'Connor, C.R., Hosseini, H.M., et al., 2009. Dung pads increase pasture production, soil nutrients and microbial biomass carbon in grazed dairy systems. Nutrient Cycling in Agroecosystems, 84(1): 81-92. Available from: https://doi.org/10.1007/s10705-008-9228-5.

Acosta-Martínez, V., Tabatabai, M.A., 2000. Enzyme activities in a limed agricultural soil. Biol. Fertil. Soils, 31(1) 85–91. Available from: https://doi.org/10.1007/s003740050628.

Acosta-Martínez, V., Waldrip, H.M., 2014. Soil enzyme activities as affected by manure types, application rates, and management practices. In: He, Z., Zhang, H. (Eds.), Applied Manure and Nutrient Chemistry for Sustainable Agriculture and Environment. Dordrecht: Springer: 99-122, Available from: https://doi.org/10.1007/978-94-017-8807-6_6.

Aggelides, S.M., Londra, P.A., 2000. Effects of compost produced from town wastes and sewage sludge on the physical properties of a loamy and a clay soil. Bioresource Technology, 71(3):253–259. Available from: https://doi.org/10.1016/S0960-8524(99)00074-7.

Ali, B., Shah, G.A.M., Traore, B., et al., 2019. Manure storage operations mitigate nutrient losses and their products can sustain soil fertility and enhance wheat productivity. Journal of Environmental Management, 241: 468–478. Available from: https://doi.org/10.1016/j.jenvman.2019.02.081.

Allison, S.D., Weintraub, M.N., Gartner, T.B., et al., 2011. Evolutionary-economic principles as regulators of soil enzyme production and ecosystem function. Soil Enzymology, 22: 229–243. Available from: https://doi.org/10.1007/978-3-642-14225-3.

Astier, M., Maass, J.M., Etchevers-Barra, J.D., et al., 2006. Short-term green manure and tillage management effects on maize yield and soil quality in an Andisol. Soil and Tillage Research, 88(1-2): 153–159. Available from: https://doi.org/10.1016/j.still.2005.05.003.

Babel, W., Biermann, T., Coners, H., et al., 2014. Pasture degradation modifies the water and carbon cycles of the Tibetan highlands. Biogeosciences, 11: 6633-6656. Available from: https://doi.org/10.5194/bg-11-6633-2014.

Bandick, A.K., Dick, R.P., 1999. Field management effects on soil enzyme activities. Soil Biology and Biochemistry, 31(11): 1471-1479. Available from: https://doi.org/10.1016/S0038-0717(99)00051-6.

Bell, C., Carrillo, Y., Boot, C.M., et al., 2014. Rhizosphere stoichiometry: Are C: N: P ratios of plants, soils, and enzymes conserved at the plant species-level? The New Phytologist, 201(2):,505–517. Available from: https://doi.org/10.1111/nph.12531.

Bell, J.M., Robinson, C.A., Schwartz, R.C., 2006. Changes in soil properties and enzymatic activities following manure applications to a rangeland. Rangeland Ecology & Management, 59(3): 314-320. Available from: https://doi.org/ 10.2111/05-172R1.1.

Brandjes, P.J., de Wit, J., van der Meer, H.G., 1996. Environmental impact of animal manure management. Food and Agriculture Organization of the United Nations. Available from: https://www.fao.org/wairdocs/lead/x6113e/x6113e00.htm#Contents.

Butler, T.J., Weindorf, D.C., Han, K.J., et al., 2013. Dairy manure compost quality effects on corn silage and soil properties. Compost Science & Utilization, 17(1): 18-24. Available from: https://doi.org/10.1080/1065657X.2009.10702395.

Cai, Y.J., Wang, X.D., Tian, L.L., et al., 2014. The impact of excretal returns from yak and Tibetan sheep dung on nitrous oxide emissions in an alpine steppe on the Qinghai-Tibetan Plateau. Soil Biology and Biochemistry, 76: 90-99. Available from: https://doi.org/10.1016/j.soilbio.2014.05.008.

Calleja-Cervantes, M.E., Fernández-González, A.J., Irigoyen, I., et al., 2015. Thirteen years of continued application of composted organic wastes in a vineyard modify soil quality characteristics. Soil Biology and Biochemistry, 90: 241-254. Available from: https://doi.org/10.1016/j.soilbio.2015.07.002.

Candemir, F., Gülser, C., 2011. Effects of different agricultural wastes on some soil quality indexes in clay and loamy sand fields. Communications in Soil Science and Plant Analysis, 42(1): 13–28. Available from: https://doi.org/10.1080/00103624.2011.528489.

Celestina, C., Hunt, J.R., Sale, P.W.G.G., et al., 2019. Attribution of crop yield responses to application of organic amendments: A critical review. Soil and Tillage Research, 186: 135–145. Available from: https://doi.org/10.1016/j.still.2018.10.002.

Celik, I., Ortas, I., Kilic, S., 2004. Effects of compost, mycorrhiza, manure and fertilizer on some physical properties of a Chromoxerert soil. Soil and Tillage Research, 78(1): 59-67. Available from: https://doi.org/10.1016/j.still.2004.02.012.

Chen, H.Q., Hou, R.X., Gong, Y.S., et al., 2009. Effects of 11 years of conservation tillage on soil organic matter fractions in wheat monoculture in Loess Plateau of China. Soil and Tillage Research, 106(1): 85–94. Available from: https://doi.org/10.1016/j.still.2009.09.009.

Chen, H.Q., Zhu, Q.A., Peng, C.H., et al., 2013. The impacts of climate change and human activities on biogeochemical cycles on the Qinghai-Tibetan Plateau. Global Change Biology, 19(10): 2940-2955. Available from: https://doi.org/10.1111/gcb.12277.

Chen, Y.S., Camps-Arbestain M, Shen Q H, et al., 2018. The long-term role of organic amendments in building soil nutrient fertility : A meta-analysis and review. Nutrient Cycling in Agroecosystems, 111(2): 103-125. Agroecosystems. Available from: https://doi.org/10.1007/s10705-017-9903-5.

Cotrufo, M.F., Ranalli, M.G., Haddix, M.L., et al., 2019. Soil carbon storage informed by particulate and mineral-associated organic matter. Nature Geoscience, 12: 989–994. Available from: https://doi.org/0.1038/s41561-019-0484-6.

Criquet, S., Braud, A., Nèble, S., 2007. Short-term effects of sewage sludge application on phosphatase activities and available P fractions in Mediterranean soils. Soil Biology and Biochemistry, 39(4): 921-929. Available from: https://doi.org/10.1016/j.soilbio.2006.11.002.

de Graaff, M.A., Hornslein, N., Throop, H.L., et al., 2019. Effects of agricultural intensification on soil biodiversity and implications for ecosystem functioning: A meta-analysis. Advances in Agronomy. Amsterdam: Elsevier: 1-44. Available from: https://doi.org/10.1016/bs.agron.2019.01.001.

Dick, W.A., Cheng, L., Wang, P., 2000. Soil acid and alkaline phosphatase activity as pH adjustment indicators. Soil Biology and Biochemistry, 32(13): 1915–1919. Available from: https://doi.org/ 10.1016/S0038-0717(00)00166-8.

Dickinson, C.H., Underhay, V.S.H., Ross, V., 1981. Effect of season, soil fauna and water content on the decomposition of cattle dung pats. New Phytologist, 88(1): 129-141. Available from: https://doi.org/ 10.1111/j.1469-8137.1981.tb04576.x.

Dinesh, R., Dubey, R.P., Prasad, G.S., 1998. Soil microbial biomass and enzyme activities as influenced by organic manure incorporation into soils of a rice-rice system. Journal of Agronomy and Crop Science, 181(3): 173-178. Available from: https://doi.org/ 10.1111/j.1439-037X.1998.tb00414.x.

Dinkelaker, B., Marschner, H., 1992. In vivo demonstration of acid phosphatase activity in the rhizosphere of soil-grown plants. Plant Soil, 144(2): 199–205. Available from: https://doi.org/10.1007/BF00012876.

Dungait, J.A.J., Bol, R., Bull, I.D., et al., 2009. Tracking the fate of dung-derived carbohydrates in a temperate grassland soil using compound-specific stable isotope analysis. Organic Geochemistry, 40(12): 1210-1218. Available from: https://doi.org/10.1016/j.orggeochem.2009.08.001.

Eden, M., Gerke H H, Houot S, 2017. Organic waste recycling in agriculture and related effects on soil water retention and plant available water: A review. Agronomy for Sustainable Development, 37(2): 11. Available from: https://doi.org/10.1007/s13593-017-0419-9.

Edmeades, D.C., 2003. The long-term effects of manures and fertilisers on soil productivity and quality: A review. Nutrient Cycling in Agroecosystems, 66(2): 165–180. Available from: https://doi.org/10.1023/A:1023999816690.

Eivazi, F., Tabatabai, M.A., 1977. Phosphatases in soils. Soil Biology and Biochemistry, 9: 167-172. Available from: https://doi.org/10.1016/0038-0717(77)90070-0.

Ekenler, M., Tabatabai, M.A., 2003. Responses of phosphatases and arylsulfatase in soils to liming and tillage systems. Journal of Plant Nutrition and Soil Science, 166(3): 281–290. Available from: https://doi.org/10.1002/jpln.200390045.

Eswaran, H., Reich, P., Padmanabhan, E., 2012. World soil resources: Opportunities and challenges. In: Lal, R., Stewart, B. (Eds.), World Soil Resources and Food Security. Boca Raton: CRC Press: 43-66.

Fan, M.S, Shen, J.B, Yuan, LX., et al., 2012. Improving crop productivity and resource use efficiency to ensure food security and environmental quality in China. Journal of Experimental Botany, 63(1): 13–24. Available from: https://doi.org/10.1093/jxb/err248.

FAO, 2003. The Yak (second edition). Available from: http://www.fao.org/docrep/006/ad347e/ad347e01. htm#bm21.9.

FAO, 2015. Intergovernmental Technical Panel on Soils. Status of the World's Soil Resources.

Fisher, K.A., Yarwood, S.A., James, B.R., 2017. Soil urease activity and bacterial ureC gene copy numbers: Effect of pH. Geoderma, 285: 1–8. Available from: https://doi.org/10.1016/j.geoderma.2016.09.012

Foster, E.J., Hansen, N., Wallenstein, M., et al., 2016. Biochar and manure amendments impact soil nutrients and microbial enzymatic activities in a semi-arid irrigated maize cropping system. Agriculture, Ecosystems & Environment, 233: 404–414. Available from: https://doi.org/10.1016/j.agee.2016.09.029.

Franco-Andreu, L., Gómez, I., Parrado, J., et al., 2017. Soil biology changes as a consequence of organic amendments subjected to a severe drought. Land Degradation & Developmen, 28(3): 897–905. Available from: https://doi.org/10.1002/ldr.2663.

Friedel, J.K., Ehrmann, O., Pfeffer, M., et al., 2006. Soil microbial biomass and activity: The effect of site characteristics in humid temperate forest ecosystems. Journal of Plant Nutrition and Soil Science, 169(2): 175–184. Available from: https://doi.org/10.1002/jpln.200521763.

Geisseler, D., Scow, K.M., 2014. Long-term effects of mineral fertilizers on soil microorganisms: A review. Soil Biology and Biochemistry, 75: 54–63. Available from: https://doi.org/10.1016/j.soilbio.2014.03.023.

Geisseler, D., Linquist, B.A., Lazicki, P.A., 2017. Effect of fertilization on soil microorganisms in paddy rice systems: A meta-analysis. Soil Biology and Biochemistry, 115: 452–460. Available from: https://doi.org/10.1016/j.soilbio.2017.09.018.

Gentsch, N., Mikutta, R., Shibistova, O., et al., 2015. Properties and bioavailability of particulate and mineral-associated organic matter in Arctic permafrost soils, Lower Kolyma Region, Russia. European Journal of Soil Science, 66(4): 722–734. Available from: https://doi.org/10.1111/ejss.12269.

German, D.P., Weintraub, M.N., Grandy, A.S., et al., 2011. Optimization of hydrolytic and oxidative enzyme methods for ecosystem studie. Soil Biology and Biochemistry, 43(7): 1387-1397. Available from: https://doi.org/10.1016/j.soilbio.2011.03.017.

Giacometti, C., Cavani, L., Baldoni, G., et al., 2014. Microplate-scale fluorometric soil enzyme assays as tools to assess soil quality in a long-term agricultural field experiment. Applied Soil Ecology, 75: 80–85. Available from: https://doi.org/10.1016/j.apsoil.2013.10.009.

Grosso, F., Bååth, E., De Nicola, F., 2016. Bacterial and fungal growth on different plant litter in Mediterranean soils: Effects of C/N ratio and soil pH. Applied Soil Ecology, 108: 1–7. DOI:10.1016/j.apsoil.2016.07.020.

Gurevitch, J., Koricheva, J., Nakagawa, S., et al., 2018. Meta-analysis and the science of research synthesis. Nature, 555(7695): 175–182. DOI:10.1038/nature25753.

Hafner, S., Unteregelsbacher, S., Seeber, E., et al., 2012. Effect of grazing on carbon stocks and assimilate partitioning in a Tibetan montane pasture revealed by $^{13}CO_2$ pulse labeling. Global Change Biology, 18(2): 528-538. Available from: https://doi.org/10.1111/j.1365-2486.2011.02557.x.

Hamid, R., Khan, M.A., Ahmad, M., et al.,, 2013. Chitinases: An update[J]. Journal of Pharmacy and Bioallied Sciences, 5(1): 21-29. Available from: https://doi.org/10.4103/0975-7406.106559.

Hatfield, J.L., Sauer, T.J., Cruse, R.M., 2017. Soil: The forgotten piece of the water, food, energy nexus. Advances in Agronomy, 143: 1-46. Available from: https://doi.org/10.1016/bs.agron.2017.02.001

Haynes, R.J., Naidu, R., 1998. Influence of lime, fertilizer and manure applications on soil organic matter. Nutr. Cycl. Agroecosystems, 51: 123–137. Available from: https://doi.org/10.1023/A:1009738307837.

He, L.Z., Gielen, G., Bolan, N.S., et al., 2015. Contamination and remediation of phthalic acid esters in agricultural soils in China: A review. Agronomy for Sustainable Developmen, 35(2): 519–534. Available from: https://doi.org/10.1007/s13593-014-0270-1.

He, Y.X., Sun, G., Liu, L., et al., 2009. Effect of yak dung on high-frigid meadow soil nutrition in northwestern Sichuan, China. Chinese Journal of Applied and Environmental Biology, 15(5): 666-671. (in Chinese with English abstract).

Huang, Y., Zou, J.W., Zheng, X.H., et al., 2004. Nitrous oxide emissions as influenced by amendment of plant residues with different C:N ratios. Soil Biology and Biochemistry, 36(6): 973–981. Available from: https://doi.org/10.1016/j.soilbio.2004.02.009.

Hunter, P.J., Teakle, G.R., Bending, G.D., 2014. Root traits and microbial community interactions in relation to phosphorus availability and acquisition, with particular reference to Brassica. Frontiers in Plant Science, 5: 27. Available from: https://doi.org/10.3389/fpls.2014.00027.

Jian, S.Y., Li, J.W., Chen, J., et al., 2016. Soil extracellular enzyme activities, soil carbon and nitrogen storage under nitrogen fertilization: A meta-analysis. Soil Biology and Biochemistry, 101: 32–43. Available from: https://doi.org/10.1016/j.soilbio.2016.07.003.

Jiang, G., Zhang, W., Xu, M., et al., 2018. Manure and mineral fertilizer effects on crop yield and soil carbon sequestration: a meta-analysis and modeling across China. Global Biogeochemical Cycles, 32(11): 1659–1672. Available from: https://doi.org/10.1029/2018GB005960.

Kallenbach, C., Grandy, A.S., 2011. Controls over soil microbial biomass responses to carbon amendments in agricultural systems: A meta-analysis. Agriculture, Ecosystems & Environment, 144(1): 241–252. Available from: https://doi.org/10.1016/j.agee.2011.08.020.

Kang, H., Kang, S., Lee, D., 2009. Variations of soil enzyme activities in a temperate forest soil. Ecological Research, 24(5): 1137-1143. Available from: https://doi.org/10.1007/s11284-009-0594-5.

Karami, A., Homaee, M., Afzalinia, S., et al., 2012. Organic resource management: Impacts on soil aggregate stability and other soil physico-chemical properties. Agriculture, Ecosystems & Environment, 148: 22-28. Available from: https://doi.org/10.1016/j.agee.2011.10.021.

Kizilkaya, R., 2008. Dehydrogenase activity in Lumbricus terrestris casts and surrounding soil affected by addition of different organic wastes and Zn. Bioresource Technology y, 99(5): 946–953. Available from: https://doi.org/10.1016/j.biortech.2007.03.004.

Kizilkaya, R., Hepsen, S., 2007. Microbiological properties in earthworm cast and surrounding soil amended with various organic wastes. Communications in Soil Science and Plant Analysis, 38(19-20): 2861–2876. Available from: https://doi.org/10.1080/00103620701663107.

Kolb, S.E., Fermanich, K.J., Dornbush, M.E., 2009. Effect of charcoal quantity on microbial biomass and activity in temperate soils. Soil Science Society of America Journal, 73: 1173. Available from: https://doi.org/10.2136/sssaj2008.0232.

Koncki, R., Ogończyk, D., Głąb, S., 2005. Potentiometric assay for acid and alkaline phosphatase. Anal. Analytica Chimica Acta, 538(1-2): 257–261. Available from: https://doi.org/10.1016/j.aca.2005.02.021.

Kuzyakov, Y., Xu, X.L., 2013. Competition between roots and microorganisms for nitrogen: Mechanisms and ecological relevance. The New Phytologist, 198(3): 656-669.

Kuzyakov, Y., Blagodatskaya, E., 2015. Microbial hotspots and hot moments in soil: Concept & review. Soil Biology and Biochemistry, 83: 184-199. Available from: https://doi.org/10.1016/j.soilbio.2015.01.025.

Larney, F.J., Hao, X., 2007. A review of composting as a management alternative for beef cattle feedlot manure in southern Alberta, Canada. Bioresource Technology, 98(17): 3221–3227. Available from: https://doi.org/10.1016/j.biortech.2006.07.005.

Le Bayon, R.C., Binet, F., 2006. Earthworms change the distribution and availability of phosphorous in organic substrates. Soil Biology and Biochemistry, 38(2): 235-246. Available from: https://doi.org/10.1016/j.soilbio.2005.05.013.

Lehmann, J., Schroth, G., 2003. Nutrient leaching, Tree, crops and Soil Fertility.

Li, L., Liang, X.Q., Li, H., et al., 2015. Phosphomonoesterase activities, kinetics and thermodynamics in a paddy soil after receiving swine manure for six years. Pedosphere, 25(2): 294-306. Available from: https://doi.org/10.1016/S1002-0160(15)60014-5.

Li, X.H., Han, X.Z., Li, H.B., et al., 2012. Soil chemical and biological properties affected by 21-year application of composted manure with chemical fertilizers in a Chinese Mollisol. Canadian Journal of Soil Science, 92: 419–428.Available from: https://doi.org/10.4141/CJSS2010-046.

Liang, Q., Chen, H.Q., Gong, Y.S., et al., 2014. Effects of 15 years of manure and mineral fertilizers on enzyme activities in particle-size fractions in a North China Plain soil. European Journal of Soil Biology, 60: 112–119. Available from: https://doi.org/10.1016/j.ejsobi.2013.11.009.

Liu, J.D., Liu, J.M., Linderholm, H.W., et al., 2012. Observation and calculation of the solar radiation on the Tibetan Plateau. Energy Conversion and Management, 57: 23-32. Available from: https://doi.org/10.1016/j.enconman.2011.12.007.

Liu, X.B., Lee Burras, C., Kravchenko, Y.S., et al., 2012. Overview of Mollisols in the world: Distribution, land use and management. Canadian Journal of Soil Science, 92(3): 383-402. Available from: https://doi.org/10.4141/CJSS2010-058.

Liu, S.B., Razavi, B.S., Su, X., et al., 2017a. Spatio-temporal patterns of enzyme activities after manure application reflect mechanisms of niche differentiation between plants and microorganisms. Soil Biology and Biochemistry, 112: 100–109. Available from: https://doi.org/10.1016/j.soilbio.2017.05.006.

Liu, Z.J., Rong, Q.L., Zhou, W., et al., 2017b. Effects of inorganic and organic amendment on soil chemical properties, enzyme activities, microbial community and soil quality in yellow clayey soil. PLoS One, 12(3): e0172767. Available from: https://doi.org/10.1371/journal.pone.0172767.

Liu, S.B., Wang, J.Y., Pu, S.Y., et al., 2020. Impact of manure on soil biochemical properties: A global synthesis. Science of the Total Environment, 745: Available from: https://doi.org/10.1016/j.scitotenv. 2020.141003.

Lovell, R.D., Jarvis, S.C., 1996. Effect of cattle dung on soil microbial biomass C and N in a permanent pasture soil. Soil Biology and Biochemistry, 28(3): 291-299. Available from: https://doi.org/10.1016/ 0038-0717(95)00140-9.

Luan, H.A., Gao, W., Huang, S.W., et al., 2019. Partial substitution of chemical fertilizer with organic amendments affects soil organic carbon composition and stability in a greenhouse vegetable production system. Soil and Tillage Research, 191: 185-196. Available from: https://doi.org/10.1016/ j.still.2019.04.009.

Luo, G.W., Li, L., Friman, V.P., et al., 2018. Organic amendments increase crop yields by improving microbe-mediated soil functioning of agroecosystems: A meta-analysis. Soil Biology and Biochemistry, 124: 105-115. Available from: https://doi.org/10.1016/j.soilbio.2018.06.002.

Maillard, É., Angers, D.A., 2014. Animal manure application and soil organic carbon stocks: A meta-analysis. Global Change Biology, 20(2): 666-679. Available from: https://doi.org/10.1111/gcb.12438.

Malcolm, R.E., 1983. Assessment of phosphatase activity in soils. Soil Biology and Biochemistry, 15(4): 403-408. Available from: https://doi.org/10.1016/0038-0717(83)90003-2.

Malik, M.A., Khan, K.S., Marschner, P., et al., 2013. Microbial biomass, nutrient availability and nutrient uptake by wheat in two soils with organic amendments. Journal of Soil Science and Plant Nutrition, 13(4): 955-966. Available from: https://doi.org/10.4067/S0718-95162013005000075.

Mancinelli, R., Marinari, S., Di Felice, V., et al., 2013. Soil property, CO_2 emission and aridity index as agroecological indicators to assess the mineralization of cover crop green manure in a Mediterranean environment. Ecological Indicators, 34: 31-40. Available from: https://doi.org/10.1016/j.ecolind.2013. 04.011.

Marschner, P., Kandeler, E., Marschner, B., 2003. Structure and function of the soil microbial community in a long-term fertilizer experiment. Soil Biology and Biochemistry, 35(3): 453-461. Available from: https://doi.org/10.1016/S0038-0717(02)00297-3.

Matilla, P.K., 2006. Ammonia emissions from pig and cattle slurry in the field and utilization of slurry nitrogen in crop production. Doctoral Dissertation. Agrifood Research Reports. Available from: http://www.mtt.fi/met/pdf/met87.pdf.

Max, J.F.J., Schurr, U., Tantau, H.J., et al., 2012. Greenhouse cover technology. Hortic. Rev. (Am. Soc. Hortic. Sci). 40, 259–396. Available from: https://doi.org/10.1002/9781118351871.ch7.

Mehta, C.M., Palni, U., Franke-whittle, I.H., et al., 2013. Compost : Its role , mechanism and impact on reducing soil-borne plant diseases. Waste Management, 34(3): 607-622. Available from: https:// doi.org/10.1016/j.wasman.2013.11.012.

Moore, J.C., Berlow, E.L., Coleman, D.C., et al., 2004. Detritus, trophic dynamics and biodiversity. Ecology Letters, 7(7): 584–600. Available from: https://doi.org/10.1111/j.1461-0248.2004.00606.x.

Nannipieri, P., 1994. The potential use of soil enzymes as indicators of productivity, sustainability and pollution. In: Pankhurst, C.E., Doube, B.M., Gupta, V.V.S.R., Grace, P.R. (Eds.), Soil Biota: Management in Sustainable Farming Systems. CSIRO, East Melbourne, 238–244.

Nannipieri, P., Ascher, J., Ceccherini, M.T., et al., 2007. Microbial diversity and microbial activity in the rhizosphere. Ciencia Del Suelo, 25(1): 89-97.

Nicolardot, B., Recous, S., Mary, B., 2001. Simulation of C and N mineralisation during crop residue decomposition: A simple dynamic model based on the C: N ratio of the residues. Plant and Soil,

228(1): 83-103. Available from: https://doi.org/10.1023/A:1004813801728.

Nicolás, C., Hernández, T., García, C., 2012. Organic amendments as strategy to increase organic matter in particle-size fractions of a semi-arid soil. Applied Soil Ecology, 57: 50-58. Available from: https://doi.org/10.1016/j.apsoil.2012.02.018.

Pascual, J.A., Moreno, J.L., Hernández, T., et al., 2002. Persistence of immobilised and total urease and phosphatase activities in a soil amended with organic wastes. Bioresource Technology, 82(1): 73-78. Available from: https://doi.org/10.1016/S0960-8524(01)00127-4.

Pausch, J., Kuzyakov, M., 2011. Photoassimilate allocation and dynamics of hotspots in roots visualized by ~(14)C phosphor imaging. Journal of Plant Nutrition and Soil Science, 174(1): 12-19. Available from: https://doi.org/10.1002/jpln.200900271.

Peacock, A.D.D., Mullen, M.D.D., Ringelberg, D.B.B., et al., 2001. Soil microbial community responses to dairy manure or ammonium nitrate applications. Soil Biology and Biochemistry, 33(7-8): 4–9. Available from: https://doi.org/10.1016/S0038-0717(01)00004-9,

Piotrowska, A., Wilczewski, E., 2012. Effects of catch crops cultivated for green manure and mineral nitrogen fertilization on soil enzyme activities and chemical properties. Geoderma, 189: 72-80. Available from: https://doi.org/10.1016/j.geoderma.2012.04.018.

Poll, C., Ingwersen, J., Stemmer, M., et al., 2006. Mechanisms of solute transport affect small-scale abundance and function of soil microorganisms in the detritusphere. European Journal of Soil Science, 57(4): 583-595. Available from: https://doi.org/10.1111/j.1365-2389.2006.00835.x.

Powers, S.M., Chowdhury, R.B., MacDonald, G.K., et al., 2019. Global opportunities to increase agricultural independence through phosphorus recycling. Earth's Future, 7(4): 370–383. Available from: https://doi.org/10.1029/2018EF001097.

Quilty, J.R., Cattle, S.R., 2011. Use and understanding of organic amendments in Australian agriculture: A review. Soil Research, 49(1): 1. Available from: https://doi.org/10.1071/sr10059.

Razavi, B.S., Zarebanadkouki, M., Blagodatskaya, E., et al., 2016. Rhizosphere shape of lentil and maize: Spatial distribution of enzyme activities. Soil Biology and Biochemistry, 96: 229-237. Available from: https://doi.org/10.1016/j.soilbio.2016.02.020.

Ren, F., Yang, X.X., Zhou, H.K., et al., 2016. Contrasting effects of nitrogen and phosphorus addition on soil respiration in an alpine grassland on the Qinghai-Tibetan Plateau. Scientific Reports, 6: 34786. Available from: https://doi.org/10.1038/srep34786.

Ren, F.L., Zhang, X.B., Sun, N., et al., 2018. A meta-analysis of manure application impact on soil microbial biomass across China's croplands. Scientia Agricultura Sinica, 51(1): 119–128. Available from: https://doi.org/10.3864/j.issn.0578-1752.2018.01.011.

Richardson, S.J., Peltzer, D.A., Allen, R.B., et al., 2004. Rapid development of phosphorus limitation in temperate rainforest along the Franz Josef soil chronosequence. Oecologia, 139(2): 267–276. Available from: https://doi.org/10.1007/s00442-004-1501-y.

Richter, D.D., Markewitz, D., Wells, C.G., et al., 1994. Soil chemical change during three decades in an old-field loblolly pine (Pinus taeda L.) ecosystem. Ecology, 75(5): 1463–1473. Available from: https://doi.org/10.2307/20166439.

Risse, L.M., Cabrera, M.L., Franzluebbers, A.J., et al., 2006. Land application of manure for beneficial reuse. In: Rice, J.M., Caldwell, D.F., Humenik, F.J. (Eds.), Animal agriculture and the environment: National Center for Manure and Animal Waste Management white papers. American Society of Agricultural and Biological Engineers, St. Joseph: 283–316.

Rodriguez-Kabana, R., Godoy, G., Morgan-Jones, G., et al., 1983. The determination of soil chitinase activity: Conditions for assay and ecological studies. Plant and Soil, 75(1): 95–106. Available from: https://doi.org/10.1007/BF02178617.

Ros, M., Pascual, J.A., Garcia, C., et al., 2006. Hydrolase activities, microbial biomass and bacterial community in a soil after long-term amendment with different composts. Soil Biology and Biochemistry, 38(12): 3443–3452.Available from: https://doi.org/10.1016/j.soilbio.2006.05.017.

Rousk, J., Bååth, E., 2007. Fungal and bacterial growth in soil with plant materials of different C/N ratios. FEMS Microbiology Ecology, 62(3): 258–267. Available from: https://doi.org/10.1111/j.1574-6941. 2007.00398.x.

Rousk, J., Brookes, P.C., Bååth, E., 2009. Contrasting soil pH effects on fungal and bacterial growth suggest functional redundancy in carbon mineralization. Applied and Environmental Microbiology, 75(6): 1589-1596. Available from: https://doi.org/10.1128/AEM.02775-08.

Saha, S., Mina, B.L., Gopinath, K.A., et al., 2008a. Organic amendments affect biochemical properties of a subtemperate soil of the Indian Himalayas. Nutrient Cycling in Agroecosystems, 80(3): 233-242. Available from: https://doi.org/10.1007/s10705-007-9139-x.

Saha, S., Mina, B.L., Gopinath, K.A., et al., 2008b. Relative changes in phosphatase activities as influenced by source and application rate of organic composts in field crops. Bioresource Technology, 99(6): 1750-1757. Available from: https://doi.org/10.1016/j.biortech.2007.03.049.

Samson, M.E., Chantigny, M.H., Vanasse, A., et al., 2020. Management practices differently affect particulate and mineral-associated organic matter and their precursors in arable soils. Soil Biology and Biochemistry, 148: 107867. Available from: https://doi.org/10.1016/J.SOILBIO.2020.107867

Sauer, D., Kuzyakov, Y., Stahr, K., 2006. Spatial distribution of root exudates of five plant species as assessed by ^{14}C labeling. Journal of Plant Nutrition and Soil Science, 169(3): 360-362. Available from: https://doi.org/10.1002/jpln.200621974.

Scotti, R., Bonanomi, G., Scelza, R., et al., 2015. Organic amendments as sustainable tool to recovery fertility in intensive agricultural systems. Journal of Soil Science and Plant Nutrition, 15(2): 333-352. Available from: https://doi.org/10.4067/S0718-95162015005000031.

Shi, R.Y., Liu, Z.D., Li, Y., et al., 2019. Mechanisms for increasing soil resistance to acidification by long-term manure application. Soil and Tillage Research, 185: 77-84. Available from: https://doi.org/10.1016/j.still.2018.09.004.

Shirani, H., Hajabbasi, M.A., Afyuni, M., et al., 2002. Effects of farmyard manure and tillage systems on soil physical properties and corn yield in central Iran. Soil and Tillage Research, 68(2): 101-108. Available from: https://doi.org/10.1016/S0167-1987(02)00110-1.

Sinsabaugh, R.L., Lauber, C.L., Weintraub, M.N., et al., 2008. Stoichiometry of soil enzyme activity at global scale. Ecology Letters, 11(11): 1252-1264. Available from: https://doi.org/10.1111/j.1461-0248.2008.01245.x.

Six, J., Frey, S.D., Thiet, R.K., et al., 2006. Bacterial and fungal contributions to carbon sequestration in agroecosystems. Soil Science Society of America Journal, 70(2): 555-569. Available from: https://doi.org/10.2136/sssaj2004.0347.

Tabbara, H., 2003. Phosphorus loss to runoff water twenty-four hours after application of liquid swine manure or fertilizer. Journal of Environmental Quality, 32(3): 1044-1052. Available from: https://doi.org/10.2134/jeq2003.1044.

Tamilselvi, S.M., Chinnadurai, C., Ilamurugu, K., et al., 2015. Effect of long-term nutrient managements on biological and biochemical properties of semi-arid tropical Alfisol during maize crop development stages s. Ecological Indicators, 48: 76–87. Available from: https://doi.org/10.1016/j.ecolind.2014. 08.001.

Tarafdar, J.C., Jungk, A., 1987. Phosphatase activity in the rhizosphere and its relation to the depletion of soil organic phosphorus. Biology and Fertility of Soils, 3(4): 199-204. Available from: https:// doi.org/10.1007/BF00640630.

Tarafdar, J.C., Claassen, N., 1988. Organic phosphorus compounds as a phosphorus source for higher plants through the activity of phosphatases produced by plant roots and microorganisms. Biology and Fertility of Soils, 5(4): 308–312.Available from: https://doi.org/10.1007/BF00262137.

Tejada, M., 2009. Application of different organic wastes in a soil polluted by cadmium: Effects on soil biological properties s. Geoderma, 153(1-2): 254–268. Available from: https://doi.org/10.1016/j. geoderma.2009.08.009.

Tejada, M., Gonzalez, J.L., Garcia-Martinez, A.M., et al., 2008. Effects of different green manures on soil biological properties and maize yield. Bioresource Technology, 99(6): 1758–1767. Available from: https://doi.org/10.1016/j.biortech.2007.03.052.

Tejada, M., Parrado, J., Hernández, T., et al., 2011. The biochemical response to different Cr and Cd concentrations in soils amended with organic wastes. Journal of Hazardous Materials, 185(1): 204–211. Available from: https://doi.org/10.1016/j.jhazmat.2010.09.019.

Thangarajan, R., Bolan, N.S., Tian, G.L., et al., 2013. Role of organic amendment application on greenhouse gas emission from soil. Science of the Total Environment, 465: 72-96. Available from: https://doi.org/10.1016/j.scitotenv.2013.01.031.

Thomsen, I.K., 2005. Crop N utilization and leaching losses as affected by time and method of application of farmyard manure. European Journal of Agronomy, 22(1): 1-9. Available from: https://doi.org/ 10.1016/j.eja.2003.10.008.

Tiquia, S.M., 2002. Evolution of extracellular enzyme activities during manure composting. Journal of Applied Microbiology, 92(4): 764-775. Available from: https://doi.org/10.1046/j.1365-2672.2002.01582.x.

Tripathy, S., Bhattacharyya, P., Equeenuddin, S.M., et al., 2008. Comparison of microbial indicators under two water regimes in a soil amended with combined paper mill sludge and decomposed cow manure. Chemosphere, 71(1): 168–175. Available from: https://doi.org/10.1016/j.chemosphere.2007.10.042.

Turner, B.L., Condron, L.M., France, C.A.M., et al., 2016. Sulfur dynamics during long-term ecosystem development. Biogeochemistry, 128(3): 281–305. Available from: https://doi.org/10.1007/s10533-016-0208-6.

Venglovsky, J., Sasakova, N., Placha, I., 2009. Pathogens and antibiotic residues in animal manures and hygienic and ecological risks related to subsequent land application. Bioresource Technology, 100(22): 5386-5391. Available from: https://doi.org/10.1016/j.biortech.2009.03.068.

Wang, Q., 2009. Prevention of Tibetan eco-environmental degradation caused by traditional use of biomass. Renewable and Sustainable Energy Reviews, 13(9): 2562-2570. Available from: https://doi.org/ 10.1016/j.rser.2009.06.013.

Webb, J., Pain, B., Bittman, S., et al., 2010. The impacts of manure application methods on emissions of ammonia, nitrous oxide and on crop response: A review. Agriculture, Ecosystems & Environment, 137(1-2): 39-46. Available from: https://doi.org/10.1016/j.agee.2010.01.001.

Xia, L.L., Lam, S.K., Yan, X.Y., et al., 2017. How does recycling of livestock manure in agroecosystems affect crop productivity, reactive nitrogen losses, and soil carbon balance? Environmental Science &

Technology, 51(13): 7450-7457. Available from: https://doi.org/10.1021/acs.est.6b06470.

Xiao. W., Chen, X., Jing, X., et al., 2018. A meta-analysis of soil extracellular enzyme activities in response to global change. Soil Biology and Biochemistry, 123: 21-32. Available from: https://doi.org/ 10.1016/j.soilbio.2018.05.001.

Xu, X.L., Ouyang, H., Kuzyakov, Y., et al., 2006. Significance of organic nitrogen acquisition for dominant plant species in an alpine meadow on the Tibet Plateau, China. Plant and Soil, 285(1): 221-231. Available from: https://doi.org/10.1007/s11104-006-9007-5.

Xu, X.L., Ouyang, H., Richter, A., et al., 2011. Spatio-temporal variations determine plant-microbe competition for inorganic nitrogen in an alpine meadow. Journal of Ecology, 99(2): 563-571. Available from: https://doi.org/10.1111/j.1365-2745.2010.01789.x.

Xu. X.N., Yan, L.M., Xia, J.Y., 2019. A threefold difference in plant growth response to nitrogen addition between the laboratory and field experiments. Ecosphere, 10(1): e02572. Available from: https://doi.org/10.1002/ecs2.2572.

Zaller, J.G., Köpke, U.. 2004. Effects of traditional and biodynamic farmyard manure amendment on yields, soil chemical, biochemical and biological properties in a long-term field experiment. Biology and Fertility of Soils, 40(4): 222-229. Available from: https://doi.org/10.1007/s00374-004-0772-0.

Zhalnina, K., Dia,s R., de Quadros, P.D., et al., 2015. Soil pH determines microbial diversity and composition in the park grass experiment. Microbial Ecology, 69(2): 395-406. Available from: https://doi.org/10.1007/s00248-014-0530-2.

Zhang, X.Y., Dong, W.Y., Dai, X.Q., et al., 2015. Responses of absolute and specific soil enzyme activities to long term additions of organic and mineral fertilizer. Science of the Total Environment, 536: 59-67. Available from: https://doi.org/10.1016/j.scitotenv.2015.07.043.

Zhang, Y.G., Yang, S., Fu, M.M., et al., 2015. Sheep manure application increases soil exchangeable base cations in a semi-arid steppe of Inner Mongolia. Journal of Arid Land, 7(3): 361-369. Available from: https://doi.org/10.1007/s40333-015-0004-5.

Zhu, K., Bruun, S., Larsen, M., et al., 2015. Heterogeneity of O_2 dynamics in soil amended with animal manure and implications for greenhouse gas emissions. Soil Biology and Biochemistry, 84: 96-106. Available from: https://doi.org/10.1016/j.soilbio.2015.02.012.

Zornoza, R., Acosta, J. A., Faz, A., et al., 2016. Microbial growth and community structure in acid mine soils after addition of different amendments for soil reclamation. Geoderma, 272: 64-72. Available from: https://doi.org/10.1016/j.geoderma.2016.03.007.

CHAPTER 4

Response of enzyme activities to metal/nanometal oxide

4.1 Effects of nanometal oxides on enzyme activity

4.1.1 Introduction

With the rapid development of nanotechnology in recent years, engineered nanometal oxide particles (ENOPs), in a size defined as 1−100nm, such as nano TiO_2, ZnO, Al_2O_3, Fe_2O_3 and CeO_2, have been widely used in various fields including electronics, biomedicine, chemical catalysis and new material development. Due to the exposure degree of ENOPs in the environment, whether they have biological toxicity has attracted extensive attention from experts and scholars in various fields (Du et al., 2011). ENOPs would inevitably be released to the environment during their production, transportation and use to generate physical/chemical/biological reactions with all components in the soil. Such migration and transformation processes will change the degree of agglomeration, dissolution rate, size or shape, surface area, surface charge and surface chemical properties and other features of nanoparticles (NPs), thereby affecting their environmental behavior and biological toxicity (Hankui et al., 2015).

Soil enzyme is a type of macromolecular active substance with biocatalytic ability and protein properties (Liu et al., 2011). Participating in various chemical reactions and biochemical processes in soil, soil enzymes are closely related to the mineralization and decomposition of organic materials, cycling of mineral nutrient elements and energy transfer (Yao et al., 2006). Enzyme activity can not only reflects the soil microbial activity, but also characterizes the transformation and migration of soil nutrients, evaluate the soil fertility and soil environmental quality, and serve as a biological index to evaluate the ENOPs pollution to soil. The small size and high specific surface area of ENOPs will enhance their diffusion and migration in soil, and promote their contact and action with the free enzymes or substrates in soil, thus possibly affecting the activity of soil enzymes (Du et al., 2011).

This section systematically reviews and summarizes relevant studies on the effects of ENOPs on enzyme activity through a large amount of literature research and analysis, and looks into the directions and priorities of future research. The dose-effect relationship between ENOPs and enzyme activity can support the research on relevant environmental benchmarks.

4.1.2 Overview of nanometal oxides

4.1.2.1 Properties and uses of nano oxides

Nanometal oxide particles have the surface effect, small size effect, quantum size effect and macroscopic quantum tunneling effect, showing completely different thermal properties, magnetic properties, electrical properties, optical properties, mechanical properties and chemical activities from micron crystalline materials with the same composition. Due to the above characteristics, ENOPs have been widely used in catalysts, sunshades, sensors, cosmetics and more fields, attracting much attention from researchers (Suresh et al., 2010). By far, commonly seen ENOPs include SiO_2, TiO_2, ZnO, CuO, CeO_2, Fe_3O_4 and Al_2O_3. For example, TiO_2 NPs are widely used in ultra violet (UV) resistant materials, textile, photocatalytic catalysts, sunscreens, coating, food packaging materials, papermaking and aerospace industries due to their special optical properties and good chemical and thermal stability (Godwin et al., 2009). ZnO NPs are typically used in the production of protective products and shading materials due to their efficient UV absorption capacity and visible light transmittance (Rousk et al., 2012). CuO NPs are widely used in semiconductor fabrication, catalysts and cell development technologies due to their high temperature superconductivity, electron correlation effect and dynamic rotation ability (Trujilloreyes et al., 2014).

4.1.2.2 Environmental behavior of ENOPs in soil

Soil is the main domicile of various pollutants including ENOPs after they enter the environment, and an important transmission link in the environmental pollution chain. Nanometal oxides can enter the soil through a variety of ways, such as atmospheric deposition of ENOPs, landfill and disposal of industrial wastes containing nanomaterials and municipal wastes, and irrigation with nanomaterial-contained wastewater and sewage. Upon entry into the environment, ENOPs migrate and transform in the environmental medium, interact with the environment and the organisms in the environment, and get absorbed and enriched, thus directly or indirectly threatening the ecosystem and human health (Klaine et al., 2008). Since ENOPs have a broad spectrum antibacterial effect and are far more toxic to organisms than to carbon nanomaterials, their biosafety has attracted increasing attention (Simondeckers et al., 2009). The main physical changes ENOPs cause in the soil are aggregation or agglomeration. Chemical transformations include the redox reactions, ion dissolution, sulfuration and phosphorylation, and modification of macromolecules or organic/inorganic ligands. In addition, the microorganisms with the highest activity and the most complex composition in the soil environment will also affect the morphology and environmental effects of NPs.

4.1.3   Effect of ENOPs on enzyme activity

Soil enzymes, involving redox enzymes, hydrolases, transferases and lyases, are derived from the secretion of soil microorganisms, secretion of plant roots, and decomposition of plant residues and soil animals, which function as an indispensable key indicator for the study of soil ecosystem (Venkataiah, 1989). Upon their entry into the soil environment, ENOPs would have physical, chemical or biological reactions with various components in the soil. Such transformation processes will affect the biochemical reaction process in the soil and the enzyme activity to varying degrees (Stone et al., 2010). According to previous reports, the effects of typical ENOPs (e.g., ZnO NPs, TiO_2 NPs, CuO NPs, CeO_2 NPs and iron nano oxides) on enzyme activity mainly manifest as the changes of urease, sucrase and catalase activities, as shown in Table 4.1.

4.1.4   Major pathways to affect enzyme activities

ENOPs affect enzyme activity mainly through three ways: Releasing toxic metal ions, changing morphological structure of enzymes or corresponding substrates in soil, and damaging biological cells, as shown in Figure 4.1. ENOPs may affect the enzyme activity through a single way, or two or three ways.

4.1.4.1   Releasing toxic metal ions

Different types of ENOPs have different physical and chemical reactions after entering the soil environment. They act upon different toxication mechanisms, and cause different changes in enzyme activity. Brunner et al. (2006) observed the cytotoxicity of a variety of nano metal oxide particles (Fe_2O_3, ZnO, CeO_2, TiO_2 and ZrO_2), and found that the metal ions released by the dissolution of NPs at least partially contributed to the

Table 4.1 Impact of ENOPs on enzyme activity

ENOPs	Soil	Organic Matter	Exposure Dosage/(mg·kg⁻¹)	Soil Enzymatic	Time/d	Action Type	Reference
ZnO NPs	Loam clay (Changshu, Jiangsu)	Organic carbon 4.6%	45.4	Catalase, peroxidase Protease	14	Significant inhibition	Du et al., 2011
ZnO NPs	Cinnamon soil (Beijing)	Total carbon 1.4%	1、5、10	Fluorescein diacetate hydrolase Dehydrogenase	7	Significant inhibition	Hou et al., 2014
CuO NPs	Silty loam (Paddy soil)	Total carbon 6.4%	1000	Urease Phosphatase Dehydrogenase	60	Significant inhibition	Xu et al., 2015
CuO NPs	Wukong soil (Changshu, Jiangsu)	0.76%	1~1000	Urease	/	Significant inhibition	Jin et al., 2010
CuO NPs	Red soil (Jiangxi Yingtan)	4.57%	1~1000	Urease	/	Very significant inhibition	Jin et al., 2010
TiO$_2$ NPs	Loam clay (Changshu, Jiangsu)	Organic carbon 4.6%	90.9	Catalase Peroxidase Urease	14	Significant inhibition Significant inhibition Significantly promoted	Du et al., 2011
TiO$_2$ NPs	Silty loam (Paddy soil)	Total carbon 6.4%	1 000 1 000 0、10、100、500、1 000	Urease Phosphatase Dehydrogenase	60	Significantly promoted Significant inhibition Concentration increased inhibition reduced	Xu et al., 2015
CuO NPs	Farmland (Poland)	Total carbon Sand: 0.6g / kg Clay: 1.2g / kg	10、100、1 000	Dehydrogenase (significantly inhibited) Urease Acid phosphatase	24	Significant inhibition	Josko et al., 2014
ZnO NPs	Farmland (Poland)	Total carbon Sand: 0.6g / kg Clay: 1.2g / kg	10、100、1 000	Dehydrogenase Urease Acid phosphatase	24	Significant inhibition	Josko et al., 2014

continued

ENOPs	Soil	Organic Matter	Exposure Dosage/ $(mg \cdot kg^{-1})$	Soil Enzymatic	Time/d	Action Type	Reference
Cr$_2$O$_3$ NPs	Farmland (Poland)	Total carbon Sand: 0.6g / kg Clay: 1.2g / kg	10、100、1 000	Urease Acid phosphatase	24	Significantly promoted	Josko et al., 2014
Fe$_3$O$_4$ NPs γ-Fe$_2$O$_3$ NPs	Vegetable garden soil (Jiangsu Yixing)	/	0、420、840、1 260	Urease Converting enzyme	30	Significantly promoted	He et al., 2011
Fe$_3$O$_4$ NPs	Loam (Jiangxi Yingtan)	/	80 000	Amylase, urease, neutral phosphatase, catalase	7	Significantly promoted	Fang et al., 2011
SiO$_2$ NPs	Corn field	/	0.42、0.84、1.26	Urease FDA hydrolase Catalase	/	Significantly promoted	He et al., 2011

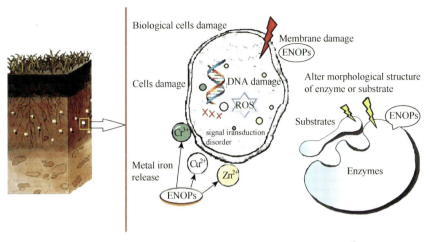

Figure 4.1 Functioning pathways of ENOPs on soil enzyme activities: ①Release toxic metal ions; ②change morphological structure of enzymes or substrates; ③damage biological cells.

cytotoxicity of nanoparticles and acted upon soil enzymes. The effects of these metal ions on soil enzymes are mainly manifested as the effects of heavy metals on proteins, which can be divided into three types: ①The addition of heavy metals can promote the coordinate bonding between the active center of enzyme and the substrate, and change the balance of the enzymic catalytic reaction and the surface charge of enzyme proteins, thus enhancing the enzyme activity, that is, imposing an activating effect. Joško et al. (2014) studied and found that nano Cr_2O_3 stimulated the urease activity and nano-ZnO stimulated and enhanced the acid phosphatase activity. ②Heavy metals occupy the active center of the enzyme, or combine with the sulfhydryl, amine and carboxyl groups of the enzyme molecules to result in the reduction of enzyme activity, that is, imposing an inhibitory effect. Xu et al. (2015a,b) studied the acting mechanism of oxide NPs on paddy soil microbial ecology, and found that different concentrations of CuO NPs treatments all significantly inhibited the enzyme activity, showing an obvious dose-effect relationship. This may be related to the metallic properties of CuO NPs, thatis, copper ions were dissolved out after CuO NPs entered the soil environment. Some studies claimed that the release of zinc ions was the main reason for the toxicity of nano-ZnO. Du et al. (2011) found in the in situ test of wheat soil that nano-ZnO dissolved after being added to the soil as a foreign source, and inhibited the activity of soil protease, catalase and peroxidase. Other studies have also shown that nano-

ZnO can inhibit the β-glucosidase, catalase, deaminase, phosphatase and superoxidase (Hankui et al., 2015). ③There is no specific correspondence between heavy metals and soil enzymes, and no impact imposed on the enzyme activity. When the ZnO NPs in soil reach a concentration of $1mg \cdot g^{-1}$, there is no significant difference between the FDAH activity and the control group ($P=0.149$).

ENOPs, especially some nanomaterials that are prone to release toxic metal ions (such as ZnO and CuO NPs), affect enzyme activity mainly through the release of metal ions. In general, metal ions will not completely dissolve out when ENOPs enter the soil environment. The contribution rate of ENOPs in nano form and released toxic metal ions to enzyme activity is not clear.

4.1.4.2 Changing the morphological structure of the enzyme or substrate
The catalytic function of soil enzymes is realized through the interaction between protein molecules. Since ENOPs are similar to proteins and molecules regarding the structure and size, they may cause confusion and abnormality of biological recognition and reaction, resulting in the disorder of the structure and function of soil enzymes.

If it's hard to dissolve metal ions or spoil soil organisms upon ENOPs' entry into the soil, they are deemed to act upon soil enzymes by changing the morphological structure of enzymes or substrates. ENOPs, such as TiO_2 NPs, CeO_2 NPs and Fe_3O_4 NPs, have good dispersion, small particle size and large specific surface area in soil. They are difficult to dissolve, highly active, and easy to combine and interact with substrates. Du et al. (2018) found through the in situ experiment if nano-TiO_2 was added from outside, the content of titanium ions in soil almost stayed unchanged; soil protease, catalase and peroxidase were inhibited; soil urease activity was significantly enhanced. Studies have shown that when TiO_2 NPs treatment concentration reaches $1000mg \cdot kg^{-1}$, the urease activity is significantly increased and phosphatase activity is severely inhibited (Xu et al., 2015a, 2015b). In the study by Liu et al. (2017a) nano-TiO_2 also had an inhibitory effect on the protease activity, and the larger the dosage added the greater the inhibitory effect. Some studies have found that the addition of nano-TiO_2 in laboratory cultivation can boost the activity of antioxidant enzymes, such as catalase, peroxidase and superoxide dismutase; the addition of nano CeO_2 can inhibit the activity of urease, catalase and FDA enzyme, and there was a dose-effect relationship (Masaki et al., 2015). Fang et al. (2018) applied nano Fe_3O_4 particles to the soil, and found the urease activity enhanced was established. The reason of this phenomenon is that the urease is a metal enzyme with a molecular weight of 480,000 Da, containing trace amounts of transition metal atoms or ions (e.g., paramagnetic nickel) that play as cofactors, coenzymes or active centers (Caldwell, 2005). When Fe_3O_4 NPs are added, urease's molecular structure phase will deform and distort under the magnetic action of Fe_3O_4 NPs. The change of their structure and function will activate its own enzyme activity.

ENOPs is easy to bind to the active center of enzyme molecules, thus hindering the recognition and interaction between enzymes and substrates. For TiO_2 NPs, CeO_2 NPs and other NPs that are uneasy to dissolve metal ions in soil but highly active, it's their chief way to impact the enzyme activity to bind to the enzyme molecules or substrate molecules in soil, change their morphology and structure, and prevent the specific recognition between the enzyme and substrate.

4.1.4.3 Damaging biological cells

ENOPs can cause membrane damage, DNA damage, and cell signal obstruction through direct contact with or induction of reactive oxygen species(ROS), which directly damage the somatic cells of microorganisms or animals/plants, and affect the secretion and release of soil enzymes by microorganisms or animals/plants.

Peroxidation damage of organisms is the toxicity mechanism most commonly accepted of ENOPs (Nel et al., 2006). Peroxidation damage is the variable of membrane permeability and the membrane damage caused by reactive oxygen. Park et al. (2008) found that the presence of nano-TiO_2 particles around the nucleus would induce the production of reactive oxygen in the area. The production of reactive oxygen was in good concert with cell activity, which can indicate the cytotoxicity. Han et al. reported that copper oxide NPs can cause oxidative stress and DNA damage in bacteria, algae, yeast, mice and human cells. Applerot et al. (2012) found that nano-CuO can promote the production of a good deal of reactive oxygen in microorganisms, rapidly consume adenosine triphosphate (ATP), and block cell signal transmission, thereby resulting in the death of a large number of microorganisms. After metal oxide particles enter the soil environment, they increase the oxidative pressure by generating ROS, and enter the cell interior through cell invagination, membrane ion channels, and cytophagocytosis to produce toxic effects on plant roots and microbial cells. Upon damage to DNA/RNA of the cell membrane or organelles, the function of cell synthase is impaired, and the synthesis and release of corresponding enzymes in soil are reduced, thus affecting the biochemical reaction process.

ENOPs can accumulate on the soil microbial membrane and make the cell wall sag, resulting in changes in membrane permeability. ENOPs may also change the membrane potential of soil microorganisms, increase the number of intracellular free radicals, and affect the metabolic process and the synthesis and secretion of enzymes in soil.

4.1.5 Main regulators of enzyme activity

The enzyme activity effect of ENOPs is affected by the particles themselves and environmental factors. Environmental factors include biotic and abiotic factors. Biotic factors mainly refer to soil microorganisms and plants. Abiotic factors principally mean the physical and chemical properties of soil, such as the organic matter content, pH and moisture content.

4.1.5.1 Properties of nano metal oxides

ENOPs have both metallic and nano properties. Numerous studies have shown that a variety of nano-oxide materials have antimicrobial effects (Lok et al., 2006; Kuang et al., 2011) that produce large perturbations in soil microbial population and community structure and affect the synthesis and secretion processes of soil enzymes.

The small size effect of ENOPs causes changes in the morphological structure of enzymes or substrates or damage to the cells of the organism, thus seriously affecting the activity of soil enzymes. In the study of Jośko et al. (2014), for dehydrogenase, the inhibition of enzyme activity by NPs in two different types of soils was much higher than that of micron particles when other testing parameters are fixed. In their study on the ecotoxicity of CuO size to urease in soil, Jin et al. found that CuO NPs had toxic effects on soil urease activity that had not been found in the micron and ionic states. This suggests that ENOPs have stronger toxic effects than micron or ionic metal oxides and greater impact on enzyme activity. Nanometal oxides, when dissolving out their metal ions in the soil, exhibit metallic properties that promote the coordinate binding of enzymes and substrates or occupy the active center of enzymes, thus affecting the activity of soil enzymes. The study by Sun et al. showed that at each sampling time during 15–60 days of culture, with the increase of nano-TiO_2 added to the black soil, the concentration of dissolved Ti ions increased, and the inhibition rate of urease activity increased significantly ($P<0.05$). The greater the number of metal ions dissolved by ENOPs in soil, the stronger the effect on enzyme activity.

The unique physical and chemical properties of ENOPs, such as the specific surface and size effect, will affect the activity of soil enzymes. However, not all nano oxides have an apparent dose-effect relationship with soil enzymes. Thus, the sensitive enzyme activity indexes of different nano oxides should be determined.

4.1.5.2 Physical and chemical properties of the soils

Since the pH, composition and content of organic matters, clay content, as well as minerals can all affect the environmental behavior of NPs and the enzyme activity itself, the physicochemical properties of soil play a regulating role in the enzyme activity of NP oxides.

When the same ENOPs are added to different types of soil, nano oxides would impose different impacts on soil enzymes due to the differences in soil physiochemical properties. Sun et al. found that when nano-TiO_2 was added to the sandy soil, black soil and peat soil, the content of microorganisms, enzymes and nitrogen in sandy soil were affected the most, and those in peat soil the least. The reason for this phenomenon is that sandy soil contains less organic matters than black soil and peat soil. TiO_2 NPs entering the soil are not easy to aggregate or adsorbed by organic matter, and free TiO_2 NPs are more likely to affect the soil microorganisms and enzyme activity. The type of electrolyte in the nanomaterial dispersion medium can modulate the antimicrobial properties of CeO_2 NPs (He et al., 2012), which vary in different soils and also affect the microbial and enzymatic effects of CeO_2 NPs in soil. When ENOPs are applied into the soil, they change physicochemical properties of the soil to some extent and indirectly affect the enzyme activity. Jin et al. found in his study that after 60 days of soil cultivation in the greenhouse, the treatment of Fe_3O_4 NPs and Fe_2O_3 NPs significantly reduced the content of ammonium nitrogen in Wushan soil and available phosphorus in red soil, and slightly reduced the pH of the soil.

Physical and chemical properties of the soil, such as pH and organic matter content, will affect the migration and transformation of ENOPs in soil, thus dictating the bioavailability and biotoxicity of NPs in soil. In the evaluation of the effects of ENOPs on enzyme activity, the physical and chemical properties of different soils shall be considered.

4.1.5.3 Biological activities in soil

Plant and microbial activities in soil will affect the aggregation state and migration process of ENOPs, indirectly changing the bioavailability and enzyme activity effect (Horst et al., 2010).

After nano oxides enter the soil, they will affect the activity and secretion of soil microorganisms and the plant root secretion, thus dictating the number of microorganisms and community structure, and then indirectly changing the activity of soil enzymes. According to the study of Fang et al., nano Fe_3O_4 has an activating effect on the bacteria and actinomycetes in red soil, and the activation effect is gradually enhanced with the increase of nano Fe_3O_4 dosage. It can inhibit the fungi in red soil, with an inhibition rate positively correlated with the dosage of Fe_3O_4, thus activating amylase, neutral phosphatase, urease and catalase in the soil. Root exudates refer to the various ions and the numerous organic substances secreted or overflowed by plant roots, whose composition and content change with the environment. Xu et al. (2015a, 2015b) considered that due to the well-developed root system of rice, the large number of plant residues in rhizosphere soil can provide abundant substrates for the hydrolysis of cellulose, thus promoting the growth of bacteria and alleviating the toxic effect of CuO NPs on microbial cells.

Specific plant or microbial agents can reduce the effect of ENOPs on enzyme activity to a certain extent and alleviate the toxic effect of ENOPs on biological cells. For soils with high ENOPs enrichment, the phytoremediation or microbial remediation mode could be considered.

4.1.6 Summary and outlook

Up to now, the research on environmental behavior and biotoxicity of ENOPs is still an emerging and active field, many parts of which are still in their infancy and lack theoretical basis. Therefore, the research breadth and depth need to be expanded and strengthened. There have been sufficient studies on the behavior and biotoxicity of ENOPs in water environment, but there are few studies on soil environment and terrestrial plants, and more need to be carried out urgently. The existing effects of ENOPs on enzyme activity are mostly inhibitory, but the specific types and extent of effects are regulated by the properties of ENOPs, soil physical and chemical properties and soil biological activities. The study of the dose-effect relationship between nano oxides and soil enzymes can provide a theoretical basis for the study of relevant environmental benchmarks and the establishment of environmental standards. Now that soil enzymes participate in almost all reactions in soil, the study on the effects of ENOPs on enzyme activity is of great significance for evaluating the ecological effects of nano oxides and the quality of soil environment. Existing studies on the biotoxicity and soil microbial effects of ENOPs have accumulated a certain basis for evaluating the environmental risks of nano oxides in soil. Existing studies focus primarily on certain types of ENOPs and rely on the results from laboratory simulations or modeling rather than actual existing engineering studies.

With the deepening of study on the biotoxicity of ENOPs, their effects on enzyme activity shall be further improved and developed in the following aspects. ①In situ experiments shall be carried out in contaminated plots to tune the differences in laboratory simulation data, and study the acting mechanism of ENOPs on soil enzymes in real environment. ②The dose-effect relationship between ENOPs and enzyme activity shall be established to correlate the dosage of different nanomaterials with the pollution degree of soil, screen out soil enzymes highly susceptible to ENOPs, formulate more scientific and reasonable indexes of soil enzymology, and more comprehensively and accurately reflect the toxicity of nano oxides on soil enzymes. ③The aggregation of ENOPs in soil will influence the change of enzyme activity, but the influential mechanism of different aggregation degrees on the enzyme activity is still unclear. It is urgent to carry out research on the aggregation characteristics of ENOPs and its effect on enzyme activity. ④Considering the composite nature of soil contamination as a sink for ENOPs released into the environment, future studies shall address the effects of composite contamination of multiple artificial nanomaterials on soil enzymes. ⑤ENOPs are an emerging soil contaminant with relatively less environmental background values, and relevant environmental benchmarks and quality standards have yet to be established.

4.2 Effect of exogenous lead contamination on microbial enzyme activity in purple soil

4.2.1 Introduction

Heavy metals in soil mainly come from natural weathering of metal-rich parent materials and human activities such as industry, mining and agriculture. Heavy metal pollution of soil is characterized by concealment, hysteresis, accumulation, non-uniformity and irreversibility. It seems impossible to restore the contaminated soil by simply cutting off the pollution source (Tabelin et al., 2018). Heavy metal pollution of soil will affect the transformation of nitrogen and phosphorus to some extent, resulting in the decrease of soil fertility. Heavy metals, when enriched in plants, will disrupt metabolism in the body and stop the cell growth and development, causing disorders of growth and development and reducing production (Rajkumar et al., 2009). In addition, heavy metals will enter the human body through exposure routes such as respiratory ingestion and skin contact. When enriched, they may cause metabolic disorders and a series of lesions. Active heavy metal elements or chemical forms in soil may migrate with the water to the surface water, groundwater or atmospheric environment, thus affecting other environmental media, and even polluting the drinking water sources (Fan et al., 2019).

The current evaluation system on soil pollution by heavy metals counts upon the total amount of heavy metals, the content of available heavy metals, and the related biological indicators to evaluate the pollution severity. Although these indicators can directly reflect the severity of soil pollution by heavy metals, some of them lack comparability and timeliness due to regional differences in soil. It is particularly necessary to warn and truly reflect the harm of heavy metals to biological system with indicators that can be used for early evaluation of soil pollution by heavy metals. A plenty of studies have pointed out that some soil enzymes are particularly sensitive to heavy metal pollution, which can be used as indicators to evaluate the degree of heavy metal pollution in soil and soil quality. Enzyme activity is used to evaluate the severity of heavy metal pollution of a single species. Gao et al. found through laboratory simulation experiments that urease and dehydrogenase, very sensitive to mercury and cadmium pollution in soil, could be used as one of the indexes to detect heavy metal pollution. Huang's research results indicated that enzyme activity could be used as an important indicator for monitoring the copper pollution. Many scholars have demonstrated that low concentration of heavy metals within a certain range can promote the enzyme activity, but with the increase of concentration, the enzyme activity would be gradually inhibited or even inactivated. Zeng et al. (2007) found that at a concentration below $500 \text{mg} \cdot \text{kg}^{-1}$, Pb could promote the enzyme activity, but at a concentration beyond $500 \text{mg} \cdot \text{kg}^{-1}$, it showed an obvious inhibitory effect. Cd can activate the soil protease activity at a low concentration ($50-100 \mu \text{nmol} \cdot \text{L}^{-1}$) (Shah and Dubey, 1998), but inhibit its activity at a high concentration. Different heavy metal ions have different types and degrees of impact on soil enzymes. Khan et al. (2007) found that Pb's impact on enzyme activity was smaller than Cd's because Cd cannot be easily adsorbed by soil colloid and can easily migrate in free state in soil. Some scholars have pointed out that Ag(I), Hg(II) and Cd(II) are more toxic to arylamidase than other heavy metals (Acostamartinez Tabatabai, 2001). The inhibition on urease activity was listed in order as Cr>Cd>Zn>Mn>Pb

(Shen et al., 2005). Different soil enzymes respond differently to the same heavy metal. For example, Cd at a certain concentration can significantly inhibit the activity of protease and alkaline phosphatase, however it leaves no significant impact on the activity of sucrase and acid phosphatase (Lorenz et al., 2006). The simulated heavy metal pollution experiments showed that the effects of heavy metals on soil enzymes were subject to the types and concentrations of heavy metals, soil properties, and varieties of soil enzymes. The longer the pollution, the weaker the inhibition (Ciarkowska et al., 2014). The effect of heavy metals on enzyme activity is a very complex issue, which is mainly subject to the types and concentrations of heavy metals, types of enzymes, soil properties and plant factors. Due to the influence of diversified factors and the difference of research methods, the studies on the influence of heavy metals on enzyme activity come to different conclusions. Therefore, the sensitive enzymes of specific heavy metals in specific soil shall be screened to evaluate the severity of soil pollution.

In this study, lead, a heavy metal widely distributed and seriously harmful, was selected as the target pollutant, and unpolluted cultivated soil and forest soil were selected as the research object. Through a two-month experiment of exogenous lead addition, soil enzymes sensitive to lead pollution in two utilization types of soil were screened out, the effects of soil properties on lead ecotoxicity were investigated, and the effects of concentration and form of heavy metals on enzyme activity were also discussed.

4.2.2   Change in content of available lead

As shown in Figure 4.2, the concentration of available Pb in Pb-contaminated soil is positively correlated with the total amount of Pb ($P<0.01$). Within the culture cycle, the available Pb in the soil treated with low concentration of Pb ($0-250$mg·kg^{-1}) underwent no significant change ($P>0.05$) with the extension of culture time, and that in the soil treated with 1000mg·kg^{-1} of Pb had a certain decrease in the short term and increased to the same level as the first day 56 days later. Under the short-term culture conditions adopted in this study, the content of available lead in purple soil was mainly subject to the amount of lead applied. Under this circumstance, Pb was not absorbed by colloids and clay particles in the soil, and 60%—98% of the exogenously applied lead could be extracted by 0.5 mol·L^{-1} EDTA.

153

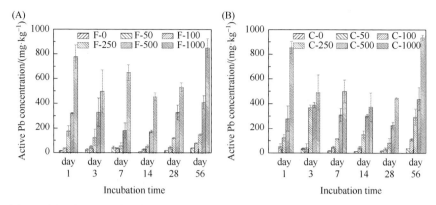

Figure 4.2 Content of available Pb in soil: (A) forest soil; (B) cultivated soil.

Figure 4.3 shows the linear fitting curve between the content of available Pb and the total amount of Pb in soil. Under the short-term simulation conditions, the content of available Pb in soils of two land use types was significantly correlated with the total Pb content ($P<0.01$), and the content of available Pb in cultivated soil was higher than that in forest soil, mainly because the organic matter content in forest soil was about 1.5 times higher than that in cultivated soil. Organic matters can absorb heavy metal ions, passivate and immobilize heavy metals, and reduce their bioavailability. However, within the culture cycle of this experiment, more than 60% of Pb could still be extracted by EDTA, indicating that less than 40% of the exogenously added heavy metal lead could be passivated in the tested soil within two months.

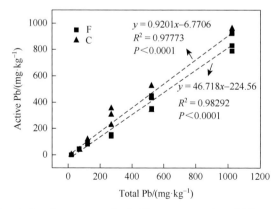

Figure 4.3 Correlation between available Pb content and total Pb content in soil. F: forest soil; C: cultivated soil.

4.2.3　Sensitivity of enzyme activity to lead contamination

After 56 days of culture, the changes of enzyme activity in the two soils were shown in Table 4.2 and Table 4.3, and the correlation between enzyme activity changes and Pb content was analyzed in Table 4.4 and Table 4.5.

Table 4.2　Enzyme activity in forest soil after 56 days incubation

Pb level/(mg·kg⁻¹)	AP/($\mu g \cdot g^{-1} \cdot h^{-1}$)	Acp/($\mu g \cdot g^{-1} \cdot h^{-1}$)	BG/($\mu g \cdot g^{-1} \cdot h^{-1}$)	NAG/($\mu g \cdot g^{-1} \cdot h^{-1}$)	LAP/($\mu g \cdot g^{-1} \cdot h^{-1}$)
0	746.93±84.98c	489.40±22.13c	194.42±16.44c	259.24±59.56a	269.67±28.48ab
50	817.16±69.50c	462.87±35.76cd	147.59±25.29c	218.73±42.16ab	313.37±70.79a
100	909.25±88.96bc	366.10±75.35d	44.58±17.73d	140.81±44.60c	199.44±32.52b
250	903±75.69bc	361.42±40.82d	41.46±2.70d	161.07±16.42bc	279.82±31.96ab
500	1226.08±126.71a	731.32±91.39b	275.58±28.61b	186±35.09abc	210.03±45.48a
1000	1063.76±57.54b	899.88±56.77a	347.37±47.52a	128.34±24c	203.34±29.27ab

Note: a, b, c and d represent significant difference at the P=0.05 level.

Table 4.3　Correlation between enzyme activity and content of heavy metal Pb in forest soil

	AP	Acp	BG	NAG	LAP	Total Pb	Available Pb
AP	1.00	0.66**	0.45*	0.55*	−0.22	0.66**	0.64**
ACP		1.00	0.90**	0.75**	−0.23	0.86**	0.87**
BG			1.00	0.78**	−0.12	0.73**	0.77**
NAG				1.00	−0.34	0.55*	0.59**
LAP					1.00	−0.39	−0.4

Note: * was significantly correlated at the 0.05 level (both sides); ** was significantly correlated at the 0.01 level (both sides).

Table 4.4　Enzyme activity in culturated soil after 56 days incubation

Pb level/(mg·kg⁻¹)	AP/($\mu g \cdot g^{-1} \cdot h^{-1}$)	Acp/($\mu g \cdot g^{-1} \cdot h^{-1}$)	BG/($\mu g \cdot g^{-1} \cdot h^{-1}$)	NAG/($\mu g \cdot g^{-1} \cdot h^{-1}$)	LAP/($\mu g \cdot g^{-1} \cdot h^{-1}$)
0	550.27±19.49a	257.69±12.37b	259.24±59.56a	108.09±9.35ab	338.44±45.10a
50	545.59±11.78a	365.21±20.38a	218.73±42.16ab	97.18±7.14b	258.18±25.64a
100	472.23±97.47ab	94.06c	140.81±44.60c	75.36±16.19b	73.52±10.79b
250	403.56±36.57ab	0e	161.07±16.42bc	159.51±12.37a	54.82±13.29b
500	458.19±19.49ab	0de	186±35.09abc	161.07±767.80a	28.33±10.19b
1000	462.87±98.97b	0d	128.34±24c	76.92±22.09b	14.31b

Note: The data in the table are the mean±standard error, n=3. a, b, c, d and e represent significant difference at the P=0.05 level.

Table 4.5 Correlation between enzyme activity and content of heavy metal Pb in culturated soil

	AP	Acp	BG	NAG	LAP	Total Pb	Available Pb
AP	1.00	0.65**	0.58*	−0.13	0.64**	−0.32	−0.34
ACP		1.00	0.60**	−0.31	0.91**	−0.64**	−0.66**
BG			1.00	−0.18	0.67**	−0.52*	−0.56*
NAG				1.00	0.18	−0.06	−0.02
LAP					1.00	−0.66**	−0.69**

Note: * was significantly correlated at the 0.05 level (both sides); ** was significantly correlated at the 0.01 level (both sides).

As can be seen from Table 4.2 and Table 4.3, except β-glucosidase, the activity of the other four enzymes in forest soil is higher than that in cultivated soil. Land use types and human activities will affect the migration and transformation of Pb in soil, and then its bioavailability and biotoxicity.

As can be seen from Table 4.2 and Table 4.3, except leucine-peptidase, the activity of the other four enzymes in forest soil was significantly positively correlated with the total Pb content and available Pb content ($P<0.05$). In other words, the phosphatase, β-glucosidase and chitinase in forest soil were more sensitive to Pb pollution when the simulated Pb pollution time lasted about two months. With the increase of Pb concentration, the microbial biomass of P decreased. A possible reason is the toxic effect of Pb on microbial cells, killing them and releasing organic matters as substrate to eventually induce the enhancement of enzyme activity. When the Pb concentration exceeds a certain level, Pb can directly bind to the enzyme or substrate and inhibit the enzyme activity.

Table 4.4 and Table 4.5 show a significant negative correlation between the activity of acid phosphatase, β-glucosidase and leucine-peptidase in cultivated soil, and the total Pb content and available Pb content ($P<0.05$). In other words, when the simulated Pb pollution last for about two months, the three enzymes in cultivated soil are more sensitive to Pb pollution. This is perhaps because the cultivated soil features a low content of organic matters and a weak absorption of metal ions. Pb^{2+} is easy to combine with free enzymes in soil and react with active proteins. For instance, it may react with sulfhydryl to produce metal sulfide, or with substrate to form complexes, thereby obscuring the binding site of enzymes and passivating/inhibiting enzyme activity. The activity of alkaline phosphatase in cultivated soil shows no correlation with the increase of lead concentration, as concluded by Angelovičová and Fazekašová (2014).

4.2.4 Dose-effect relationship between enzyme activity and lead concentration

The enzyme activity in forest soil was positively correlated with the Pb concentration. The dose-effect relationship between the four sensitive enzymes and the Pb concentration could be well characterized by the proportional function under the Pb concentration set for experiments ($P<0.05$). It can be calculated from Table 4.6 that when Pb concentration is $767\text{mg}\cdot\text{kg}^{-1}$, $417\text{mg}\cdot\text{kg}^{-1}$, and $622\text{mg}\cdot\text{kg}^{-1}$, the activity of alkaline phosphatase, β-glucosidase and leucine-peptidase are doubled. The sensitivity of enzyme activity to Pb in forest soil is as follows: β-glucosidase > alkaline phosphatase > leucine-peptidase > acid phosphatase.

Table 4.6 Fitting equation between enzyme activity in forest soil and Pb.

Enzyme	Fitting equation	r	F value	P
AP	$Y=391.09+0.51X$	0.731	47.08	<0.001
ACP	$Y=839.28+0.33X$	0.403	12.48	0.002
BG	$Y=100.08+0.24X$	0.489	17.23	<0.001
NAG	$Y=111.97+0.18X$	0.239	6.34	0.023

The simulated Pb pollution generally had an inhibitory effect on the enzyme activity in cultivated soil. Logistic model can well characterize the dose-effect relationship between the three sensitive enzymes and the concentration of Pb ($P<0.05$, Table 4.7). According to the fitting equation, the ED_{20} values of acid phosphatase, β-glucosidase and leucine-peptidase under simulated Pb pollution were $91.93\text{mg}\cdot\text{kg}^{-1}$, $47.97\text{mg}\cdot\text{kg}^{-1}$, and $29.07\text{mg}\cdot\text{kg}^{-1}$, respectively; their ED_{50} values were $96.58\text{mg}\cdot\text{kg}^{-1}$, $51.06\text{mg}\cdot\text{kg}^{-1}$, and $124.09\text{mg}\cdot\text{kg}^{-1}$. The results showed that the sensitivity of enzyme activity to Pb in cultivated soil was: leucine-peptidase > β-glucosidase > acid phosphatase.

Table 4.7 Fitting equation between enzyme activity in culturated soil and Pb

Enzyme	Fitting equation	R	F	P	$ED_{20}/$ $(\text{mg}\cdot\text{kg}^{-1})$	$ED_{50}/$ $(\text{mg}\cdot\text{kg}^{-1})$
ACP	$Y=12.25+299.2/(1+(X/96.58)^{28.07})$	0.826	39.77	<0.001	91.93	96.58
BG	$Y=154.06+105.18/(1+(X/51.06)^{22.23})$	0.467	88.99	<0.001	47.97	51.06
LAP	$Y=-109.18+453.63/(1+(X/24.09)^{0.96})$	0.819	33.21	<0.001	29.07	124.09

4.2.5 Summary

Under the two land use types of purple soil, the microbial enzyme activity in forest soil was generally higher than that in cultivated soil, and the available Pb was slightly lower than that in cultivated soil. The results indicated that due to the frequent human interference, the cultivated soil featured a reduced content of organic matter and a weakened absorption of lead, thus enhancing its inhibition of microbial and enzyme activity. Except LAP, the activity of the other four enzymes in forest soil was significantly or very significantly positively correlated with the amount of exogenous lead and available lead. Exogenous lead pollution could promote the mineralization and recycling process of C and P in forest land, but had no significant effect on the recycling process of N. The activity of ACP, β-glucosidase and LAP in cultivated soil was significantly or very significantly negatively correlated with the amount of exogenous lead and available lead. Lead pollution would inhibit the nutrient mineralization and cycling in cultivated land to a certain extent, doing harm to the crop growth.

4.3 Toxicity of nano-CuO particles to maize and microbial community largely depends on its bioavailable fractions

4.3.1 Introduction

The use of engineered nanomaterials, especially Cu-based NPs, has extensively increased in recent decades (Anjum et al., 2015a). They have been broadly applied in many industries, such as medicine, catalysts, energy and environment, electronics and optics, which have experienced worldwide increases of 200 tons per year (Keller et al., 2013). These NPs present a critical risk to plants and soil microbes which are the preliminary media accumulating Cu-based NPs and initiating their impact on humans and the environment (Gao et al., 2018; Nath et al., 2018; Rawat et al., 2018).

Soils provide nutrients and habitats for plants and microbes. The antimicrobial nature of nano-CuO particles allows their application to protect wood products from being biodegraded by fungi (Abramova et al., 2013; Evans et al., 2008). Though Cu is an essential micronutrient, a critical level of nano-CuO particles also impairs plant growth (Anjum et al., 2015b). The elevated ROS and protein dysfunction are considered to be the mechanisms inducing the depression and inhibition of microbial and plant growth (Anjum et al., 2015a; Kohen and Nyska, 2002; Yruela, 2005). In addition, nano-CuO toxicity to plants and microbial communities also depends on the extent of Cu-ion release, the forms of Cu, soil pH, soil types and particle sizes (McKee and Filser, 2016; Frenk et al., 2013). The extent and speciation of heavy metals generally governs both microbial and plant responses to heavy metal contamination (Abdu et al., 2017; He et al., 2019; Shahid et al., 2017). For instance, the bioavailable fraction of Cu (e.g., soluble and exchangeable Cu or

DTPA-extractable Cu) contributes to the toxicity of nano-CuO particles, and reducible Cu may also be responsible for toxicity by inducing nano-accrued oxidative stress. However, there is insufficient information regarding the responses of plants and microbes (e.g., microbial activities and microbial community compositions) to different chemical species of Cu (mainly referring to exchangeable Cu and reducible Cu) in nano-CuO-treated soils.

Enzymes, mainly released by microbes and plant roots, enable the release of labile carbon (C), nitrogen (N) and phosphorus (P) compounds from complex soil organic matter and drive the nutrient cycling process (Sinsabaugh et al., 2008; Waring et al., 2014). Their activities have been frequently recommended as early indicators for plants and microbes in response to environmental stress (e.g., nano-CuO particles accumulation). Especially around the roots, the interactions between microbes and plants are fairly drastic. The responses of various enzymes can be the main mechanisms underlying the impact of nano-CuO particles on plant growth, microbial activities and communities. Therefore, the investigation and visualization of enzyme activities, especially on the rhizoplane, is a critical issue that must be addressed to fully recognize the effects of nano-CuO particles on plants and microbes (Liu et al., 2017a,b; Razavi et al., 2016).

In this study, emerging in situ zymography was applied to investigate the distribution of enzyme activities on the rhizoplane in response to the application of nano-CuO particles, as this technique avoids the shortcomings of previous destructive plant sampling methods (Razavi et al., 2017). Three enzymes related to C, N and P cycles were considered: β-glucosidase (β-glu), N-acetyl-glucosaminidase (NAG) and phosphomonoesterase (Phos). The soil microbial communities were analyzed using the 16S rRNA high-throughput sequencing technique to elucidate their responses to nano-CuO particles. CuSO$_4$, commonly used to evaluate soil Cu availability and phytotoxicity, was included for comparison (Ginocchio et al., 2006). Maize (*Zea mays* L.) was grown for 42 days, and three treatments were included: ①control soil without Cu addition; ②soil treated with nano-CuO particles (50nm diameter); ③soil treated with CuSO$_4$. Plant characteristics (e.g., height, Cu content in shoots and roots, shoot and root biomasses) were measured to reveal the impacts of nano-CuO particles on plant growth and their accumulation in different tissues. Cu chemical speciation variations were identified using DTPA extractions and BCR sequential extraction procedures. This study aims to ①visualize the distribution of enzyme activities in soil profiles as affected by nano-CuO particles and CuSO$_4$; ②demonstrate the contribution of Cu ions released from the nano-CuO particles to diverse Cu chemical speciation; ③fascertain the mechanisms underlying the influence of nano-CuO particles on plant growth, soil microbial activity and community composition.

4.3.2 Response of plants to nano-CuO and CuSO$_4$

Plants play a key role in maintaining ecosystem stability and soil fertility. The accumulation of Cu ions or nano-CuO in plants reduces Cu concentrations in the soil but also poses a threat to humans via food production (Reddy et al., 2016). However, the mechanisms of the responses of plants and soil microbes with respect to chemical speciation and the extractability of Cu from nano-CuO are yet to be investigated. Thus, it is essential to investigate and compare the accumulation and phytotoxicity of nano-CuO and CuSO$_4$ to plants (Jośko, 2019). The results showed that the shoot biomass and plant heights in CuSO$_4$ were only slightly lower than for the nano-CuO ($P>0.05$) treatment but were much lower than the control soil after 42 days of incubation (Figure 4.4A and Figure 4.4B). This indicated that CuSO$_4$ and nano-CuO had similar negative impacts on maize growth. Even though the phytotoxicity of both materials was similar, the mechanisms were not the same according to the chemical speciations of Cu. In CuSO$_4$-treated soils, exchangeable Cu was the most dominant form (Figure 4.5A, approx 5 times greater than nano-CuO), which was directly bioavailable to plants and microbes. However, in nano-CuO-treated soils, the dominant form was reducible Cu (Figure 4.5B), suggesting that part of the toxicity may also be attributed to nano-accrued oxidative stress. In addition, significantly higher oxidizable Cu levels were also found in the nano-CuO-treated soils (Figure 4.5C) and were more than 5 times greater compared to CuSO$_4$. Oxidizable Cu is generally bound to soil organic matter (Alan and Kara, 2019). This result is consistent with Jośko (2019), who recognized that the concentrations of Cu bound to organic matter were highest in soils treated with nano-CuO particles. They attributed this to the small size of nano-CuO, which allows easier soil penetration and interactions with organic matter. Gradual changes of DTPA-extractable Cu and exchangeable Cu levels in nano-CuO-treated soils were consistently confirmed (Figure 4.5 and Figure 4.6), indicating a gradual release of Cu^{2+} ions. This gradually released Cu^{2+} also explained the gradual decrease in soil pH in the nano-CuO-treated soil (Figure 4.7), because released Cu^{2+} may react with water to form Cu(OH)$_2$ and leave H$^+$ in soil solutions. Root uptake of Cu^{2+} may also induce the release of H$^+$ ions from roots to soil. Moreover, the DTPA-extractable Cu in the nano-CuO particles was only half of that in CuSO$_4$ at day 42 (Figure 4.6). This gradual release and significantly reduced DTPA-extractable Cu in the nano-CuO-treated soils may also have contributed to the slightly lower impact of nano-CuO on maize growth.

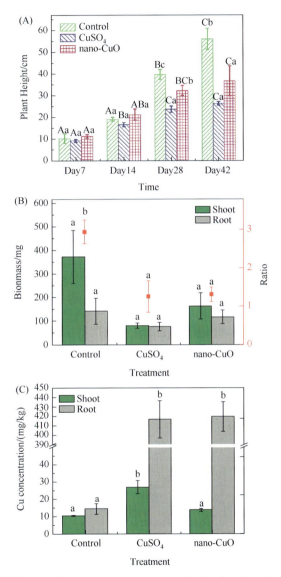

Figure 4.4 (A) Plant height among three treatments during the whole incubation period; (B) Plant biomass and shoot/root biomass ratio among three treatments; (C) Cu concentration in the shoot and root biomass of three treatments. Control, CuSO₄ and nano-CuO represent the control, CuSO₄-treated and nano-CuO-treated soil. Error bars stand for standard errors (SE). Credit: Reprinted with permission from Pu et al., 2019.

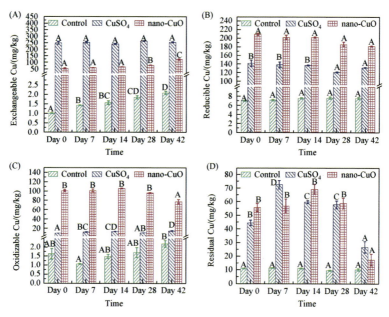

Figure 4.5 Variation of chemical speciation of Cu with time: (A) Exchangeable Cu, (B) Reducible Cu, (C) Oxidizable Cu and (D) Residual Cu. Control, CuSO$_4$ and nano-CuO represent the control, CuSO$_4$-treated and nano-CuO-treated soil. Different letters show the significant change of chemical forms with time ($P<0.05$). Error bars stand for standard errors (SE). Credit: Reprinted with permission from Pu et al., 2019.

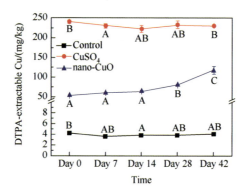

Figure 4.6 DTPA-extractable Cu among three treatments during the whole incubation period. Capital letters indicate significant differences in different time ($P < 0.05$). Control, CuSO$_4$ and nano-CuO represent the control, CuSO$_4$-treated and nano-CuO-treated soil. Error bars represent standard error (SE). Credit: Reprinted with permission from Pu et al., 2019.

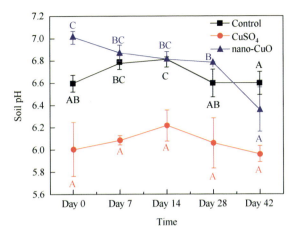

Figure 4.7 Response of soil pH to different treatments during the whole incubation period. Capital letters indicate significant differences in different time ($P < 0.05$). Control, CuSO$_4$ and nano-CuO represent the control, CuSO$_4$-treated and nano-CuO-treated soil. Error bars represent standard error (SE). Credit: Reprinted with permission from Pu et al., 2019. Copyright 2019 Elsevier.

Cu accumulations in the root biomass showed no differences between the nano-CuO and CuSO$_4$ treatments (Figure 4.4C). However, significantly higher concentrations of Cu were found in the shoot biomass under the CuSO$_4$ treatment. The control soil and the nano-CuO-treated soil had similar Cu concentrations in their shoots. The absorbance peak presented at the optimal wavelength that was specific to nano-CuO (Figure 4.8). This revealed that the shoots in the nano-CuO-treated soils contained nano-Cu particles, according to El-Trass et al. (2012). These results indicated that in the nano-CuO-treated soils, Cu mainly accumulated in the roots but rarely transferred to the shoots. Our results were different from a previous study showing that the Cu concentrations in the shoots of nano-CuO-treated maize were 1.2 times higher compared to Cu^{2+}-treated plants (Wang et al., 2012). The lower accumulations of Cu in the shoots of the nano-CuO-treated maize could be induced by the heterogeneity and complexity of the soil, which enhanced the chemical transformation of Cu. However, in a study by Wang et al. (2012), a hydroponic culture was used, where relatively weak interactions between Cu and the environment may have occurred due to the homogeneous and simple environment.

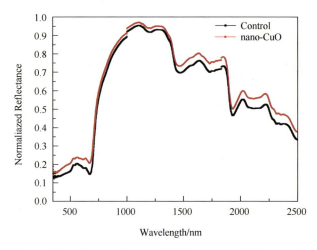

Figure 4.8 Spectral absorbance of oven-dried aboveground biomass for control and nano-CuO treatment Control and nano-CuO represents the control and nano-CuO- treated soil. Due to the low shoot biomass in CuSO₄ treatment, the spectrometer was not able to recognize the spectra of the biomass. Therefore, only the spectra of shoots from control and nano-CuO treatment were presented. Credit: Reprinted with permission from Pu et al., 2019.

The DTPA-extractable Cu in the $CuSO_4$-treated soil was four times greater than in the nano-CuO-treated soil (Figure 4.6). Similarly, the exchangeable Cu in the $CuSO_4$-treated soil was approximately five times higher compared to the nano-CuO-treated soil (Figure 4.5A). However, the differences in Cu content in the shoots for both treatments were much smaller, indicating that the maximum Cu content that a plant could hold was not determined by the soil bioavailable Cu, but rather by plant species and growth stage.

4.3.3 Response of microbial community compositions to nano-CuO and $CuSO_4$

According to Chao1 and the OUT number, the abundance of the microbial community in the control group was the highest, followed by the nano-CuO and $CuSO_4$- treated soils (Table 4.8). In addition, the control soil also had the highest Shannon and Simpson indexes when compared to the nano- CuO and $CuSO_4$- treated soils, indicating that species diversity decreased after Cu application. Both indexes were only slightly smaller in the nano-CuO-treated soil compared to the control, suggesting that the toxicity of nano- CuO is somewhat less than CuSO$_4$ and the slight impact may be due to the Cu ions released from the nano-CuO-particles (Rousk et al., 2012).

Table 4.8 Alpha-diversity indexes of Control, Nano-CuO and CuSO$_4$ soils

Treatments [a]	Chao1	OUT number	Shannon index	Simpson index
CuSO$_4$	269.11	510	3.33	0.88
Nano-CuO	317.67	604	4.69	0.98
Control	329.67	634	4.94	0.99

[a] Control, CuSO$_4$ and nano-CuO represent the control, CuSO$_4$-treated and nano-CuO-treated soil.

Five major phyla (Proteobacteria, Acidobacteria, Firmicutes, Chloroflexi and Gemmatimonadetes) were present at relative abundances of over 2% in all samples in this study (Figure 4.9A). The most abundant phylum was the Proteobacteria (42% average of all samples). Proteobacteria increased 25% and 20% in CuSO$_4$-treated and nano-CuO-treated soils, respectively, in comparison with the control. This means that Proteobacteria had the strongest tolerance to Cu ions and to nano-CuO contamination (Zhao et al., 2019). Moreover, the increase was mainly provided by Gammaproteobacteria (Figure. 4.6B, +26.68% in CuSO$_4$ and +14.71% in nano-CuO). This result is in accordance with Li et al. (2014), showing that Gammaproteobacteria were the dominant Proteobacteria. Zhao et al. (2019) found that Gammaproteobacteria were positively correlated with total Cu. Mendes et al. (2011) found that more serious ecological disturbances induced higher abundances of Gammaproteobacteria. Our study found higher abundances of Gammaproteobacteria (47%) in the CuSO$_4$-treated soil compared to the nano-CuO-treated soil (34%) and in the control group (19%), demonstrating that nano-CuO had less impact on the bacteria. The second most abundant phylum was Acidobacteria (36%) in the control group and nano-CuO-treated soils, but its abundance decreased to 7% in the CuSO$_4$-treated soil (Figure 4.9A). A positive relationship between soil pH and the relative abundance of Acidobacteria was also found, which may be provided by Acidobacteria subgroup 6 because it was previously found to be positively related to soil pH (Liu et al., 2016; Rousk et al., 2010). A previous study also concluded that the Acidobacteria subgroup 6 was very sensitive to environmental stress (Barns et al., 2007), also supporting the lower impact of nano-CuO particles on the bacteria that was observed. Compared to the nano-CuO-treated and control soils, Firmicutes reached 27% under the CuSO$_4$ treatment due to a sharp increase of Clostridia. This, together with the Proteobacteria, suggests that both bacteria can survive in Cu ion-contaminated environment.

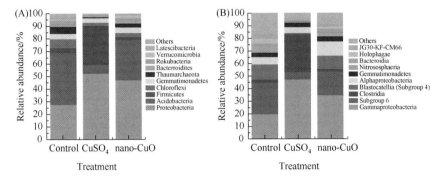

Figure 4.9 (A) Horizontal histogram of bacterial gate classification based on 16s rRNA; (B) a histogram based on the 16s rRNA-based bacterial class. Credit: Reprinted with permission from Pu et al., 2019. Copyright 2019 Elsevier.

4.3.4 Responses of enzyme activities on the rhizoplane to nano-CuO and CuSO₄

Enzymes are generally considered as early indicators for soil microbial activities for revealing the influences of environmental stress (Sinsabaugh et al., 2008; Asadishad et al., 2018). Especially for the regions around and on the root surfaces, the enzyme activities are very prominent because of the co-release of enzymes by microbes and roots. The novel zymography technique provided an opportunity to visualize the impacts of heavy metal contamination on enzyme activities in the roots (including the rhizosphere and rhizoplane in this study). The results of image processing show that enzyme activities on the rhizoplane in CuSO₄-treated soils were barely recognizable, indicating that CuSO₄ application significantly depressed the Beta-glu, NAG and Phos activities on the rhizoplane (Figure 4.10–Figure 4.12). The antimicrobial activity of Cu^{2+} may have been responsible for this strong inhibition (Asadishad et al., 2018; Bondarenko et al., 2012). In general, enzyme activities on the rhizoplane were all found to be positively related to soil pH but were negatively correlated with DTPA-extractable Cu and exchangeable Cu (Figure 4.13). This means that the inhibition of nano-CuO to plant and microbial activities was weaker than that of CuSO₄ because of its lower bioavailable Cu content and higher soil pH (Qiu and Smolders, 2017).

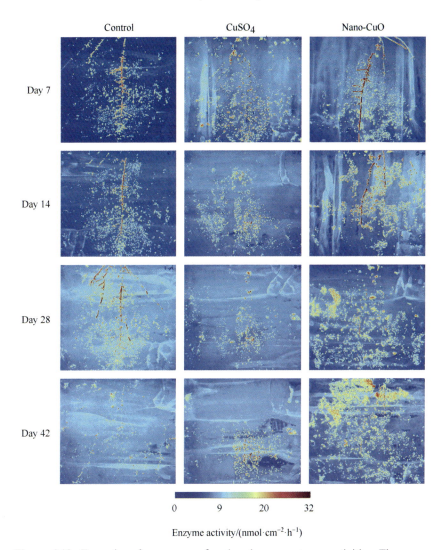

Figure 4.10 Examples of zymograms for phosphomonoesterase activities. Three rows represent response of activities of three treatments: ①control, ②CuSO₄, ③nano-CuO. Figures from top to bottom are the measurements at days 7, 14, 28 and 42. Control, CuSO₄ and nano-CuO represent the control, CuSO₄-treated and nano-CuO-treated soil. The color bar corresponds to phosphomonoesterase activity ($nmol \cdot cm^{-2} \cdot h^{-1}$). Credit: Reprinted with permission from Pu et al., 2019.

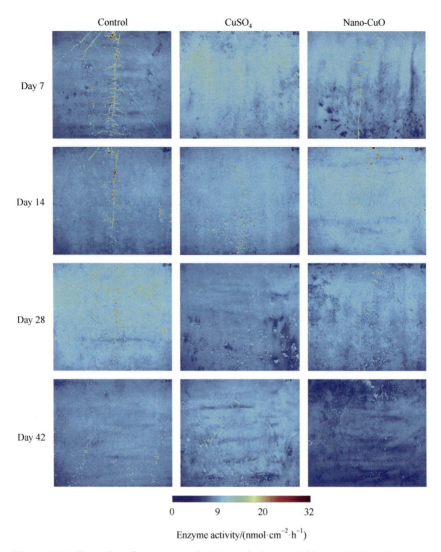

Figure 4.11 Examples of zymograms for N-acetyl-glucosaminidase activities. Three rows represent response of activities of three treatments: ①control, ②CuSO₄, ③nano-CuO. Figures from top to bottom are the measurements at day 7, 14, 28 and 42. The color bar corresponds to N-acetyl-glucosaminidase activity ($nmol \cdot cm^{-2} \cdot h^{-1}$). Credit: Reprinted with permission from Pu et al., 2019.

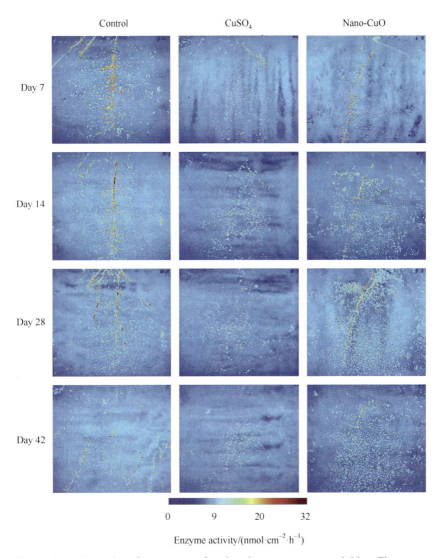

Figure 4.12 Examples of zymograms for phosphomonoesterase activities. Three rows represent response of activities of three treatments: ①control, ②CuSO₄, ③nano-CuO. Figures from top to bottom are the measurements at day 7, 14, 28 and 42. The color bar corresponds to phosphomonoesterase activity (nmol·cm^{-2} h^{-1}). Credit: Reprinted with permission from Pu et al., 2019.

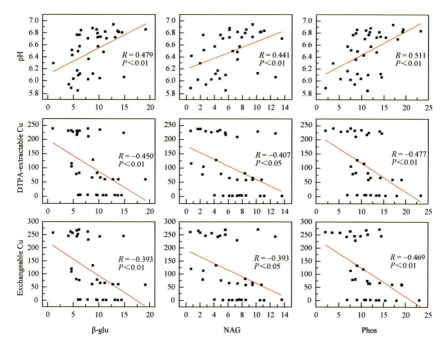

Figure 4.13 Correlation between enzyme activities on the rhizoplane and exchangeable Cu and DTPA-extractable Cu. R is the correlation coefficient and P shows the significance of the relationship. β-glu represents the activity of β-glucosidase on the rhizoplane; NAG represents the N-acetyl-glucosaminidase activity on the rhizoplane; Phos represents the phosphomonoesterase activity on the rhizoplane. Credit: Reprinted with permission from Pu et al., 2019.

Due to the low enzyme activities in the rhizoplane in the $CuSO_4$-treated soil, only the images in nano-CuO-treated soil and the control soil are presented (Figure 4.14). Enzyme activities on the rhizoplane all decreased with time. The control soil had slightly larger NAG activity compared to the nano-CuO-treated soil (Figure 4.14A). This enzyme is mainly responsible for catalyzing the hydrolysis of chitin, which is a major constituent in the cell walls of fungi. The relatively low NAG activity on the rhizoplane may be induced by the antimicrobial function of nano-CuO because nano-CuO is a typical fungicide. In addition, similar phosphomonoesterase activities on the rhizoplane between the nano-CuO-treated soil and the control soil were found (Figure 4.14B). This is in accordance with Jośko et al. (2014), who concluded that application of nano-CuO particles at a rate of 1000mg·kg^{-1} did not alter phosphatase activity in comparison with the control soil.

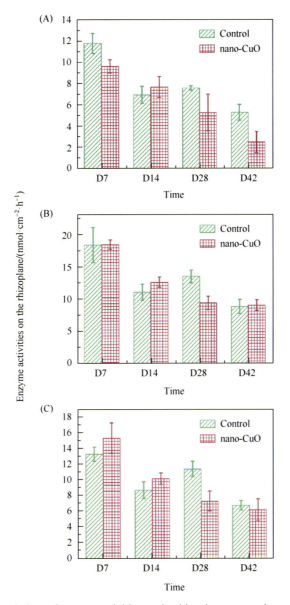

Figure 4.14 Variations of enzyme activities on the rhizoplane among three treatments: (A) NAG, (B) Phos, (C) β-glu. Error bars represent standard error (SE). β-glu represents the activity of β-glucosidase on the rhizoplane; NAG represents the N-acetyl-glucosaminidase activity on the rhizoplane; Phos represents the phosphomonoesterase activity on the rhizoplane. Credit: Reprinted with permission from Pu et al., 2019.

4.3.5　Agricultural implications

This study demonstrated some potential benefits of using nano-CuO particles as a fungicide for use in agriculture, rather than the soluble Cu salt. First, nano-CuO particles were much less toxic to plants and microbes than $CuSO_4$. With equivalent applications of Cu, the bioavailability of nano-CuO particles was much lower than that of $CuSO_4$. The gradual release of Cu ions from nano-CuO particles not only gave plants and microbes an opportunity to adapt to the environmental stress but also provided a persistent antimicrobial effect (Gao et al., 2018). Second, Cu in nano-CuO-treated soil mainly accumulated in the roots instead of in the shoot portions, indicating that nano-CuO applications had a lower probability of severely toxifying the grains.

4.3.6　Summary

In this study, variations in chemical speciation of Cu in nano-CuO-treated soil over time were investigated to understand the mechanisms for the toxicity of nano-CuO particles to soil microbes and maize. Gradual increases of DTPA-extractable Cu and exchangeable Cu were found in nano-CuO-treated soil, but the final concentrations for both forms at day 42 were only half of those in $CuSO_4$-treated soil. Enzyme activities on the rhizoplane, as visualized by the zymography technique, were positively related to soil pH and negatively correlated with DTPA-extractable Cu and exchangeable Cu. Similarly, lower microbial community diversity was also found in $CuSO_4$-treated soil compared to nano-CuO-treated soil, indicating that inhibition of $CuSO_4$ for microbes was stronger than for nano-CuO. Cu accumulations in shoot biomass were the highest and plant biomass was lower in $CuSO_4$-treated soil. These results demonstrated that bioavailable Cu had a more important role in the toxicity of Cu to plants and microbes. The gradually released bioavailable Cu in nano-CuO-treated soil explained its lower toxicity compared to that of Cu ions. This study showed the greater importance of bioavailable Cu content for toxicity modulation rather than the scale of the Cu particles.

4.4　Influences of nano-ZnO particles on plant and microbes grown in Pb-contaminated soil

4.4.1　Impact of ZnO nanoparticles on soil lead bioavailability and microbial properties

4.4.1.1　Introduction

Nano-ZnO particles have been frequently used in many industries in recent decades, e.g., biofertilizers, paints, coatings and personal-care products

(Rizwan et al., 2017; Singh et al., 2018). The global annual production of nano-ZnO particles ranges from 550t to 33400t and has become the third most commonly used metal-based nanomaterial (Rajput et al., 2018). The huge amounts of nano-ZnO utilization increased the opportunities for its accumulation in edaphic and aqueous systems (Osmond and McCall, 2010; Spisni et al., 2016).

Nanotechnology has been gaining increasing attention in the field of environmental remediation, and nanosized materials have been used as the main sorbents and immobilizers (Liu and Zhao, 2007; Hao et al., 2010; Gil-díaz and Lobo, 2014; Mahdavi and Afkhami, 2015; Zhang et al., 2019). Nano-ZnO particles are considered a promising material to detoxify some heavy metals (e.g., Cd, V and Pb) from soil or aqueous solutions (Mahdavi and Afkhami, 2015; Hussain et al., 2018; Yin et al., 2018). For instance, the addition of nano-ZnO particles can remediate soil by mitigating oxidative stress symptoms of *Leucaena leucocephala* induced by other heavy metals (Venkatachalam et al., 2017). Reduction, physical sorption, chemical sorption and coprecipitation were previously proposed as the main mechanisms by which nano-ZnO remediates heavy metal-contaminated soil (Mahdavi and Afkhami, 2015). Soil pH, organic matter content and clay content were also the proposed main factors influencing the impact of nano-ZnO particles (Gil-díaz and Lobo, 2014; Stefanowicz et al., 2020). For instance, Waalewijn-Kool et al. (2014) found that the high organic matter content and low pH is related to the higher concentration of soluble Zn. This suggests that the impact of nano-ZnO particles on heavy metals is influenced by multiple factors and that the mechanisms are complicated. Nevertheless, the behaviors of nano-ZnO particles in heavy metal-contaminated soil are still scarce and require a deeper understanding (Hussain et al., 2018).

At present, most research has focused on the effect of nano-ZnO on plants, and the results have been inconsistent. For example, proper utilization of nano-ZnO particles can mitigate crop Zn deficiency, stimulate germination and suppress plant disease (Singh et al., 2018). However, when very high concentrations of nano-ZnO particles are applied in low Cd-contaminated soil, plant growth is depressed by taking up more Cd. The shoot biomass or root-to-shoot ratio of carrots (*Daucus catota*) was not affected when nano-ZnO particles were applied at rates of $0.5-500 \text{mg} \cdot \text{Zn} \cdot \text{kg}^{-1}$ soil (Ebbs et al., 2016). By applying nano-ZnO particles at rates of $0-1000 \mu\text{g} \cdot \text{mL}^{-1}$, the root length of cabbage (*Brassica oleracea* var. *capitate* L.) decreased with increasing nano-ZnO concentrations (Pokhrel and Dubey, 2013) These results indicate that the effect of nano-ZnO particles on other heavy metals is dose-specific

(Zhang et al., 2019). Suman et al. (2015) investigated the effect of nano-ZnO particles on marine algae *Chlorella vulgaris* and found increased levels of lipid peroxidation as the nano-ZnO particle dose increased from $50mg \cdot L^{-1}$ to $300mg \cdot L^{-1}$. They concluded that the concentration of particles and exposure time are important for the response. Compared with a large number of studies on plants, studies on the effects of nano-ZnO on soil microbes are relatively scarce. The presence of nano-ZnO particles may also induce an influence on soil microbes (Zuverza-Mena et al., 2017). For example, the response of soil microbial biomass and diversity to nano-ZnO particles was evaluated, and all of these factors were found to be reduced with nano-ZnO particles (Ge et al., 2011). Enzyme activities, as an early indicator of plant and microbe responses to environmental stresses, were found to be strongly inhibited by high concentrations of nano-ZnO particles (García-Gómez et al., 2018; Kwak et al., 2017). However, these rare reports mainly focused on the impact of nano-ZnO on microbes without considering environmental factors (such as heavy metals). Thus, a more comprehensive understanding of the impact of nano-ZnO particles on soil microbes and biochemical properties, especially in soil contaminated with bulk heavy metals (e.g., Pb), needs to be studied.

Here, we investigated the effect of nano-ZnO particles on soil microbial properties and Pb bioavailability in Pb-contaminated soil. Our aims are to ①elucidate the effect of nano-ZnO particles on reducing Pb bioavailability and ②evaluate the coeffect of both nano-ZnO particles and Pb particles on soil microbial properties.

4.4.1.2 Impact of nano-ZnO on Zn and Pb bioavailability

Bioavailable Pb and Zn concentrations are closely related to the concentration of added nano-ZnO and incubation time (Figure 4.15). The concentrations of bioavailable Pb significantly decreased during the 60-day experiment. The concentration of bioavailable Pb treated with $300mg \cdot kg^{-1}$ nano-ZnO was the lowest at day 30 of the experiment among all treatments ($P<0.05$). The effect of nano-ZnO particles on the bioavailable Pb concentration on day 60 was more significant than that on day 30. On day 60 of the experiment, the bioavailable Pb concentration was reduced to $196.94mg \cdot kg^{-1}$ and $175.33mg \cdot kg^{-1}$ by adding $10mg \cdot kg^{-1}$ and $150mg/kg$ nano-ZnO, respectively. The effect of $300mg \cdot kg^{-1}$ nano-ZnO was similar to that of $150mg \cdot kg^{-1}$ nano-ZnO ($P>0.05$).

The concentration of bioavailable Zn also changes over time (Figure 4.16). The bioavailable Zn increased with increasing nano-ZnO particle concentration. Similar to bioavailable Pb, the concentration of bioavailable Zn in soil

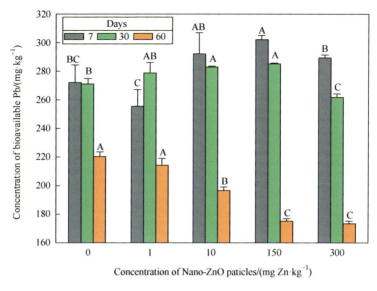

Figure 4.15 Change in bioavailable Pb with different doses of nano-ZnO. Error bars stand for standard errors (SE). Capital letters represent significant differences between different treatments (*P*<0.05). Credit: Reprinted with permission from Huang et al., 2022.

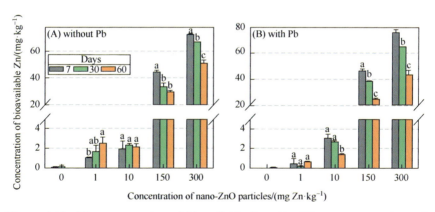

Figure 4.16 The concentration of bioavailable Zn with different concentrations of nano-ZnO. Error bars stand for standard errors (SE). Lowercase letters represent significant differences at different times. (*P*<0.05). Credit: Reprinted with permission from Huang et al., 2022.

decreased over time in the 150mg·kg^{-1} and 300mg·kg^{-1} ZnO treatments. However, the addition of 400mg·kg^{-1} Pb^{2+} had no effect on the bioavailable Zn concentration.

The results of this experiment suggested that the addition of nano-ZnO to soil increased the concentration of bioavailable Zn in soil, and the concentration of Zn decreased over time. This phenomenon is consistent with many studies showing that nano-ZnO in the soil will gradually dissolve into Zn^{2+} over time and increase the content of exchangeable and weakly bound Zn, thus increasing the DTPA-extractable Zn content in soil (Kool et al., 2011; Scheckel et al., 2010; Waalewijn-Kool et al., 2013), which demonstrated that the concentration of bioavailable Zn is positively correlated with the concentration of nano-ZnO (Table 4.9).

Table 4.9 Correlation between different variables in Pb-contaminated and Pb-free soil

	β-glu	MBC	SBR	q_{CO_2}	pH	Bio-Zn	Bio-Pb
β-glu	1	0.021	0.172	−0.16	**−0.623**[*]	0.129	0.466
MBC	0.14	1	**0.923**[**]	**−0.936**[**]	**−0.559**[*]	−0.054	**0.848**[**]
SBR	−0.082	**0.651**[**]	1	**−0.813**[**]	**−0.681**[**]	−0.009	**0.843**[**]
q_{CO_2}	−0.278	**−0.945**[**]	**−0.549***	1	**0.589**[*]	0.127	**−0.884**[**]
pH	−0.27	−0.156	0.105	0.238	1	−0.285	**−0.748**[**]
Bio-Zn	**−0.55**[**]	0.018	**0.536***	0.143	0.025	1	0.075
Conc-Zn	**−0.576**[*]	−0.093	0.428	0.260	−0.009	**0.979**[**]	-

Note: The upper part of the table represents the correlation of the indicators of lead-contaminated soil, and the lower part represents the lead-free soil. β-glu represents the β-glu activity. Bio-Zn and Bio-Pb represent bioavailable zinc and bioavailable lead, respectively. Conc-Zn represents the concentration of nano-ZnO. MBC represents the microbial biomass carbon. SBR represents the soil basal respiration. q_{CO_2} represents the microbial metabolic quotient. Significant correlations were shown in Bold. *Significant at $P<0.05$. **Significant at $P<0.01$.

Nano-ZnO interacts with soil particles and Pb through gravity and Brownian motion (Tourinho et al., 2012). When nano-ZnO was added to Pb-contaminated soil, both nano-ZnO and Zn^{2+} competitively adsorbed Pb, resulting in a change in the bioavailable Pb concentration in the soil (Wei et al., 2019; Zhang et al., 2020). Compared with metal ions, soil colloids have a stronger adsorption capacity for nano-ZnO (Zhao et al., 2012). Therefore, the bioavailable Pb slightly increased by the addition of nano-ZnO on day 30. Due to the fixation of Pb by organic matter, soil colloids and iron manganese oxide in soil, the concentration of available Pb in all treatments decreased with time. In addition, on day 60 of the experiment, the bioavailable Pb concentration decreased with increasing nano-ZnO concentration, which demonstrates that nano-ZnO also has a certain fixation effect on Pb in soil. Zhang et al. (2020) also found that nano-ZnO significantly reduced the concentration of DTPA-extractable Cd.

4.4.1.3 Impact of nano-ZnO on the soil microbial activity of Pb-contaminated soil

For the Pb-free soil treatment, the soil MBC content gradually decreased after adaptive fluctuations at the early stage (Figure 4.17A and Figure 4.17B). The MBC of the 300mg·kg^{-1} nano-ZnO treatment (ranging from 107.72–362.99µg C·g^{-1} soil) was slightly lower than that of the control (ranging from 127.2–500.22µg C·g^{-1} soil). At the beginning of the experiment, the MBC content in Pb-contaminated soil was lower than that of Pb-free soil except for the 300mg·kg^{-1} treatment. The MBC of Pb-contaminated soil also decreased with time after adaptive fluctuations, but no significant difference ($P>0.05$) was detected among the ZnO treatments in Pb-contaminated soil.

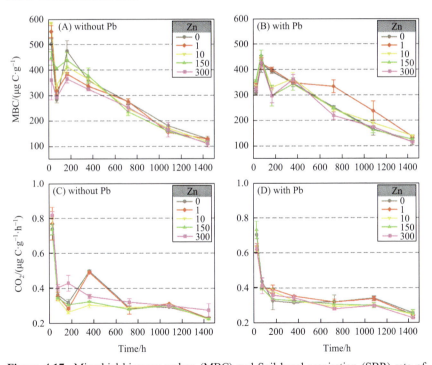

Figure 4.17 Microbial biomass carbon (MBC) and Soil basal respiration (SBR) rate of different treatments during 60 days incubation period. Error bars stand for standard errors (SE). Credit: Reprinted with permission from Huang et al., 2022.

The SBR rate in all treatments decreased sharply and then stabilized (Figure 4.17C and Figure 4.17D). The respiration rate of the 300mg·kg^{-1} nano-ZnO treatment was slightly higher than that of the other treatments in the Pb-free treatment group. However, there was no difference among treatments of Pb-contaminated soils.

The inhibitory effect of high-concentration nano-ZnO on soil MBC has been reported by other studies (Ge et al., 2011). Although nano-ZnO reduced the MBC of lead-free soil (3 days and 7 days) and lead-contaminated soil (7 days) at some points, the MBC value of lead-contaminated soil or lead-free soil was almost the same at the end of the experiment. Therefore, in summary, nano-ZnO does not seem to cause regular changes in MBC.

SBR represents the overall activity of microorganisms and the rate of soil carbon turnover. According to the correlation analysis, SBR is positively correlated with bioavailable Zn in Pb-free soil and bioavailable Pb in Pb-contaminated soil and negatively correlated with pH value in Pb-contaminated soil (Table 4.9). SBR is closely related to the quality of the soil environment and reflects the stress of the soil environment (Zhang et al., 2010). High organic matter content and low pH will lead to an increase in the microbial SBR rate (Romero-Freire et al., 2016). Although some reports claim that soil pH has a significant effect on SBR, the results of this experiment show that Zn and Pb in soil are the main reasons for the changes in SBR. From the results of this experiment, nano-ZnO had no significant effect on the pH value of lead-free soil (Figure 4.18A), but decreased the pH value of lead-contaminated soil during the first 7 days

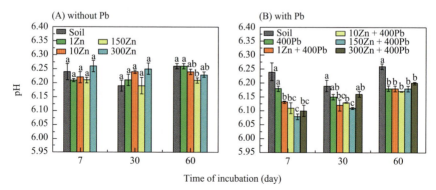

Figure 4.18 Changes of pH value in different treatments. Error bars stand for standard errors (SE). Lowercase letters represent significant differences between different treatments. ($P<0.05$). Credit: Reprinted with permission from Huang et al., 2022.

(Figure 4.18B). This shows that Zn may be the main reason that affects SBR in Pb-free soil, which can be explained by nano-ZnO stimulating the respiration of soil microorganisms. Ge et al. (2011) reported that low concentrations of nano-ZnO slightly increased the basal respiration of soil microorganisms. However, in lead-contaminated soil, correlation analysis shows that SBR has a significant relationship with available lead and pH. This indicated that the impact of soil pH changes and effective lead concentration changes on SBR is also not negligible. Thus, we speculated that Zn is responsible for the increase in SBR in a direct way to stimulate microbial respiration because Zn is an essential nutrient for living organisms and in an indirect way by lowering the pH (along with Pb).

The soil metabolic quotient (q_{CO_2}) represents the energy required to convert organic matter into biomass through respiration per unit of microbial mass in a unit of time. It is used to evaluate the impact of environmental factors and heavy metal pollution on microbial stress. Zhang et al. (2010) reported that the more serious the soil pollution was, the higher was the soil microbial metabolic quotient. We found that the addition of high concentrations of nano-ZnO increased the q_{CO_2} in Pb-free soil compared with the control, which indicated that nano-ZnO increased the microbial energy demand to convert C sources into biomass (Anderson and Domsch, 2010). In Pb-contaminated soil, we found that nano-ZnO decreased the q_{CO_2} value ($P<0.05$) after 30 days, which indicated that the microbial energy demand to convert C sources into biomass decreased and indicated that the environmental stress on microorganisms was relieved. However, this phenomenon was related to the concentration of nano-ZnO. The q_{CO_2} values of the 150 and 300mg·kg^{-1} nano-ZnO treatments were higher than those of the 1 and 10mg·kg^{-1} treatments in Pb-contaminated soil. This is because a high concentration of nano-ZnO will also stress the microorganisms, thus increasing the q_{CO_2} value. Therefore, we concluded that the appropriate concentration of nano-ZnO could reduce the stress of Pb on microbial metabolism.

According to the results of q_{CO_2}, in the Pb-free treatment group (Figure 4.19A), the q_{CO_2} of each treatment strongly decreased and then gradually increased. In Pb-free soil, the 300mg·kg^{-1} nano-ZnO treatment had a higher q_{CO_2} (1.1–2.59) than the control treatment (0.69–1.83), while the 10mg·kg^{-1} and 150mg·kg^{-1} nano-ZnO treatments had the lowest q_{CO_2} (0.65–1.91 and 0.87–2.03, respectively) between 72 and 720 hours. Except for the 300mg·kg^{-1} nano-ZnO treatment group, the q_{CO_2} values of all treatment groups were essentially the same at the end of the experiment. In the

Pb-contaminated treatment, the 150mg·kg^{-1} nano-ZnO treatment first reached the lowest point of q_{CO_2} (0.82mg C·g^{-1} biomass·h^{-1}) (Figure 4.19B). However, the blank group of Pb-contaminated soil (400mg·kg^{-1} Pb only) was the last group, reaching the lowest point of q_{CO_2}. The q_{CO_2} of the blank group in Pb-contaminated soil (0.82−2.27) was higher than that of the other treatment after 30 days. Treatment with 1mg·kg^{-1} and 10mg·kg^{-1} nano−ZnO showed the lowest q_{CO_2} after 30 days (0.95−1.82 and 0.90−1.74, respectively).

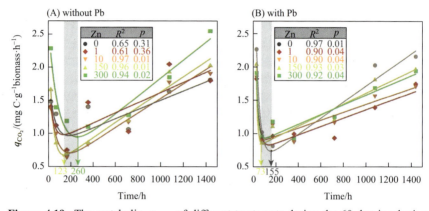

Figure 4.19 The metabolic q_{CO_2} of different treatments during the 60-day incubation period. The curve in the figure represents the nonlinear fitting result of each group of data. Credit: Reprinted with permission from Huang et al., 2022.

4.4.1.4 Impact of nano-ZnO on the soil enzyme activity of lead-contaminated soil

The enzyme activity was fitted from the Michaelis-Menten equation, and then V_{max} was recognized. The addition of nano-ZnO to the Pb-free soil significantly decreased the β-glucosidase activity in the soil, but the trend receded on day 60; only the highest ZnO concentration decreased this enzymatic activity (Figure 4.20A). Only the 300mg·kg^{-1} nano-ZnO-treated group had reduced β-glucosidase activity by 24.39% on day 60. On the other hand, the changes in β-glucosidase activity in Pb-contaminated soil were dependent on the concentration of nano-ZnO and incubation time (Figure 4.20B). The addition of 400mg·kg^{-1} Pb visibly decreased the β-glucosidase activity compared with the control. However, the β-glucosidase activity significantly increased with the addition of 10mg·kg^{-1}, 150mg·kg^{-1} and 300mg·kg^{-1} nano-ZnO after 7 days and 10mg·kg^{-1} after 30 days ($P<0.01$). Notably, nano-ZnO significantly promoted β-glucosidase activity in

Pb-contaminated soil at 7 and 30, while the promoting effect of nano-ZnO after 60 days was not significant.

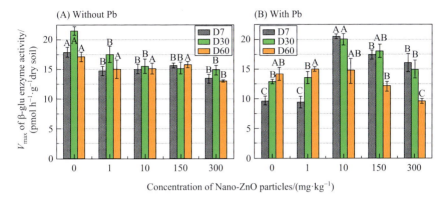

Figure 4.20 Activity of β-glucuronidase in different treatments. Error bars stand for standard errors (SE). Capital letters represent significant differences between different treatments, and lowercase letters represent significant differences at different times (*P*<0.05). Credit: Reprinted with permission from Huang et al., 2022.

Soil enzyme activity is recommended as one of the standard biochemical indicators for evaluating the quality of metal-contaminated soil (Merino et al., 2016). Glucose is an important carbon source of soil microorganisms, which can be produced by the decomposition of cellulose oligosaccharides and cellobiose by β-glucosidase in the process of cellulose saccharification. In the Pb-free group, we found that the concentrations of bioavailable Zn and nano-ZnO were negatively correlated with β-glucosidase activity (Table 4.9). Additionally, the available Zn increases with the concentration of nano-ZnO (Figure 4.16), while the β-glu activity remains stable regardless of the nano-ZnO dose (Figure 4.20A). Therefore, we speculate that the addition of nano-ZnO significantly reduced the activity of β-glucosidase in Pb-free soil due to the Zn content (not only the DTPA-extractable Zn) in soil. Some studies also confirmed that metal ions can chelate with the substrate or react with the enzyme-substrate complex to inactivate the enzymes (Mikanova, 2006). The addition of Pb significantly reduced the β-glucosidase activity compared with the control, while the addition of nano-ZnO increased the β-glucosidase activity of the Pb-contaminated soil during the first 7 days. According to the correlation analysis, we found that β-glucosidase activity was only negatively correlated with pH in Pb-contaminated soil, which was consistent with previous studies, showing that β-glucosidase activity was strongly correlated with pH value (Stark et al., 2014; Xiao et al., 2006). Changes in soil pH may be an important factor affecting β-glucosidase activity because studies have reported that β-glucosidase activity is negatively correlated with pH, and the data in this experiment also confirm the correlation between β-glucosidase activity and pH (Table 4.9). However, for lead-contaminated soil, the reasons leading to changes in enzyme activity may be diverse. On the one hand, the addition of nano-ZnO changed the EC value of the soil, resulting in a change in bioavailable lead content during the first 7 days, which is reflected in Figure 4.15. Changes in the content of lead and zinc together lead to a temporary decrease in pH, which leads to changes in enzyme activity. According to the correlation analysis, we also found that there was a significant negative correlation between bioavailable lead and pH (Table 4.9). On the other hand, zinc ions have lower biological toxicity than Pb. The addition of nano-ZnO increases the concentration of zinc ions to compete with lead ions for binding sites, thereby increasing enzyme activity. Numerous studies have shown that nanometals affect enzyme activity in various pathways (Deng et al., 2017; Du et al., 2018). Although this experiment found a correlation between pH changes and β-glucosidase activity, the mechanism of the toxicity of nano-ZnO in reducing Pb still needs to be further studied.

4.4.1.5 Impact of nano-ZnO on the soil microbial community of Pb-contaminated soil

16S rRNA sequence analysis was used for soil samples on day 60 to investigate the soil microbial community structure. The microbial diversity indices are presented in Table 4.10. Chao1 is usually considered an indicator of microbial richness, and the Shannon and Simpson indices are indicators of microbial diversity. The number of Chao1 and the Shannon index in the nano-ZnO treatment was the highest, followed by the joint treatments in nano-ZnO and Pb, and the control soil was the lowest. However, there was no significant difference in the Simpson index among all treatments.

Table 4.10 Indices of microbial diversity.

Sample	Nano-ZnO concn./(mg·kg^{-1})	Pb concn./(mg·kg^{-1})	Chao1	Shannon	Simpson
S1	0	0	718.97	4.79	0.97
S2	300	0	910.35	5.44	0.99
S3	0	400	813.98	4.97	0.97
S4	300	400	865.04	5.32	0.98

The relative abundance of bacteria at the phylum level is presented in Figure 4.21. In total, 29 identified phyla were observed, Proteobacteria, Thaumarchaeota, Acidobacteria, Chloroflexi, Bacteroidetes, Gemmatimonadetes and Firmicutes were the dominant phyla, which varied between different treatments. Compared with the control (S1: 43.22%), the abundance of Proteobacteria in S2/S3/S4 decreased by 32.3%, 10.57% and 15.04%, respectively. The abundance of Bacteroidetes (S1: 11.38%) showed a similar trend, decreasing by 36.82%, 43.32% and 39.02% in S2/S3/S4, respectively. In contrast, the abundances of Thaumarchaeota (S1: 9.63%, S2: 13.94%, S3: 16.86%, S4: 13.75%), Acidobacteria (S1: 6.4%, S2: 12.07%, S3: 8.59%, S4: 9.51%), Gemmatimonadetes (S1: 2.85%, S2: 6.21%, S3: 3.91%, S4: 5.08%) and Chloroflexi (S1: 10.51%, S2: 14.62%, S3: 12.61%, S4: 11.96%) increased with the addition of nano-ZnO and Pb. The addition of Pb inhibited the abundance of Firmicutes (3.72%) compared with the control (5.08%). However, nano-ZnO slightly facilitated Firmicutes abundance, i.e., 5.59% in the nano-ZnO alone treatments and 4.39% in the combined treatments.

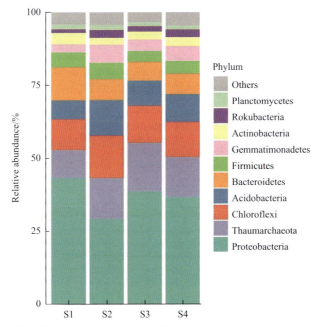

Figure 4.21 Relative abundance of microbial community structure at the phylum level. S1: 0mg·kg^{-1} nano-ZnO + 0mg·kg^{-1} Pb; S2: 300mg·kg^{-1} nano-ZnO + 0mg·kg^{-1} Pb; S3: 0mg·kg^{-1} nano-ZnO + 400mg·kg^{-1} Pb; S4: 300mg·kg^{-1} nano-ZnO + 400mg·kg^{-1} Pb. Credit: Reprinted with permission from Huang et al., 2022.

Heavy metal is an important factor affecting the microbial community in the soil environment (Guo et al., 2017; Song et al., 2018). These effects are generally reflected in the diversity and richness of the microbial community. Our results showed that the addition of nano-ZnO and/or Pb both increased the diversity of the microbial community compared with the control, and the diversity of the microbial community in the treatment with nano-ZnO was higher than that of the Pb treatment. Moreover, the Shannon index and Simpson index, which are usually used as indicators of diversity, also showed the same trend. Low concentrations of Pb ($175mg \cdot kg^{-1}$) have been found to increase the abundance and diversity of soil microorganisms (An et al., 2018). Similar results were observed in our study. The effect of nano-ZnO on microbial diversity is obvious (Table 4.10). After 60 days, the addition of nano-ZnO reduced the concentration of bioavailable Pb in soil, which could induce higher microbial diversity in Pb-contaminated soil.

The graphic of the relative abundance of microbial community structure at the phylum level intuitively shows the differences in microbial community structure among all treatments (Figure 4.21). Because of the better tolerance to heavy metals (Bouskill et al., 2010), Proteobacteria was the most abundant phylum, and the changes were mainly provided by Gammaproteobacteria (Figure 4.22). This is in accordance with Zhang et al. (2020), showing that Gammaproteobacteria is the dominant class in metal-contaminated soil. As the main genus of Gammaproteobacteria, the relative abundance of Lysobacter changed obviously (Figure 4.22). Our results show that Pb enhanced the relative abundance of Lysobacter and that nano-ZnO reduced the abundance of Lysobacter in Pb-contaminated soils. Lysobacter is sensitive to Pb pollution, and the relative abundance of Lysobacter increases with Pb pollution (Niu et al., 2020). The decrease in Lysobacter indicated that nano-ZnO alleviated the environmental pressure caused by Pb. Acidobacteria and Gemmatimonadetes have strong tolerance to heavy metals in acidic soil (Li et al., 2020). The main class of Acidobacteria is Subgroup 6 (Figure 4.22), but we did not find a genus associated with Acidobacteria, or the relative abundance detected was very low. Acidobacteria Subgroup 6 has been reported to be highly sensitive to environmental stress and has a high relative abundance in heavy metal-contaminated soil (Barns et al., 2007; Fatimawali et al., 2020). We found that nano-ZnO increased the relative abundance of Acidobacteria Subgroup 6 in Pb-contaminated soil, but the relative abundance was lower than that in the treatment with nano-ZnO only. There is also a similar trend in Gemmatimonadetes. The changes in the phylum Gemmatimonadetes were dominated by the class Gemmatimonadetes (Figure 4.22), and Gemmatimonas was the main genus we detected. Lur and Higley (2015) reported that the abundance of Gemmatimonadetes increased with heavy metal pollution levels. Hemmat-Jou et al. (2018) found that the abundance of Gemmatimonas increased with the concentration of Zn and Pb. Both Acidobacteria Subgroup 6 and Gemmatimonadetes are heavy metal-sensitive classes. The changes in Acidobacteria Subgroup 6 and Gemmatimonas indicated that nano-ZnO increased environmental stress to soil microorganisms, thus changing the abundance of metal-tolerant bacteria. However, in Pb-contaminated soil, the disturbance of nano-ZnO to microorganisms was influenced by Pb. The changes in microorganisms related to Pb pollution and sensitivity to heavy metals indicate that the effects of nano-ZnO and Pb on microorganisms are mutually restrictive.

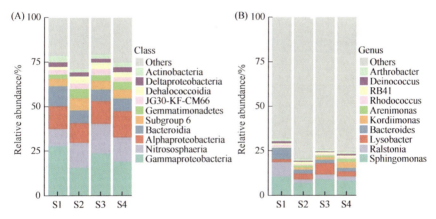

Figure 4.22 Relative abundance of microbial community structure at the class (A) and genus (B) levels. S1: 0mg·kg^{-1} nano-ZnO + 0mg·kg^{-1} Pb; S2: 300mg·kg^{-1} nano-ZnO + 0mg·kg^{-1} Pb; S3: 0mg·kg^{-1} nano-ZnO + 400mg·kg^{-1} Pb; S4: 300mg·kg^{-1} nano-ZnO + 400mg·kg^{-1} Pb. Credit: Reprinted with permission from Huang et al., 2022.

4.4.1.6 *Relationship between different variables based on redundancy analysis (RDA)*

To clarify the relationship between various variables, the redundancy analysis (RDA) method was used to analyze the corresponding relationship between various environmental variables and microbial indicators in Pb-contaminated soil. Environmental variables (Bio-Zn, Bio-Pb, Conc-Zn, pH) were used as explanatory variables, and microbial indicators (MBC, SBR, q_{CO_2}, enzyme activity) were used as response variables. Before performing the RDA, the gradient length was determined by DCA analysis. The result shows that the gradient length of each constraint axis is 0.21 at most, so it is suitable to use RDA for the next analysis.

From the RDA results in Table 4.11, it can be seen that axis 1 and axis 2 explain 75.82% of the total variance of all variables, and the adjusted explanation rate is 66.2%. The analysis results of the explanatory variables show that the bioavailable lead concentration has the greatest impact on the microbial indicators, with an explanatory rate of 69.2%, accounting for 91.2% of the weight of all explanatory variables. The second is pH and bioavailable zinc, and Zn-conc has little effect on microbial indicators.

Table 4.11 Results of RDA analyses between environmental variables and microbial indicators

Summary	Axis 1	Axis 2	Axis 3	Axis 4
Eigenvalues	0.7035	0.0547	0.0002	0
Pseudo-canonical correlation	0.9446	0.5155	0.1642	0.1128
Explained fitted variation (cumulative)	92.77	99.98	100	100
Environment variables	**Explains/%**	**Contribution/%**	**pseudo-F**	**P**
Bio-Pb	69.2	91.2	29.2	0.002
pH	4.1	5.3	1.8	0.18
Zn-conc	0.1	0.2	<0.1	0.924
Bio-Zn	2.4	3.2	1	0.344

In the RDA ordination diagram (Figure 4.23), the angle between the explanatory variables and the response variables or between the response variables indicates the correlation between them. The higher the correlation between the variables, the smaller the angle between the arrows and the same direction. The arrow in the same direction indicates a positive correlation, and the arrow in the opposite direction indicates a negative correlation. It can be seen from the figure that the bioavailable lead has varying degrees of influence on the microbial indicators of SBR, MBC, q_{CO_2} and enzyme activity, especially closely related to the three indicators of SBR, MBC and q_{CO_2}. This shows that changes in the content of bioavailable lead in lead-contaminated soil and changes in pH are the main reasons for inducing changes in microbial SBR, MBC and q_{CO_2}, and the effect of bioavailable lead is greater than that of pH. This is consistent with our previous analysis results. In addition, the figure also shows that pH, bioavailable Pb and bioavailable Zn dominate the changes in β-glucosidase activity in lead-contaminated soil. Among them, pH had the strongest correlation, and bioavailable Pb and bioavailable Zn were also involved in the changes in enzyme activity. This result supports our previous analysis of changes in enzyme activity in lead-contaminated soil.

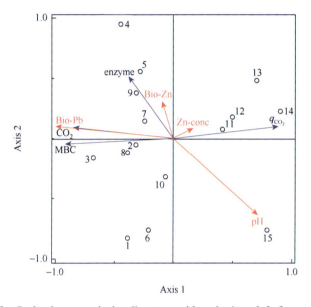

Figure 4.23 Redundancy analysis, diagrams with axis 1 and 2 for environmental variables (Bio-Zn, Bio-Pb, Zn-Conc and pH value) and microbial indicators (MBC, SBR, q_{CO_2}, enzyme activity) in Pb-contaminated soil. Credit: Reprinted with permission from Huang et al., 2022.

4.4.2 Effects of nano ZnO addition on metal morphology and plant growth in lead-contaminated soil

4.4.2.1 Introduction

With the continuous expansion of the production and application of NPs, more and more engineering NPs are discharged into the soil. NPs released into the soil may interact with pollutants to alter the bioaccumulation of heavy metals in plants and the toxicity of heavy metals to plants. Therefore, the study on the response of plants to NPs in polluted soil can provide reference for the practical application of NPs.

ZnO NPs are a kind of common metal-based NPs, which are confirmed by recent studies as an adsorption material to alleviate the phytotoxicity of heavy metals. Therefore, the application of ZnO NPs in soil has attracted wide attention. For example, it has been shown that ZnO NPs can be used as a coating for removing toxic chemicals and biological substances (including heavy metals) (Behnajady et al., 2006). ZnO NPs have also been reported to be able to reduce the oxidative stress of *leucaena glauca* to the heavy metal lead (Venkatachalam et al., 2017). Moreover, experiments have shown that the average root length of corn increases with the addition of ZnO NPs at $0-1000\mu g\cdot mL^{-1}$, while the root length of cabbage decreases with the increase of ZnO NPs concentration (Pokhrel and Dubey, 2013). These studies indicate that plant growth is widely accepted as an index of toxicity evaluation, which reveals the importance of the amount and time of ZnO NPs exposure in the studies. Therefore, it is necessary to evaluate the response of ZnO NPs to plants in lead-contaminated soil from the perspective of ZnO NPs concentration and culture time.

In addition, the toxicity of heavy metals in soil to plants is also affected by factors such as the morphological characteristics of heavy metals. The NPs of different forms of heavy metals in soil usually affect the absorption of heavy metals by plants and the toxicity of metal-based NPs to plants (Giller et al., 2009; He et al., 2019). For example, the toxicity of ZnO NPs and CuO NPs is mainly determined by the bioavailable components of zinc and copper in the soil. Also, the exchangeable and reducible heavy metals can increase the oxidative pressure of cells and thus produce toxicity to microorganisms. However, there are few reports on the response of plants to different chemical forms of lead in lead-contaminated soil treated by ZnO NPs.

Therefore, in the experiment of this chapter, sampling BCR extraction technique was used to reveal the variation characteristics of different forms of lead and zinc in soil exposed to different concentrations of ZnO NPs. The effects of ZnO NPs on plant growth in lead-contaminated soil were analyzed based on the changes of Pb and Zn content in different forms in soil and the accumulation of Pb and Zn in different parts of plants. The biological effects of ZnO NPs at different concentrations on plants in lead-contaminated soil were elucidated.

4.4.2.2 Changes of Pb and Zn in different forms
4.4.2.2.1 Zn
The determination of different forms of zinc in the soil show (Figure 4.24A) that ZnO NPs will quickly transform into the residual and exchangeable state after being added to the soil. The proportion of exchangeable Zn and reducible

Zn is second only to the residual state, and the proportion of oxidizable Zn is the lowest, not exceeding 10% of the total. The proportion of different forms of zinc is affected by the concentration of ZnO NPs. As the concentration of ZnO NPs increases, the proportion of exchangeable Zn gradually increases. When the concentration of ZnO NPs exceeds 150mg·kg^{-1}, the proportion of exchangeable Zn increases obviously. With the increase of ZnO NPs, the proportion of residual Zn decreases obviously, reducible Zn increases slightly, and oxidized Zn decreases slightly.

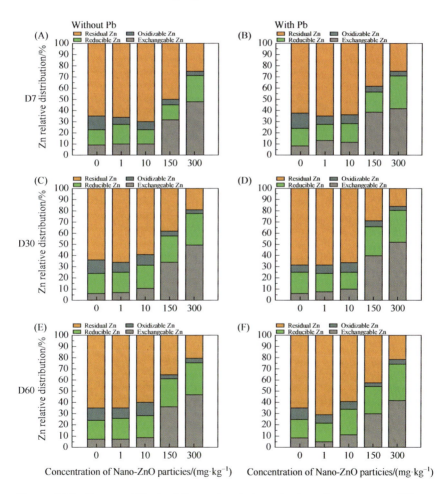

Figure 4.24 Content of Zn in different chemical species in each treatment at different time periods.

When the metal-based NPs like CuO NPs and ZnO NPs enter the soil environment, corresponding metal ions will be released with the dissolution of the NPs, and different kinds of heavy metals exist in different forms in the soil. In this experiment, the concentration of exchangeable Zn in the soil increases gradually as the ZnO NPs dissolve in the soil to produce Zn^{2+}. When the concentration of ZnO NPs exceeds $150mg \cdot kg^{-1}$, the Zn^{2+} arising from the dissolution of ZnO NPs significantly increases the proportion of exchangeable Zn in soil. Since free Zn^{2+} cannot stably combine with iron and manganese oxides or soil organic matter to form reducible oxides in a short time, the ratio of reducible Zn in soil does not change significantly (Martin-Barranco et al., 2020). The transition of Zn^{2+} dissolved out of ZnO NPs into the residual state is a process where the soil consolidates and stabilizes Zn^{2+}. Although the concentration of residual Zn also increases with the increase of ZnO NPs added, its increase rate is smaller than that of ZnO NPs dissolving out Zn^{2+}. As a result, when ZnO NPs addition increases, the proportion of exchangeable Zn increases gradually while that of residual Zn decreases correspondingly. Compared with the more stable residual Zn, the exchangeable Zn is more easily absorbed and utilized by plants. Therefore, ZnO NPs can increase the concentration of exchangeable Zn in soils contaminated and uncontaminated by lead, thus increasing the accumulation risk of heavy metal Zn in plants.

4.4.2.2.2 Pb

The morphological changes of lead in soil are shown in Figure 4.25. According to the results, lead exists in soil mainly in the reducible state and residual state, accounting for 55%—70% and 24%—37% of the total amount, respectively. The proportion of exchangeable Pb is small, only 3%—5% of the total amount. The proportion of oxidable Pb is the lowest, accounting for not more than 3%. With the increase of ZnO NPs addition, the proportion of reducible Pb increases gradually, and that of residual Pb decreases gradually. When ZnO NPs addition exceeds $10mg \cdot kg^{-1}$, it can be found that the concentration of reducible lead increases and that of the residual lead decreases. The concentration of oxidable Pb and exchangeable Pb don't change significantly.

Different from the occurrence form of zinc, lead exists in the soil environment mainly in a reducible state, which is consistent with the study findings by Xu et al. (2018). Plants have different difficulties in absorbing different forms of heavy metals, among which the exchangeable heavy metals can be directly absorbed by plants. Although the reducible state has better stability than the exchangeable, it can still be used by plants. More stable than other forms, the residual Pb can hardly be absorbed and utilized by plants (Zhu et al., 2011). The experimental results show that with the increase of ZnO NPs concentration, the proportion of residual Pb decreases, reducible Pb increases, and exchangeable Pb does not change significantly. These results indicate that ZnO NPs can increase the instability of lead and transform residual Pb into reducible Pb in soil, thus increasing the risk of lead entering into plants to some extent.

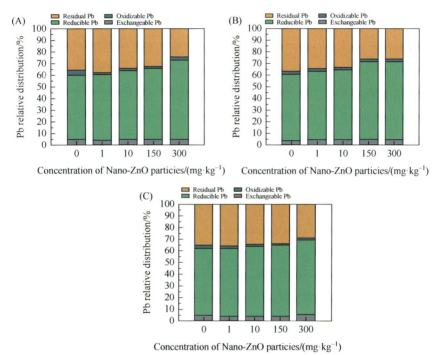

Figure 4.25 Content of Pb in different chemical species in each treatment at different time periods (A)Day 7, (B)Day 30, (C)Day 60.

4.4.2.3 *Changes of heavy metal content in plants*
4.4.2.3.1 Changes of Zn content in plants
Plants play an important role in element cycling in the soil. Plants growing in soils contaminated by heavy metals will accumulate a certain amount of heavy metals inside during the growth process. Therefore, the heavy metal content in different parts of plants can reflect the threat of heavy metals to some extent. The experimental results show that ZnO NPs at 150mg·kg^{-1} and 300mg·kg^{-1} concentrations can significantly increase the zinc content in the above-ground and underground parts of the corn fields (Figure 4.26A).

ZnO NPs at low concentration ($0-10\text{mg}\cdot\text{kg}^{-1}$) have no significant impact on the zinc content in above-ground and underground parts of the corn fields. Corn roots obviously have a better enrichment effect on zinc than the above-ground part.

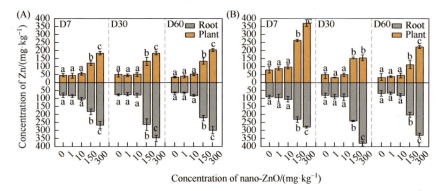

Figure 4.26 Concentrations of Zn in aboveground and belowground biomass for each treatment. (A) Soil not contaminated by Pb; (B) Pb-contaminated soil. Lower case letters represent the significant differences between different treatment groups at the same time ($P<0.05$). The error line is the mean standard errors ($\pm$SE) of the three parallel samples.

Through correlation analysis (Table 4.12), it was found that the content of exchangeable Zn in soil was positively correlated with the content of zinc in aboveground and underground parts of plants. Therefore, the increase of exchangeable Zn in soil was the main reason for the increase of zinc content in the corn. The study of Sungur et al. (2014) also showed that the variation of zinc concentration in plants was mainly determined by the proportion of exchangeable and reducible heavy metal components in soil. Combined with the change of the concentration of exchangeable Zn in this experiment, it can be concluded that with the concentration increase of the addition of ZnO NPs, a rate greater than $150\text{mg}\cdot\text{kg}^{-1}$ can significantly increase the content of exchangeable zinc in the soil, thus increasing the accumulation of zinc in the corn. In addition, the experiment also found that the underground part of the corn contained more Zn than the aboveground part, which may be resulted from multiple mechanisms of the conversion and accumulation of NPs in soil and plants (Capaldi Arruda et al., 2015). Compared with granular ZnO NPs, ionic Zn^{2+} can be well absorbed by the root system of corn, and gradually transferred to the aboveground part of corn alongside the growth and life activity of corn. Due to its small particle size, granular ZnO NPs may also directly penetrate into the walls and membranes of corn epidermal cells and root cortex cells in the form of NPs, and the xylem along with a series of complex processes, and then be transported by the xylem (Rajput et al., 2018). Unlike ionic Zn, plants' direct absorption of ZnO NPs is dominated by a variety of cellular transport processes (Etxeberria et al., 2009). Therefore, zinc entering the corn in the form of ZnO NPs is more difficult to transfer to the aboveground part than that in the form of ions, thereby resulting in a higher zinc content in corn roots than in the aboveground part.

Table 4.12 Correlation among indexes based on pearson correlation coefficient

	Exchangeable Zn	Plant height	AGB	BGB
Shoot Zn	**0.99****	0.14	0.203	0.469
Root Zn	**0.978****	0.176	0.166	0.457
Shoot Pb	-	**−0.911****	**−0.685****	**−0.609***
Root Pb	-	**−0.848****	**−0.625****	**−0.585***

Note: Bold fonts indicate significant correlations. $*P<0.05$; $**P<0.01$. Aboveground Zn/Pb content and underground Zn/Pb content represent the Zn/Pb content in aboveground and underground parts of the corn fields, respectively.

4.4.2.3.2 Changes of Pb content in plants

The measurement results of lead content in plants showed (Figure 4.27) that lead concentration in the underground part of corn was higher than that in the aboveground part, and ZnO NPs had a more significant effect on lead concentration in the underground part of corn than the aboveground part. With the increase of ZnO NPs addition, the lead concentration in the underground part of corn increased, but the lead concentration in the aboveground part increased only on the 7th day of the experiment, and showed little impact on day 30 and day 60. Overall, ZnO NPs can increase lead accumulation in the corn, especially in the roots.

Due to their adsorption of pollutants, NPs in soil environment will affect the toxicity and environmental behavior of co-existing pollutants (Deng et al., 2017; Li et al., 2020). The high affinity of NPs to metal ions will promote the migration of metal ions in the aqueous phase (Vitkova et al., 2018; Zhang et al., 2019), which may lead to increased uptake of metal ions by plant roots. According to analysis of the morphology of lead, ZnO NPs can promote the transformation of residual Pb into reducible Pb in soil, and reducible Pb and exchangeable Pb can be absorbed by plant roots (Sungur et al., 2014). Therefore, adding ZnO NPs to lead-contaminated soil can increase lead absorption of the corn to a certain

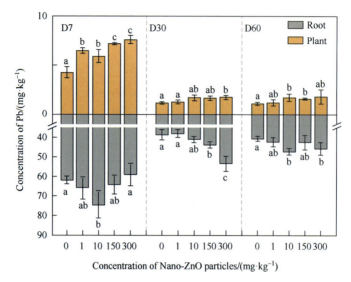

Figure 4.27 Pb Concentrations in aboveground and underground biomass at each treatment. Lower case letters represent the significant differences between different treatment groups at the same time ($P<0.05$). The error line is the mean standard errors ($\pm$SE) of the three parallel samples.

extent, and further cause ecological risks. In addition, some studies have found that nanoparticles can also act as pollutant carriers in soil, facilitating the entry of other pollutants into organic bodies (Naddafi et al., 2011; Shen et al., 2012). For example, Jia et al. (2017) found that ZnO NPs can increase the probability of pollutants entering the organic body and increase the liver damage by Pb^{2+} in mice. Therefore, in this experiment, ZnO NPs may also act as the carrier of Pb^{2+} to promote the absorption of Pb^{2+} by the corn. In conclusion, ZnO NPs added to lead-contaminated soil can increase the uptake of lead by plants, and particularly the accumulation of lead in plant roots, indicating that ZnO NPs applied to lead-contaminated soil have higher environmental risks.

4.4.2.3.3 Changes in plant growth

The changes of plant height and root length reflect the effects of heavy metals on plant growth rate. In soil uncontaminated by lead, ZnO NPs of a concentration above 10mg·kg^{-1} significantly promoted the growth of aboveground parts of corn on the 30th day of the experiment (Figure 4.28A), and the corn's plant height increased by 26%—32%. However, ZnO NPs had no significant effect on root length. In lead-contaminated soil (Figure 4.28B), ZnO NPs had little effect on the corn's plant height, increasing it by only 5%—13% on the day 30 of the experiment. Compared with changes in the plant height and root length, those in plant biomass can more specifically reflect the effects of heavy metals on plant growth. According to the results shown in Figure 4.28A, ZnO NPs of a concentration above $10mg \cdot kg^{-1}$ can significantly increase the biomass of aboveground part (11%—67%) and underground part (69%—146%) of corn in soil without lead contamination. However, in lead-contaminated soil (Figure 4.29B), ZnO NPs decreased the biomass in aboveground (10%—23%) and underground (19%—63%) parts of the corn.

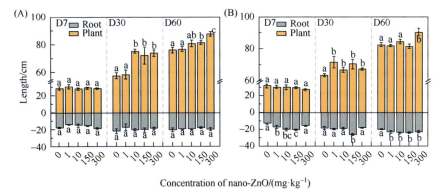

Figure 4.28 Lengths of shoot and root at each treatment. (A) Soil not contaminated by Pb; (B) Pb-contaminated soil. Lower case letters represent the significant differences between different treatment groups at the same time ($P<0.05$). The error line is the mean standard errors ($\pm$SE) of the three parallel samples.

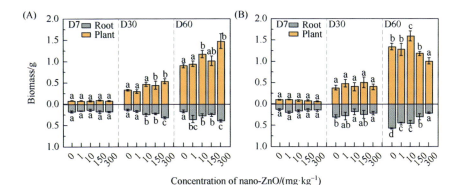

Figure 4.29 Biomasses of aboveground and belowground at each treatment. (A) Soil not contaminated by Pb; (B) Pb-contaminated soil. Lower case letters represent the significant differences between different treatment groups at the same time ($P<0.05$). The error line is the mean standard errors ($\pm$SE) of the three replicates.

Because manganese, iron, copper, zinc and boron are essential trace elements for plant growth (Welch and Graham, 1999), and trace elements in soil can promote the plant growth and increase the plant's dry weight (Torun et al., 2001). In this experiment, the addition of ZnO NPs can provide trace elements for the growth of corn, and increase the concentration of zinc in corn, thus promoting the growth of aboveground part of corn and increasing the biomass of aboveground and underground parts of corn. There have been similar results reported. When the concentration of ZnO NPs added is below $400mg \cdot kg^{-1}$, its addition can provide necessary trace elements for plants and promote plant growth (Lina et al., 2008; Liu et al., 2015; Mertens et al., 2004). In lead-contaminated soil, ZnO NPs didn't significantly promote the growth of the aboveground part of corn, but reduced the biomass of aboveground part and underground part of the corn, thus inhibiting the growth of corn. This is perhaps because the addition of ZnO NPs increased the zinc concentration in corn, and enhanced the corn's absorption of lead, and the presence of lead stimulated the toxicity of ZnO NPs. Overall, it demonstrated an inhibitory effect on the growth of corn. Related studies have also shown that the coexisting metal ions and nanoparticles can produce synergistic toxicity. For example, Cu^{2+} promotes the generation of ROS induced by ZnO NPs, thus increasing the toxicity of ZnO (Moussa et al., 2016). In addition, it can be found from the correlation analysis results in Table 4.12 that lead concentration in both underground and aboveground parts of corn is significantly negatively correlated with its plant height and biomass, indicating that the increase of Pb absorbed by the corn is the main reason for its growth inhibition.

4.4.3 Summary

In this study, the effects of ZnO NPs on plant growth and soil microbial activity under lead contamination were studied through soil pot experiment and sealed soil culture experiment. The environmental behavior of ZnO NPs and lead in soil and their response to plants and microorganisms were investigated to provide reference for the biosafety of ZnO NPs in soil. The main conclusions are as follows.

(1) The chemical form of zinc and lead in soil is affected by ZnO NPs. ZnO NPs mainly exist in soil in the residual state and exchangeable state. Because ZnO dissolves and releases Zn^{2+} in soil, increasing the concentration of ZnO NPs will reduce the proportion of the residual state and increase that of the exchangeable state of soil zinc. Lead exists in soil mainly in the reducible and residual states, its concentration subject to ZnO NPs. With the

increase of ZnO NPs concentration, the proportion of reducible lead gradually increases, and that of residual lead gradually decreases.

(2) ZnO NPs can promote the growth of corn by increasing the concentration of exchangeable zinc in soil and the content of zinc in underground and aboveground parts of corn. In lead-contaminated soil, ZnO NPs can promote the lead absorption and inhibit the growth of corn by increasing the reducible lead in soil. For lead entering the corn, a large part accumulates in the underground part of plant, and a small amount is transferred to the aboveground part as the corn grows.

(3) ZnO NPs will dissolve in soil and produce Zn^{2+}, thus increasing the concentration of available zinc in soil. Meanwhile, due to ZnO's fixation of Pb^{2+}, ZnO NPs at a concentration above $10mg \cdot kg^{-1}$ significantly decreased the concentration of available lead in soil 60 days later.

(4) ZnO NPs can reduce the soil microbial biomass carbon and accelerate the soil microbial respiration rate in lead-free soil, but leave no significant impact on the MBC and SBR in lead-free soil. The high concentration of ZnO NPs increases the energy demand of microorganisms and puts a strain on their metabolism. In the lead-contaminated soil, ZnO NPs can reduce the environmental pressure on microorganisms by reducing the concentration of available Pb in soil, eventually lowering the value of q_{CO_2}. The results of soil metabolic entropy show that ZnO NPs at an appropriate concentration can lower the energy requirements during the soil microbial metabolism, thus dropping the environmental pressure caused by lead.

(5) Different enzymes respond to ZnO NPs and lead differently, among which chitinase and phosphatase have weak responses to ZnO NPs, while ZnO NPs have significant effects on β-glucosidase. The Zn^{2+} and Pb^{2+} dissolved from ZnO NPs and lead can stimulate chitinase and boost its activity. However, in lead-contaminated soil, ZnO NPs decreased the by decreasing the concentration of available Pb and the proportion of nitrogen cycling-related bacteria (*Thaumarchaeota*). Since Zn^{2+} dissolved from ZnO NPs marks a trace element for microbial growth, ZnO NPs at a concentration above $10mg \cdot kg^{-1}$ can significantly improve the phosphatase activity in soil contaminated or uncontaminated by lead. ZnO NPs at a concentration of $10-150mg \cdot kg^{-1}$ can improve the β-glucosidase activity in lead-contaminated soil, with the activating effect decreasing with the increase of ZnO NPs concentration. The influence of heavy metals or nano-metal particles on enzyme activity in soil may be a biochemical collaborative process, which is not only dependent on the concentration of available metal ions and physicochemical properties of soil, but also closely related to the abundance of microorganisms in soil.

(6) ZnO NPs can improve the microbial diversity of lead-contaminated soil, but compared with the control group, the improvement effect of ZnO NPs on microbial diversity is subject to the lead pollution. Some lead/zinc-related changes in microbial relative abundance indicate the impact on microorganisms by the interference between ZnO NPs and lead.

Due to the limitation of laboratory experiments, the research content of this subject still needs to be supplemented and improved. Some suggestions are hereby put forward.

(1) In the experiment to explore the effect of ZnO NPs on the growth of plants in lead-contaminated soil, because the plants were cultured for 60 days only and it was an indoor pot experiment, the experiment space had certain constraints on the growth of plants. Therefore, a suitable site can be selected for outdoor simulated planting in the future, and on this basis the culture time can be extended to observe the changes of plants in a complete growth cycle.

(2) During the experiment, it was found that soil properties were a very crucial factor that had an impact on the environmental behavior of ZnO NPs and lead. Therefore, the responses of plants and microorganisms in soil with different properties could be compared in the follow-up experiment.

(3) The high-throughput sequencing data of this experiment lacked biological replicates, since insufficient volume had been reserved for sampling. In the follow-up study, it is thus recommended to add the detection of parallel samples during the detection so as to enhance the reliability of experimental data in terms of statistical significance.

4.5 Effects of acid rain on heavy metal release and enzyme activity in contaminated soil

4.5.1 Introduction

Metal resources act as an important foundation for the economic development in China. However, with the acceleration of our industrialization process, the heavy metal pollution problem accumulated has gradually been revealed in recent years. A large number of heavy metal pollutants have entered the ecosystem to widely distribute in our living environment. Major heavy metal pollution events occurred frequently in some river basins and regions (Cao and Huang, 2017; Mkhinini et al., 2020). Over the past 50 years, about 30000t of Cr and 800000t of Pb have been released into the global environment, with a large portion pooled in the soil, causing serious heavy metal pollution to the soil (Chen et al., 2020; Ren et al., 2015).

In recent years, acid rain has become a serious regional and even global environmental problem. Acid rain does serious harm to both aquatic and terrestrial ecosystems, leading to surface water acidification, forest & crop degradation, corrosion of buildings, changes of the physical and chemical properties of soil, as well as direct and indirect harm to human health (Han et al., 2017). China is the third largest acid rain region in the world, second only to Europe and North America. According to the Bulletin of the Environmental Status of China, acid rain is mainly seen in the south of the Yangtze River and the east of the Qinghai-Xizang Plateau, among which the southwest region is a traditional acid rain precipitation area in China that is listed as one of the four major acid rain distribution regions. The forest area in the Sichuan basin suffers the most from acid rain, about 280,000 hectares, accounting for 32% of the forest land area. Acid rain is principally seen in Zigong and Luzhou in the southern Sichuan economic zone, Deyang in the Chengdu economic zone, and Panzhihua in the Panxi economic zone. These areas happen to be places where heavy metal contaminated sites are relatively concentrated. In some polluted sites, especially areas where heavy metal waste residues are stored, acid rain will not only speed up the leaching of heavy metals in the waste residues, but also interact with heavy metal ions to cause more serious harm. In the process of soil acidification, in addition to the decrease of soil pH, it also apparently boosts the release intensity of toxic heavy metal ions such as manganese, chromium, copper, lead and zinc, and aggravates the toxicity of contaminated soil, thus affecting the growth of crops and endangering human health (Zhu et al., 2018). Soil acidified by the leaching of acid rain will inhibit the activity of microorganisms and enzymes, change the soil microbial population, and destroy the transformation and recycling process of nutrient elements in the soil. However, in areas where heavy metal residues are stored, acid rain will not only accelerate the leaching of heavy metal ions in heavy metal residues, allowing more heavy metals to enter the soil, but also interact with heavy metal ions to cause more serious harm to the soil (Xu and Liu, 2005).

Soil enzymes are biocatalysts in the soil, which are used to catalyze biochemical reactions in the soil. Most soil enzymes come from microorganisms, while a small part from plants or animals in the soil (Zhang et al., 2013). Soil enzymes are widely used as the evaluation indicators of soil quality. The distribution and activity of soil enzymes are related to the physicochemical properties of soil, microbial community structure, vegetation coverage, and soil pollution, among others. A variety of enzymes exist in soil to support the complex biochemical reactions there. It is estimated that there are at least 500 enzymes that play a key role in C, N, P and S cycles in soil. However, only a small part has been recognized and tested to date. The diversity of enzymes in soil is related to the diversity of soil community, the complexity of organic matters in soil and the difference of physical matrices in soil (Li et al., 2005). Reflecting the rate of important bio-mediated processes in the soil, the activity of soil enzymes can thus be taken as a good indicator of biochemical processes such as biological activities, energy conversion, detoxification, heavy metal fixation, plant production and nutrient turnover in the soil.

The purpose of this section is to explore the release characteristics and migration rules of anions and cations in contaminated soil and to establish an effective numerical model, by simulating the migration of lead and chromium in soil under the acid rain through HYDRUS-2D. Combined with the in situ enzyme spectrometry, the spatial and temporal distribution characteristics of enzyme activity reflected the compound pollution of soil by heavy metals and acid rain, providing the basis for the effect of acid rain on heavy metals in soil environment.

4.5.2 Effect of dynamic simulated acid rain on Cr(VI) release characteristics and enzyme activity in contaminated soil

4.5.2.1 Leaching characteristics of hexavalent chromium under acid rain of different pH values

With advance of the leaching process, the concentration of hexavalent chromium in the leaching solution at the bottom of the soil is shown in Figure 4.30, and the measured and simulated leaching amounts are shown in Table 4.13. The total leaching amount of hexavalent chromium in the acid rain treatment group with pH=3 is greater than that in the treatment groups with pH=5 and pH=7. This indicates that compared with the neutral rain, the acid rain can increase the cumulative leaching amount of hexavalent chromium in the soil. The lower the pH, the greater the cumulative leaching amount. By comparing the simulated and measured leaching amounts, it can be seen that the model has a good simulation effect, but the simulation result is slightly higher than the actual one. That's because a small amount of hexavalent chromium will still be reduced to trivalent chromium in the soil. The lower the pH, the higher the reaction amount. The leaching amount and proportion of hexavalent chromium in the elution process are shown in Table 4.14. In the elution of the first 900mL acid rain, the hexavalent chromium accounts for more than 60% of the total leaching amount. In other words, the migration of hexavalent chromium caused by precipitation mainly occurs in the first 8 months at the contaminated site, during which hexavalent chromium causes the greatest pollution risks to soil and groundwater. Treatment and prevention measures shall be taken for the contaminated sites as soon as possible to prevent expansion of the pollution (Figure 4.31 and Figure 4.32).

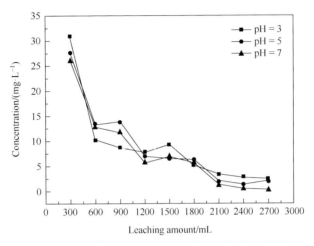

Figure 4.30 Cr(VI) content in leaching solution under acid rain of different pH values.

Table 4.13 Cr(VI) Leaching amount under acid rain of different pH values.

Treatment	pH=3		pH=5		pH=7	
Leaching amount/mg	Measured	Simulated	Measured	Simulated	Measured	Simulated
1st leaching	14.94	14.80	16.46	16.74	15.41	16.15
2nd leaching	6.66	9.00	5.93	6.82	5.74	5.84
3rd leaching	2.57	3.85	1.57	2.32	0.91	1.92
Cumulated leaching	24.17	27.65	23.96	25.88	22.06	23.91

Table 4.14 Amount and proportion of Hexavalent Chromium during the leaching process.

Treatment	pH=3		pH=5		pH=7	
Leacing process	Amount/mg	Proportion/%	Amount/mg	Proportion/%	Amount/mg	Proportion/%
1-1	9.29	38.42	8.29	34.62	7.90	35.83
1-2	3.03	12.55	4.03	16.81	3.91	17.75
1-3	2.62	10.83	4.14	17.28	3.59	16.28
2-1	2.34	9.70	2.07	8.64	1.80	8.17

continued

Treatment	pH=3		pH=5		pH=7	
Leacing process	Amount/mg	Proportion/%	Amount/mg	Proportion/%	Amount/mg	Proportion/%
2-2	2.76	11.42	1.98	8.26	2.16	9.80
2-3	1.56	6.43	1.88	7.84	1.77	8.04
3-1	1.01	4.17	0.57	2.39	0.48	2.16
3-2	0.83	3.43	0.40	1.69	0.25	1.13
3-3	0.74	3.05	0.59	2.47	0.18	0.82

Note: 1-1, 1-2, and 1-3 represent the leaching amounts of hexavalent chromium and its proportions to the total leaching amount in the leaching solution obtained from the first, second and third sampling in the first leaching process, which are the same as that of 2-1, 2-2, 2-3, 3-1, 3-2, and 3-3.

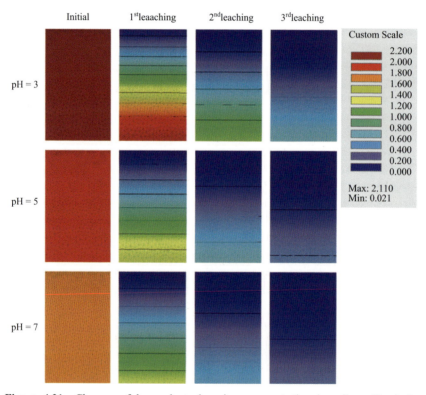

Figure 4.31 Changes of hexavalent chromium concentration in soil profile during hydrus-2d simulated leaching process($\mu mol \cdot cm^{-3}$).

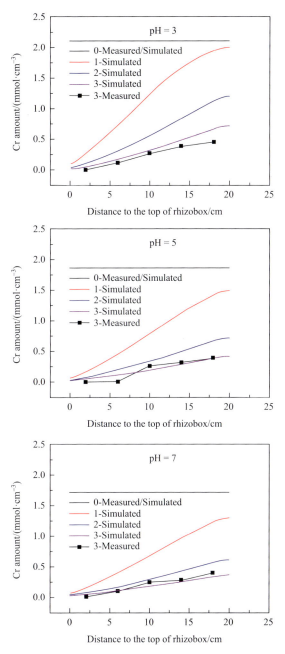

Figure 4.32 Measured and simulated concentrations of hexavalent chromium in soil profile. 0 represents the initial state of leaching, and 1, 2, 3 represent the end of the first, second, third leaching, respectively.

4.5.2.2　HYDRUS-2D simulation results and measured hexavalent chromium content in soil profile

Comparison suggests that the HYDRUS-2D model has a better simulation effect. With the decrease of pH of the leaching acid rain, the content of hexavalent chromium in soil solution at the initial state increased, indicating that the acidic conditions were more favorable for the leaching of hexavalent chromium in soil. However, with advance of the leaching process, the migration of hexavalent chromium in soil slowed down with the decrease of pH of leaching acid rain. According to the results of isothermal adsorption experiment, under a low pH, the soil had a greater adsorption capacity of hexavalent chromium to prevent the downward migration of hexavalent chromium and slow down the migration of hexavalent chromium. This was because chromium ions existed in a variety of forms in soil solutions in a pH range of 1 to 7, such as $Cr_2O_7^-$, $HCrO_4^-$, $Cr_3O_{10}^{2-}$ and $Cr_4O_{13}^{2-}$, with $HCrO_4^-$ being the dominant ion. The mechanism of hexavalent chromium adsorption is the electrostatic attraction between $HCrO_4^-$ ion and the soil surface. With the decrease of solution pH value, the electrostatic attraction between the positive charged group and $HCrO_4^-$ ion on the soil surface increased, thus enhancing the adsorption.

The simulated value of the treatment group (pH=7) is slightly lower than the measured value, because the HYDRUS-2D simulated value only contains the hexavalent chromium adsorbed by soil particles and the hexavalent chromium in soil solution. In contrast, the flame atomic absorption spectrometry method used in alkali solution extraction in the actual measurement method can extract a wider range of hexavalent chromium, making the measured content slightly higher. The treatment group (pH=3) is higher than the measured value. In addition to the above reasons, the acidic solution environment provides a large number of H^+ ions. Cr(VI) is easy to be reduced under acidic conditions due to its high redox potential (1.3V in the standard state), and part of the hexavalent chromium is reduced to trivalent chromium. According to the chemical reaction in Eqs. (4.1), the higher the acid, the more the reaction tends to approach the right side, and the more $HCrO_4^-$ ions reduced to trivalent chromium. According to the relationship between the adsorption precipitation of trivalent chromium and pH value, in this experiment, due to the pH range of 3−7, the deposition is not the dominating process. When generated, trivalent chromium ions mainly exist in the form of Cr^{3+} and $Cr(OH)^{2+}$. The soil contains some iron and aluminum oxides, and the resulting trivalent chromium reacts with the iron and aluminum oxides to be fixed in the soil.

HYDRUS is a physical model that can well simulate the physical changes of water and solute in soil. Because the chemical processes in soil cannot be calculated, the simulated values are lower than the measured values. The treatment group (pH=5) was equally affected by both factors.

$$HCrO_4^- + 7H^+ + 3e^- \longrightarrow Cr^{3+} + 4H_2O \qquad (4.1)$$

4.5.2.3 Distribution of phosphatase in soil profile before and after acid rain leaching

In situ enzyme spectrometry was used to test the distribution of phosphatase in soil, and the enzyme spectrum images of phosphatase before and after acid rain leaching at different pH values were obtained, as shown in Figure 4.33. The number of pixels representing phosphatase in the rectangular subimage and their proportion in the region were calculated, and the distribution density of phosphatase in soil was described accordingly, as shown in Figure 4.34.

Before start of the leaching experiment, the phosphatase activity density was close to zero, indicating that the initial enzyme activity of Cr(VI)-contaminated soil was severely inhibited by hexavalent chromium. After the first acid rain leaching, Cr(VI) in the upper part of the soil profile migrated downward under the leaching effect of rain, the content of hexavalent chromium in the upper part of the soil was reduced, and the inhibition effect on phosphatase was weakened. A small amount of phosphatase appeared in the upper part of the soil profile, indicating that the enzyme activity began to recover to a certain extent, though it was still under the strong inhibition effect of residual hexavalent chromium and acid rain. With the increase of leaching times, the hexavalent chromium in the soil gradually migrated downward and leached at the bottom. In the vertical direction, the hexavalent chromium gradually decreased from the top of the soil profile, leading to the decrease of its inhibitory effect on phosphatase activity and the increase of enzyme activity. Finally, after the end of the third leaching, the enzyme activity of the whole soil profile was largely restored compared with the initial state.

Because part of the hexavalent chromium adsorbed on soil particles still existed in the soil of the treatment group supplemented with exogenous Cr(VI) after the acid rain leaching, the enzyme activity was significantly lower than that of the blank soil control group after the same acid rain leaching. In the distribution image of phosphatase activity at the same time, the enzyme activity density of the acid rain treatment group (pH=7) was higher than that of the acid rain treatment group (pH=5), while that of the acid rain treatment group (pH=3) was much lower than the other two. This indicated that acid rain had a great inhibitory effect on phosphatase activity,

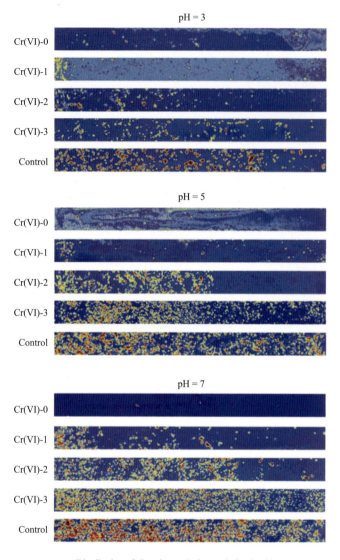

Distribution of phosphatase before and after leaching

Figure 4.33 Distribution of phosphatase activity at soil profile. On the left is the top surface of the soil; Cr(VI)-0 represents the enzyme activity before leaching in chromium-contaminated soil. Cr(VI)-1, Cr(VI)-2, Cr(VI)-3 represent the enzyme activity after the first, second, third leaching in chrome-contaminated soil. Control represents the enzyme activity after the end of leaching in blank soil. During the entire acid rain leaching process, the phosphatase activity in blank soil did not change significantly. Only the distribution of enzyme activity after the entire leaching process is shown here.

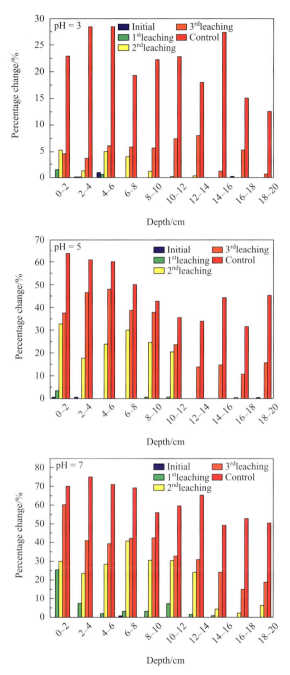

Figure 4.34 Percentage changes of phosphatase pixel with soil depth.

and the lower the pH, the more obvious the inhibitory effect. The recovery of phosphatase activity in the longitudinal soil profile was accelerated with the increase of pH of the leaching acid rain, because the increase of pH accelerated the transport and leaching of hexavalent chromium in the soil, and the corresponding phosphatase activity led to different enzyme activity distributions with the vertical variation of hexavalent chromium content.

4.5.3 Effect of dynamic simulated acid rain on the release characteristics and enzyme activity of Pb in contaminated soil

4.5.3.1 Leaching characteristics of lead under elution acid rain of different pH values

With advance of the elution process, the concentration of lead in the leaching solution at the bottom of the soil is shown in Figure 4.35, and the measured and simulated leaching amounts are shown in Table 4.15. The total leaching amount of lead in the acid rain treatment group with pH=3 is greater than that in the treatment groups with pH=5 and pH=7. This indicates that compared with the neutral rain, the acid rain can increase the cumulative leaching amount of lead in the soil. The lower the pH, the greater the cumulative leaching amount. By comparing the simulated value and the measured value of leaching amount, it can be seen that the model has a good simulation effect, but the result is slightly lower than the actual value, because under acidic conditions, a small amount of lead in carbonate bound state will be converted into exchangeable state in the soil, and the lower the pH, the more converted. The concentration of lead in the leaching solution decreased with the increase of leaching amount, but compared with the hexavalent chromium ion, the concentration of lead in the leaching solution changed slightly, the concentration decreased slowly, and the overall stability was relatively high. The leaching amount and proportion of lead ions in the leaching process are shown in Table 4.16. Among them, the first leaching features the largest concentration, and the amount of lead leached from the first 200mL of leaching solution accounts for about 20% of the total amount. In the initial state, the molar concentration of lead ion in the soil solution is much lower than that of hexavalent chromium ion. In the leaching solution, the molar concentration of lead ion is also much lower than that of hexavalent chromium ion, because metal cations exist in the soil in various states. Among them, exchangeable lead only accounts for a small proportion (Halim et al., 2005). Thus, lead ions pose a smaller threat to the ecological environment than hexavalent chromium ions.

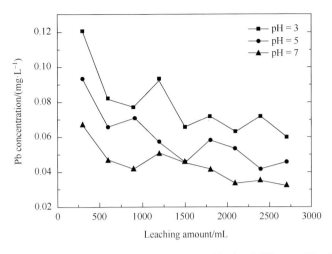

Figure 4.35 Pb content in leaching solution under acid rain of different pH values.

Table 4.15 Pb leaching amount under acid rain of different pH values.

Item	pH=3		pH=5		pH=7	
Leaching amount/mg	Measured	Simulated	Measured	Simulated	Measured	Simulated
1st leaching	0.132	0.101	0.103	0.078	0.057	0.052
2nd leaching	0.103	0.086	0.06	0.043	0.047	0.046
3rd leaching	0.081	0.052	0.049	0.031	0.024	0.019
Cumulated leaching	0.316	0.239	0.212	0.152	0.128	0.117

Table 4.16 Amount and proportion of lead leached during the leaching process.

Treatment	pH=3		pH=5		pH=7	
Leaching process	Amount/mg	Proportion/%	Amount/mg	proportion/%	Amount/mg	Proportion/%
1-1	0.061	19.20%	0.044	20.74	0.028	21.86
1-2	0.037	11.79%	0.028	13.11	0.016	12.50
1-3	0.034	10.84%	0.031	14.49	0.013	10.17
2-1	0.044	13.88%	0.022	10.61	0.019	14.48
2-2	0.028	8.75%	0.015	7.27	0.015	11.78
2-3	0.031	9.89%	0.023	10.76	0.013	10.00
3-1	0.026	8.17%	0.02	9.54	0.008	6.35
3-2	0.031	9.89%	0.013	6.23	0.009	7.15
3-3	0.024	7.60%	0.016	7.34	0.007	5.73

Note: 1-1, 1-2, and 1-3 represent the leaching amounts of lead and its proportions to the total leaching amount in the leaching solution obtained from the first, second and third sampling in the first leaching process, which are the same as that of 2-1, 2-2, 2-3, 3-1, 3-2 and 3-3.

Compared with hexavalent chromium ions, the leaching process of lead ions is relatively gentle and the leaching concentration is rather uniformly distributed, but there is still the phenomenon that the initial leaching concentration is much higher than the subsequent ones. Therefore, certain water and seepage prevention measures shall still be taken for lead-contaminated soil. The release of lead ions is a prolonged and continuous process. With the elapse of time, more and more lead ions will be leached. In a long run, huge threats will be brought to the ecological environment. Thus, in addition to the water and seepage prevention measures, actions shall be taken to treat and restore the lead-contaminated soil as soon as possible.

4.5.3.2 HYDRUS-2D simulation results and measured lead content in soil profile

The simulated image of the soil profile is shown in Figure 4.36, and the measured and simulated concentrations of lead in the soil are shown in Figure 4.37.

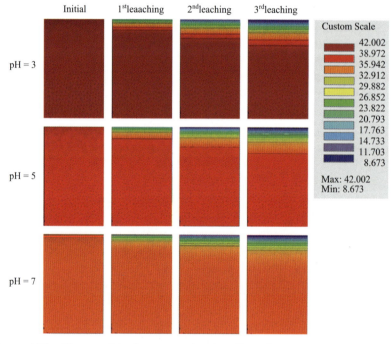

Figure 4.36 Change of lead concentration in soil profile during the HYDRUS-2D simulated leaching (nmol·cm^{-3}). Initial represents the initial state of leaching, while 1, 2, 3 refer to the end of the first, second, third leaching, respectively.

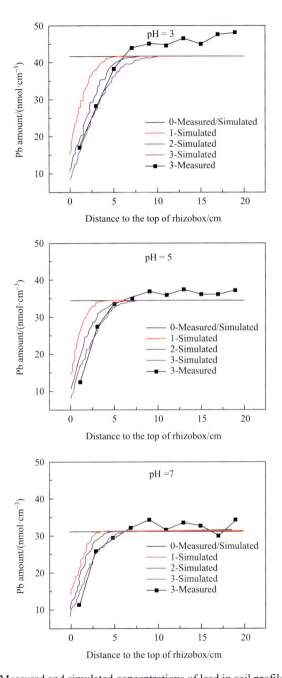

Figure 4.37 Measured and simulated concentrations of lead in soil profile.

Comparison proves that the HYDRUS-2D model has a good simulation effect. As the pH of leaching acid rain decreased, the lead content in the soil solution at the initial state increased, indicating that the acidic conditions were more conducive to the leaching of lead from the soil. This is because lead exists in a variety of forms in the soil, among which the carbonate bound lead is the element formed through coprecipitation with carbonate minerals in the soil. This form of lead is very sensitive to the change of the surrounding soil environment, and especially susceptible to the change of soil pH value. The increase of pH value can help transform the free-state lead into the carbonate-state lead in the soil, while the decrease of pH value is conducive to the release of the free-state lead into the environment again, thereby changing the lead leaching amount under different pH values (Wu and Kim, 2017).

With progress of the elution, the transport of lead in the soil is accelerated with the decrease of the pH of the leaching acid rain. According to results of the isothermal adsorption experiment, compared with low pH values, the soil has a greater adsorption capacity for lead ions at high pH values, circumstances that hinder the downward transport of lead ions to finally slow down the transport of lead ions. The adsorption mechanism of exchangeable lead is the electrostatic attraction between lead ions and the soil surface. With increase of pH value of the solution, the number of negatively charged groups on the soil surface increases, and the electrostatic attraction with metal cations of the lead increases, thereby enhancing the adsorption.

Among the various forms of lead in soil, exchangeable lead accounts for a small proportion of the total amount. It is an existential form with the greatest impact on the soil ecological environment, groundwater and human health, because of its strongest transport and transformation ability and its ability to be absorbed and utilized by plants (Lee, 2006). In the HYDRUS-2D simulation process, only the transport of exchangeable lead in the soil was simulated. According to the comparison between simulation results and the measured results of the soil profile, it can be concluded that the model established by HYDRUS-2D has a good simulation effect on the transport of lead ions in the soil, but the measured value is slightly greater than the simulated value, and the lower the pH, the greater the deviation. This is because in the elution process, there is still the transformation of carbonate bound lead into free lead in the soil. Moreover, the lower the pH, the more free lead released in the elution process. As a physical model, HYDRUS can better simulate the physical changes of water flow and solute in the soil. Unable to calculate the chemical process in the soil, it thus can not calculate the release process of carbonate bound lead. In addition, in the measuring of the content of free lead in the soil, due to the change of soil structure during the sampling process, part of the exchangeable lead existing in the non-mobile area was released.

4.5.3.3 *Distribution of soil phosphatase before and after acid rain leaching*

In situ enzyme spectrometry was used to test the distribution of phosphatase in soil, and the enzyme spectrum images of phosphatase before and after acid rain leaching at different pH values were obtained, as shown in Figure 4.38. The number of pixels representing phosphatase in the rectangular sub-image and their proportion in the region were calculated, and the distribution density of phosphatase in soil was described accordingly, as shown in Figure 4.39.

Before the leaching experiment, the density of phosphatase activity was low, indicating that the initial enzyme activity of Pb-contaminated soil was strongly inhibited by lead, though much higher than the initial enzyme activity of hexavalent chromium contaminated soil profile, indicating that hexavalent chromium is far more toxic to soil than lead. After the first acid rain leaching, the exchangeable Pb in the upper part of the soil profile moved downward under the leaching effect of rainwater, the exchangeable Pb content in the upper soil decreased, its inhibitory effect on phosphatase was weakened, and its enzyme activity density increased slightly. With the increase of leaching times, the exchangeable lead in the soil gradually migrated downward and leached at the bottom. In the vertical direction, the exchangeable lead gradually decreased from the top of the soil profile, leading to the decrease of its inhibitory effect on phosphatase activity and the increase of enzyme activity. Finally, after the end of the third leaching, the enzyme activity of the whole soil profile was largely restored compared with the initial state.

In the soil treatment group with exogenous Pb addition, some lead still existed in the soil after the acid rain leaching as adsorbed on soil particles in the carbonate bound state, iron and manganese oxide bound state, organic bound state and residual state, with a small proportion of exchangeable lead leached. Therefore, compared with the experimental group with hexavalent chromium, the activity of phosphatase changed less, but it could still show the change of lead content in exchangeable state to some extent. The enzyme activity of the treatment group with exogenous Pb was significantly lower than that of the blank control group after leaching with the same acid rain, indicating that lead in the soil could impose a certain inhibitory effect, far

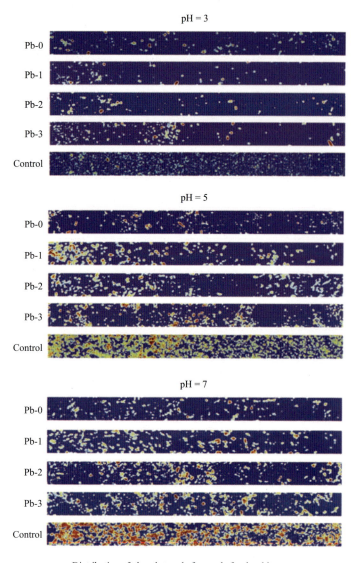

Distribution of phosphatase before and after leaching

Figure 4.38 Distribution of phosphatase activity at soil profile. On the left is the top surface of the soil; Pb-0 represents the enzyme activity before leaching in chrome-contaminated soil. Pb-1, Pb-2, Pb-3 represent the enzyme activity after the first, second and third leaching in chrome-contaminated soil. Control represents the enzyme activity after the end of leaching in blank soil. During the entire acid rain leaching process, the phosphatase activity in blank soil did not change significantly. Only the distribution of enzyme activity after the entire leaching process is shown here.

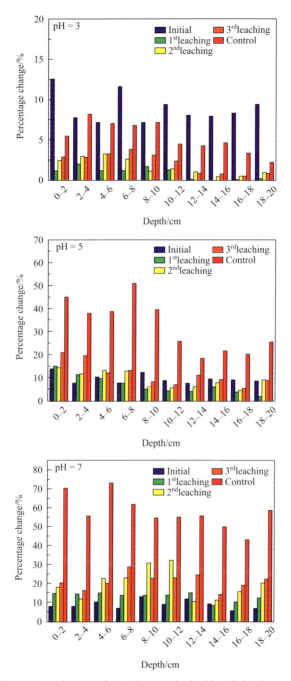

Figure 4.39 Percentage changes of phosphatase pixel with soil depth.

smaller than that of hexavalent chromium, on the phosphatase activity. In the distribution image of phosphatase activity at the same time, the enzyme activity density of the acid rain treatment group (pH=7) was slightly higher than that of the acid rain treatment group (pH=5), while that of the acid rain treatment group (pH=3) was much lower than the other two. This indicated that acid rain had a great inhibitory effect on phosphatase activity, and the lower the pH, the more obvious the inhibitory effect.

4.5.4 Summary

In this work, hexavalent chromium and lead ions are selected as the representatives of the anion and cation of heavy metals. The transport performance of the two heavy metal ions is studied by dynamically simulating the soil contaminated by hexavalent chromium and lead with acid rain of different pH values. Also, it measured the adsorption performance parameters of the soil to hexavalent chromium and lead ions under different pH values, used HYDRUS-2D to simulate the transport process of heavy metals, and took in situ zymogram technology to detect the changes of enzyme activity during leaching, thus establishing the transport model of hexavalent chromium and lead in soil as well as the response mechanism of enzyme activity under acid rain leaching conditions. The main conclusions are:

(1) The acid rain leaching will promote the leaching of hexavalent chromium and lead ions in soil, and the lower the pH of acid rain, the more heavy metal ions leached. In the low pH environment, the soil has a greater adsorption capacity for hexavalent chromium ions, but a smaller one for lead ions. As a result, leaching with acid rain of a low pH will speed up the transport of lead ions and slow down the transport of hexavalent chromium. However, in a long run, the soil contaminated by hexavalent chromium with acid rain of a low pH features a higher cumulative leaching amount.

(2) The content of hexavalent chromium and lead in soil has a great inhibitory effect on phosphatase activity; the lower the pH, the higher the content of hexavalent chromium and lead, the more obvious the inhibitory effect. The inhibitory effect of hexavalent chromium on phosphatase activity was much greater than that of lead, indicating that hexavalent chromium had a stronger toxic effect on soil. As the leaching process continues, the phosphatase in the contaminated soil gradually recovered its activity, in a rate accelerated with the increase of the pH of the leaching acid rain. However, since lead ions exist in the soil in various forms, the exchangeable lead moved downward, while other forms of lead stayed relatively stable

and still existed in the soil in large quantities to inhibit the phosphatase activity. As a result, the phosphatase activity did not change significantly compared with the hexavalent chromium treatment group before and after leaching. Acid rain also inhibited the phosphatase activity; the lower the pH, the more obvious the inhibition effect.

(3) HYDRUS-2D has a good simulation effect on the migration of anions and cations in the soil profile, and the redox reaction occurring in the leaching process will make the simulated value slightly higher than the actual value. Since hexavalent chromium is more easily reduced to trivalent chromium at a low pH; the lower the pH, the greater the deviation between the simulated value and the actual value. The difference between the simulated and measured lead ion values is subject to the transformation from carbonate bound lead to exchangeable lead. The lower the pH, the greater the transformation. As a result, the simulated value is lower than the measured value.

(4) For the soil of the site actually contaminated by chromium, the leaching of hexavalent chromium caused by precipitation mainly occurs in the first 8 months, during which the hexavalent chromium brings the largest risks to the soil and groundwater. However, in lead-contaminated sites, the leaching of lead is a slow and long-term process, and the most is leached in the early stage of rainwater leaching. Therefore, for the two contaminated sites, it is necessary to well adopt waterproof and seepage prevention measures, and complete the treatment as soon as possible to control the diffusion range of heavy metals in the soil.

Limited by energy and time, as well as experimental conditions in the early stage, there are still some parts not detailed enough in the research of this topic, and some work needs to be improved, which can be further discussed in subsequent experiments. Here, we put forward the following three tips.

(1) To explore the transport behavior of anions and cations of heavy metals in the soil under the acid rain leaching, this chapter only selects two representative anions and cations as the research objects. In the later stage, more types of anions and cations shall be added and the differences in transport characteristics among different anions and cations shall be compared.

(2) HYDRUS-2D is primarily used for main modules in this study. For the combined application of additional HP2 module, C-Ride-2D module and DualPerm-2D module, a thorough discussion of more factors such as the transport of colloidal solute and chemical reactions of major ions will make the simulation results more accurate.

(3) This study only discusses the ion transport and its influence on enzyme activity in soil contaminated by a single heavy metal, while soil contaminated by multiple heavy metal ions is more often the case. It is thus of great significance to study the transport behavior and toxicity of different heavy metal ions under the acid rain leaching.

Main References

Abdu, N., Abdullahi, A.A., Abdulkadir, A., 2017. Heavy metals and soil microbes. Environmental Chemistry Letters, 15(1): 65–84. Available from: https: doi.org/10.1007/s10311-016-0587-x.

Abramova, A., Gedanken, A., Popov, V., et al.,, 2013. A sonochemical technology for coating of textiles with antibacterial nanoparticles and equipment for itsimplementation. Materials Letters, 96: 121–124. Available from: https: doi.org/ 10.1016/j.matlet.2013.01.041.

Acosta-Martínez, V., Tabatabai, M.A., 2001. Arylamidase activity in soils: Effect of trace elements and relationships to soil properties and activities of amidohydrolases. Soil Biology and Biochemistry, 33(1): 17–23. Available from: https: doi.org/ 10.1016/S0038-0717(00)00109-7.

Alan, M., Kara, D., 2019. Comparison of a new sequential extraction method and the BCR sequential extraction method for mobility assessment of elements around boron mines in Turkey. Talanta, 194: 189–198. Available from: https: doi.org/ 10.1016/j.talanta.2018.10.030.

An, F.Q., Diao, Z., Lv, J.L., 2018. Microbial diversity and community structure in agricultural soils suffering from 4 years of Pb contamination. Canadian Journal of Microbiology, 64(5): 305–316. Available from: https: doi.org/ 10.1139/cjm-2020-0242.

Anderson, T.H., Domsch, K.H., 2010. Soil microbial biomass: The eco-physiological approach. Soil Biology and Biochemistry, 42(12): 2039–2043. Available from: https: doi.org/ 10.1016/j.soilbio.2010.06.026.

Angelovičová, L., Fazekašová, D., 2014. Contamination of the soil and water environment by heavy metals in the former mining area of Rudňany (Slovakia). Soil and Water Research, 9(1): 18–24. Available from: https: doi.org/ 10.17221/24/2013-SWR.

Anjum, N.A., Adam, V., Kizek, R., et al.,, 2015a. Nanoscale copper in the soil–plant system–toxicity and underlying potential mechanisms. Environmental Research, 138: 306–325. Available from: https: doi.org/ 10.1016/j.envres.2015.02.019.

Anjum, N.A., Singh, H.P., et al.,, 2015b. Too much is bad: An appraisal of phytotoxicity of elevated plant-beneficial heavy metal ions. Environmental Science and Pollution Research International, 22(5): 3361–3382. Available from: https: doi.org/10.1007/s11356-014-3849-9.

Applerot, G., Lellouche, J., Lipovsky, A., et al.,, 2012. Understanding the antibacterial mechanism of CuO nanoparticles: Revealing the route of induced oxidative stress. Small, 8(21): 3326–3337. Available from: https: doi.org/10.1002/smll.201200772.

Asadishad, B., Chahal, S., Akbari, A., et al.,, 2018. Amendment of agricultural soil with metal nanoparticles: Effects on soil enzyme activity and microbial community composition. Environmental Science & Technology, 52(4): 1908–1918. Available from: https: doi.org/ 10.1021/acs.est.7b05389.

Barns, S.M., Cain, E.C., Sommerville, L., et al.,, 2007. Acidobacteria Phylum sequences in uranium-contaminated subsurface sediments greatly expand the known diversity within the Phylum.

Applied and Environmental Microbiology, 73(9): 3113–3116. Available from: https: doi.org/ 10.1128/AEM.02012-06.

Bondarenko, O., Ivask, A., Käkinen, A., et al.,, 2012. Sub-toxic effects of CuO nanoparticles on bacteria: Kinetics, role of Cu ions and possible mechanisms of action. Environmental Pollution, 169: 81–89. Available from: https: doi.org/ 10.1016/j.envpol.2012.05.009.

Bouskill, N.J., Barker-Finkel, J., Galloway, T.S., et al.,, 2010. Temporal bacterial diversity associated with metal-contaminated river sediments. Ecotoxicology, 19(2): 317–328.

Brunner, T.J., Wick, P., Manser, P., et al.,, 2006. *In vitro* cytotoxicity of oxide nanoparticles: Comparison to asbestos, silica, and the effect of particle solubility. Environmental Science & Technology, 40(14): 4374–4381. Available from: https: doi.org/ 10.1021/es052069i.

Caldwell, B.A., 2005. Enzyme activities as a component of soil biodiversity: A review. Pedobiologia, 49(6): 637–644. Available from: https: doi.org/ 10.1016/j.pedobi.2005.06.003.

Cao, Q.Y., Huang, Z.H., 2017. Review on speciation analysis of heavy metals in polluted soils and its influencing factors. Ecologic Science,36(6): 222–232. Available from: https: doi.org/ 10.14108/j.cnki. 1008-8873.2017.06.030.

Caporaso J G, Lauber C L, Walters W A, et al.,, 2011. Global patterns of 16S rRNA diversity at a depth of millions of sequences per sample. Proceedings of the National Academy of Sciences of the United States of America, 108(Suppl 1): 4516–4522. Available from: https: doi.org/ 10.1073/pnas.1000080107.

Caporaso, J.G, Lauber, C.L., Walters, W.A., et al.,, 2011. Global patterns of 16S rRNA diversity at a depth of millions of sequences per sample. Proceedings of the National Academy of Sciences of the United States of America, 108(Suppl 1): 4516–4522. Available from: https: doi.org/ 10.1073/pnas.1000080107.

Chai, H.K., Yao, J., Sun, J.J., et al., 2015. The effect of metal oxide nanoparticles on functional bacteria and metabolic profiles in agricultural soil. Bulletin of Environmental Contamination and Toxicology, 94(4): 490–495. Available from: https: doi.org/ 10.1007/s00128-015-1485-9.

Chen Z M, Wang Q, Ma J W, et al.,, 2020. Fungal community composition change and heavy metal accumulation in response to the long-term application of anaerobically digested slurry in a paddy soil. Ecotoxicology and Environmental Safety, 196: 110453. Available from: https: doi.org/ 10.1016/j. ecoenv.2020.110453.

Ciarkowska K, Sołek-Podwika K, Wieczorek J. 2014. Enzyme activity as an indicator of soil-rehabilitation processes at a zinc and lead ore mining and processing area. Journal of Environmental Management, 132: 250–256. Available from: https: doi.org/ 10.1016/j.jenvman.2013.10.022.

Deng R, Lin D.H, Zhu L.Z, et al, 2017. Nanoparticle interactions with co-existing contaminants: joint toxicity, bioaccumulation and risk. Nanotoxicology, 11(5): 591–612. Available from: https: doi.org/ 10.1080/17435390.2017.1343404.

Dimkpa C.O, McLean J.E, Latta D.E, et al, 2012. CuO and ZnO nanoparticles: Phytotoxicity, metal speciation, and induction of oxidative stress in sand-grown wheat. Journal of Nanoparticle Research, 14(9):1125. Available from: https: doi.org/ 10.1007/s11051-012-1125-9.

Du, W.C., Sun, Y.Y., Ji R., et al., 2011. TiO_2 and ZnO nanoparticles negatively affect wheat growth and soil enzyme activities in agricultural soil. Journal of Environmental Monitoring: JEM, 13(4): 822–828. Available from: https: doi.org/ 10.1039/c0em00611d.

Du, W.C., Xu, Y.W., Yin, Y., et al., 2018. Risk assessment of engineered nanoparticles and other contaminants in terrestrial plants. Current Opinion in Environmental Science & Health, 6: 21–28. Available from: https: doi.org/ 10.1016/j.coesh.2018.07.010.

Ebbs, S.D., Bradfield, S.J., Kumar, P., et al., 2016. Accumulation of zinc, copper, or cerium in carrot (Daucus carota) exposed to metal oxide nanoparticles and metal ions. Environmental Science: Nano, 3(1):114–126. Available from: https: doi.org/ 10.1039/c5en00161g.

El-Trass, A., ElShamy, H., El-Mehasseb, I., et al., 2012. CuO nanoparticles: Synthesis, characterization, optical properties and interaction with amino acids. Applied Surface Science, 258(7): 2997–3001. Available from: https: doi.org/ 10.1016/j.apsusc.2011.11.025.

Evans, P., Matsunaga, H., Kiguchi, M., 2008. Large-scale application of nanotechnology for wood protection. Nature Nanotechnology, 3(10): 577. Available from: https: doi.org/ 10.1038/nnano.2008.286.

Fan, Z.X., Zhang, Q., Gao, B., et al., 2019. Removal of hexavalent chromium by biochar supported nZVI composite: Batch and fixed-bed column evaluations, mechanisms, and secondary contamination prevention. Chemosphere, 217: 85–94. Available from: https: doi.org/ 10.1016/j.chemosphere.2018. 11.009.

Fang, G.D., Si, Y.B., 2011 Effects of Nanoscale Fe_3O_4 on microbial communities, enzyme activities and 2,4-D degradation in red soil. Scientia Agricultural Sinica, 44(6): 1165–1172. (in Chinese)

Fatimawali M., Kepel, B.J., Gani, M.A., et al., 2020. Comparison of bacterial community structure and diversity in traditional gold mining waste disposal site and rice field by using a metabarcoding approach. International Journal of Microbiology, 2020: 1858732. Available from: https: doi.org/ 10.1155/2020/1858732.

Gao, X.L., Chen, C.T.A., Wang, G., et al., 2010. Environmental status of *Daya* Bay surface sediments inferred from a sequential extraction technique. Estuarine, Coastal and Shelf Science, 86(3): 369–378. Available from: https: doi.org/ 10.1016/j.ecss.2009.10.012.

Gao, X.Y., Avellan, A., Laughton, S., et al., 2018. CuO nanoparticle dissolution and toxicity to wheat (*Triticum aestivum*) in rhizosphere soil. Environmental Science & Technology, 52(5): 2888–2897. Available from: https: doi.org/ 10.1021/acs.est.7b05816.

Gao, X.Y., Rodrigues, S.M., Spielman-Sun E., et al., 2019. Effect of soil organic matter, soil pH, and moisture content on solubility and dissolution rate of CuO NPs in soil. Environmental Science & Technology, 53(9): 4959–4967. Available from: https: doi.org/ 10.1021/acs.est.8b07243.

García-Gómez, C., Fernández, M.D., García, S., et al., 2018. Soil pH effects on the toxicity of zinc oxide nanoparticles to soil microbial community. Environmental Science and Pollution Research, 25(28): 28140–28152. Available from: https: doi.org/ 10.1007/s11356-018-2833-1.

Ge. Y., Schimel, J.P., Holden, P.A., 2011. Evidence for negative effects of TiO_2 and ZnO nanoparticles on soil bacterial communities. Environmental Science & Technology, 45(4): 1659–1664. Available from: https: doi.org/ 10.1021/es103040t.

Gil-díaz, M.M., Lobo, M.C., 2014. Immobilisation of Pb and Zn in Soils Using Stabilised Zero-valent Iron Nanoparticles : Effects on Soil Properties. 42:1776–1784. Available from: https: doi.org/ 10.1002/ clen.201300730.

Ginocchio, R., Sánchez, P., de la Fuente, L.M., et al., 2006. Agricultural soils spiked with copper mine wastes and copper concentrate: Implications for copper bioavailability and bioaccumulation. Environmental Toxicology and Chemistry, 25(3): 712–718. Available from: https: doi.org/ 10.1897/05-105R.1

Godwin, H.A., Chopra, K., Bradley, K.A., et al., 2009. The university of California center for the environmental implications of nanotechnology. Environmental Science & Technology, 43(17): 6453–6457.

Guber, A., Kravchenko, A., Razavi, B.S., et al., 2018. Quantitative soil zymography: Mechanisms, processes of substrate and enzyme diffusion in porous media. Soil Biology and Biochemistry, 127: 156–167. Available from: https: doi.org/ 10.1016/j.soilbio.2018.09.030.

Guo, H.H., Nasir, M., Lv, J.L., et al., 2017. Understanding the variation of microbial community in heavy metals contaminated soil using high throughput sequencing. Ecotoxicology and Environmental Safety, 144: 300–306. Available from: https: doi.org/ 10.1016/j.ecoenv.2017.06.048.

Halim, C.E., Scott, J.A., Amal, R., et al., 2005. Evaluating the applicability of regulatory leaching tests for assessing the hazards of Pb-contaminated soils. Journal of Hazardous Materials, 120(1–3): 101–111. Available from: https: doi.org/ 10.1016/j.jhazmat.2004.12.039.

Han, W., Sun, C.X., Su, J., 2017. Effects of Elevated Temperature and Simulated Acid Rain on Enzyme Activity and Temperature Sensitivity of Paddy Soil. Journal of Ecology and Rural Environment,33(12): 1117–1124.

Hao, Y.M., Man, C., Hu, Z.B., 2010. Effective removal of Cu (II) ions from aqueous solution by amino-functionalized magnetic nanoparticles. Journal of Hazardous Materials, 184(1–3): 392–399. Available from: https: doi.org/ 10.1016/j.jhazmat.2010.08.048.

He, E.K., Qiu, H., Huang, X.Y., et al., 2019. Different dynamic accumulation and toxicity of ZnO nanoparticles and ionic Zn in the soil sentinel organism *Enchytraeus crypticus*. Environmental Pollution, 245: 510–518. Available from: https: doi.org/ 10.1016/j.envpol.2018.11.037.

He, S.Y., Feng, Y.Z., Ren, H.X., et al., 2011. The impact of iron oxide magnetic nanoparticles on the soil bacterial community. Journal of Soils and Sediments, 11(8): 1408–1417. Available from: https: doi.org/ 10.1007/s11368-011-0415-7.

He, X., Kuang, Y.S., Li, Y.Y., et al., 2012. Changing exposure media can reverse the cytotoxicity of ceria nanoparticles for *Escherichia coli*. Nanotoxicology, 6(3): 233–240. Available from: https: doi.org/ 10.3109/17435390.2011.569097.

Hemmat-Jou, M.H., Safari-Sinegani, A.A., Mirzaie-Asl, A., et al., 2018. Analysis of microbial communities in heavy metals-contaminated soils using the metagenomic approach. Ecotoxicology, 27(9): 1281–1291. Available from: https: doi.org/ 10.1007/s10646-018-1981-x.

Horst, A.M., Neal, A.C., Mielke, R.E., et al., 2010. Dispersion of TiO₂ nanoparticle agglomerates by *Pseudomonas aeruginosa*. Applied and Environmental Microbiology, 76(21): 7292–7298. Available from: https: doi.org/ 10.1128/AEM.00324-10.

Hou, Z., Chen, Z., Shen, Z.Y., et al., , 2014. Effects of Zinc oxide nanoparticles on enzyme activities of Soil microorganisms. Journal of Agro-Environment Science, 33(6):1153–1158. (in Chinese)

Huang, H.Y., Chen, J.S., Liu, S.B., et al., 2022. Impact of ZnO nanoparticles on soil lead bioavailability and microbial properties. Science of the Total Environment, 806: 150299. Available from: https: doi.org/10.1016/j.scitotenv.2021.150299.

Hussain, A., Ali, S., Rizwan, M., et al., 2018. Zinc oxide nanoparticles alter the wheat physiological response and reduce the cadmium uptake by plants. Environmental Pollution, 242: 1518–1526. Available from: https: doi.org/ 10.1016/j.envpol.2018.08.036.

Jin, S.Y., Wang, Y.J., Wang, P., et al.,, 2010. Comparative Secotoxicity of nanometer-micrometer-sized CuO and Ionic Copper to soil urease. Asian Journal of Ecotoxicology, (6):835–841.

Jośko, I., 2019. Copper and zinc fractionation in soils treated with CuO and ZnO nanoparticles: The effect of soil type and moisture content. Science of the Total Environment, 653: 822–832. Available from: https: doi.org/ 10.1016/j.scitotenv.2018.11.014.

Jośko, I., Oleszczuk, P., Futa, B., 2014. The effect of inorganic nanoparticles (ZnO, Cr_2O_3, CuO and Ni) and their bulk counterparts on enzyme activities in different soils. Geoderma, 232: 528–537. Available from: https: doi.org/ 10.1016/j.geoderma.2014.06.012.

Keller, A.A., McFerran, S., Lazareva, A., et al., 2013. Global life cycle releases of engineered nanomaterials. Journal of Nanoparticle Research, 15(6): 1692. Available from: https: doi.org/ 10.1007/s11051-013-1692-4.

Khan, S., Cao, Q., Hesham, A.E.L., et al., 2007. Soil enzymatic activities and microbial community structure with different application rates of Cd and Pb. Journal of Environmental Sciences, 19(7): 834–840. Available from: https: doi.org/ 10.1016/S1001-0742(07)60139-9.

Klaine S J, Alvarez P J J, Batley G E, et al., 2008. Nanomaterials in the environment: Behavior, fate, bioavailability, and effects. Environmental Toxicology and Chemistry, 27(9): 1825–1851.

Kohen, R., Nyska, A., 2002. Oxidation of biological systeme. Toxicol. Pathol. 30, 620–650. Available from: https: doi.org/ 10.1080/0192623029016672.

Kool, P.L., Ortiz, M.D., van Gestel, C.A.M., 2011. Chronic toxicity of ZnO nanoparticles, non-nano ZnO and $ZnCl_2$ to *Folsomia candida* (Collembola) in relation to bioavailability in soil. Environmental Pollution, 159(10): 2713–2719. Available from: https: doi.org/10.1016/j.envpol.2011.05.021.

Kuang, Y.S., He, X., Zhang, Z.Y., et al., 2011. Comparison study on the antibacterial activity of nano- or bulk-cerium oxide. Journal of Nanoscience and Nanotechnology, 11(5): 4103–4108. Available from: https: doi.org/ 10.1166/jnn.2011.3858.

Kwak, J.I., Yoon, S.J., An, Y.J., 2017. Long-term effects of ZnO nanoparticles on exoenzyme activities in planted soils. Environmental Engineering Research, 22(2): 224–229. Available from: https: doi.org/ 10.4491/eer.2016.103.

Lee, E., 2006. Long-term Leaching Characteristics of Lead Contaminated Soils Treated with Soluble Phosphate. Journal of Korea Academia-Industrial cooperation Society,7(3): 453–457.

Li ,Y.Y., Chen, L.Q., Wen, H.Y., et al., 2014. 454 pyrosequencing analysis of bacterial diversity revealed by a comparative study of soils from mining subsidence and reclamation areas. Journal of Microbiology and Biotechnology, 24(3): 313–323. Available from: https: doi.org/ 10.4014/jmb.1309.09001.

Li, C.Z., Wang, X.F., Huang, H., et al., 2020. Effect of multiple heavy metals pollution to bacterial diversity and community structure in farmland soils. Human and Ecological Risk Assessment: an International Journal, 27: 724–741. Available from: https: doi.org/10.1080/10807039.2020.1752143.

Li, J., Shu, W.Q., Chen, J.A., et al, 2005. Relationships between enzyme activities and chemical property and microbial quantity in landfill soil. Chinese Journal of Ecology, 24(9): 1043–1047.

Liu, C., Yao, M.J., Stegen, J.C., et al., 2017. Long-term nitrogen addition affects the phylogenetic turnover of soil microbial community responding to moisture pulse. Scientific Reports, 7: 17492. Available from: https: doi.org/ 10.1038/s41598-017-17736-w.

Liu, J.J., Sui, Y.Y., Yu, Z.H., et al., 2016. Diversity and distribution patterns of acidobacterial communities in the black soil zone of Northeast China. Soil Biology and Biochemistry, 95: 212–222. Available from: https: doi.org/10.1016/j.soilbio.2015.12.021.

Liu, R.Q., Zhao, D.Y., 2007. Reducing leachability and bioaccessibility of lead in soils using a new class of stabilized iron phosphate nanoparticles. Water Research, 41(12): 2491–2502. Available from: https: doi.org/ 10.1016/j.watres.2007.03.026.

Liu, S.B., Razavi, B.S., Su, X., et al., 2017. Spatio-temporal patterns of enzyme activities after manure application reflect mechanisms of niche differentiation between plants and microorganisms. Soil

Biology and Biochemistry, 112: 100–109. Available from: https: doi.org/ 10.1016/j.soilbio.2017. 05.006.

Liu, S.J., Xia, X., Chen, G.M., et al, 2011. Study Progress on Functions and Affecting Factors of Soil Enzymes. Chinese Agricultural Science Bulletin, 27(21): 1–7.

Lok, C.N., Ho, C.M., Chen, R., et al., 2006. Proteomic analysis of the mode of antibacterial action of silver nanoparticles. Journal of Proteome Research, 5(4): 916–924. Available from: https: doi.org/ 10.1021/pr0504079.

Lorenz, N., Hintemann, T., Kramarewa, T., et al., 2006. Response of microbial activity and microbial community composition in soils to long-term arsenic and cadmium exposure. Soil Biology and Biochemistry, 38(6): 1430–1437. Available from: https: doi.org/ 10.1016/j.soilbio.2005.10.020.

Lur, G., Higley, M.J., 2015. Glutamate receptor modulation is restricted to synaptic microdomains. Cell Reports, 12(2): 326–334. Available from: https: doi.org/ 10.1016/j.celrep.2015.06.029.

Mahdavi, S., Afkhami, A., Jalali, M., 2015. Reducing leachability and bioavailability of soil heavy metals using modified and bare Al_2O_3 and ZnO nanoparticles. Environmental Earth Sciences, 73(8): 4347–4371. Available from: https: doi.org/ 10.1007/s12665-014-3723-6.

Marx, M.C., Kandeler, E., Wood, M., et al., 2005. Exploring the enzymatic landscape: Distribution and kinetics of hydrolytic enzymes in soil particle-size fractions. Soil Biology and Biochemistry, 37(1): 35–48. Available from: https: doi.org/ 10.1016/j.soilbio.2004.05.024.

Masaki, S., Shiotsu, H., Ohnuki, T., et al., 2015. Effects of CeO_2 nanoparticles on microbial metabolism. Chemical Geology, 391: 33–41. Available from: https: doi.org/ 10.1016/j.chemgeo.2014.10.026.

McKee, M.S., Filser, J., 2016. Impacts of metal-based engineered nanomaterials on soil communities. Environmental Science: Nano, 3(3): 506–533. Available from: https: doi.org/ 10.1039/c6en00007j.

Mendes, R., Kruijt, M., de Bruijn, I., et al., 2011. Deciphering the rhizosphere microbiome for disease-suppressive bacteria. Science, 332(6033): 1097–1100. Available from: https: doi.org/ 10.1126/science. 1202007.

Merino, C., Godoy, R., Matus, F., 2016. Soil microorganisms and enzyme activity at different levels of organic matter stability. Journal of Soil Science and Plant Nutrition, 16:14–30 . Available from: https: doi.org/ 10.4067/S0718-95162016005000002.

Mganga, K.Z., Razavi, B.S., Kuzyakov, Y., 2016. Land use affects soil biochemical properties in Mt. Kilimanjaro region. CATENA, 141: 22–29. Available from: https: doi.org/ 10.1016/j.catena.2016. 02.013.

Mikanova, O., 2006. Effects of heavy metals on some soil biological parameters. Journal of Geochemical Exploration, 88(1–3): 220–223. Available from: https: doi.org/ 10.1016/j.gexplo.2005.08.043.

Mkhinini, M., Boughattas, I., Alphonse, V., et al., 2020. Heavy metal accumulation and changes in soil enzymes activities and bacterial functional diversity under long-term treated wastewater irrigation in East Central Region of Tunisia (Monastir governorate). Agricultural Water Management, 235: 106150. Available from: https: doi.org/ 10.1016/j.agwat.2020.106150.

Moscatelli, M.C., Lagomarsino, A., Marinari, S., et al., 2005. Soil microbial indices as bioindicators of environmental changes in a poplar plantation. Ecological Indicators, 5(3): 171–179. Available from: https: doi.org/ 10.1016/j.ecolind.2005.03.002.

Nath, J., Dror, I., Landa, P., et al., 2018. Synthesis and characterization of isotopically-labeled silver, copper and zinc oxide nanoparticles for tracing studies in plants. Environmental Pollution, 242: 1827–1837. Available from: https: doi.org/ 10.1016/j.envpol.2018.07.084.

Nel, A., Xia, T., Mädler, L., et al., 2006. Toxic potential of materials at the nanolevel. Science, 311(5761): 622–627.

Niu, X.Y., Zhou, J., Wang, X.N., et al., 2020. Indigenous bacteria have high potential for promoting *Salix integra* thunb. remediation of lead-contaminated soil by adjusting soil properties. Frontiers in Microbiology, 11: 924. Available from: https: doi.org/ 10.3389/fmicb.2020.00924.

Osmond, M.J., McCall, M.J., 2010. Zinc oxide nanoparticles in modern sunscreens: An analysis of potential exposure and hazard. Nanotoxicology, 4(1): 15–41. Available from: https: doi.org/ 10.3109/17435390903502028.

Park, E.J., Yi, J., Chung, K.H., et al., 2008. Oxidative stress and apoptosis induced by titanium dioxide nanoparticles in cultured BEAS-2B cells. Toxicology Letters, 180(3): 222–229. Available from: https: doi.org/ 10.1016/j.toxlet.2008.06.869.

Peng, C., Xu, C., Liu, Q.L., et al., 2017. Fate and transformation of CuO nanoparticles in the soil-rice system during the life cycle of rice plants. Environmental Science & Technology, 51(9): 4907–4917. Available from: https: doi.org/ 10.1021/acs.est.6b05882.

Pokhrel, L.R., Dubey, B., 2013. Evaluation of developmental responses of two crop plants exposed to silver and zinc oxide nanoparticles. Science of the Total Environment, 452: 321–332. Available from: https: doi.org/ 10.1016/j.scitotenv.2013.02.059.

Pu, S.Y., Yan, C., Huang, H.Y., et al., 2019. Toxicity of nano-CuO particles to maize and microbial community largely depends on its bioavailable fractions. Environmental Pollution, 255: 113248. Available from: https: doi.org/10.1016/j.envpol.2019.113248.

Qiu, H., Smolders, E., 2017. Nanospecific phytotoxicity of CuO nanoparticles in soils disappeared when bioavailability factors were considered. Environmental Science & Technology, 51(20): 11976–11985. Available from: https: doi.org/ 10.1021/acs.est.7b01892.

Rajkumar, M., Vara Prasad, M.N., Freitas, H., et al., 2009. Biotechnological applications of serpentine soil bacteria for phytoremediation of trace metals. Critical Reviews in Biotechnology, 29(2): 120–130. Available from: https: doi.org/ 10.1080/07388550902913772.

Rajput, V.D., Minkina, T.M., Behal, A., et al., 2018. Effects of zinc-oxide nanoparticles on soil, plants, animals and soil organisms: A review. Environmental Nanotechnology, Monitoring & Management, 9: 76–84. Available from: https: doi.org/ 10.1016/j.enmm.2017.12.006.

Rauret, G., López-Sánchez, J.F., Sahuquillo, A., et al., 1999. Improvement of the BCR three step sequential extraction procedure prior to the certification of new sediment and soil reference materials. Journal of Environmental Monitoring: JEM, 1(1): 57–61. Available from: https: doi.org/ 10.1039/a807854h.

Rawat, S., Pullagurala, V.L.R., Hernandez-Molina, M., et al., 2018. Impacts of copper oxide nanoparticles on bell pepper (*Capsicum annum* L.) plants: A full life cycle study. Environmental Science: Nano, 5(1): 83–95. Available from: https: doi.org/ 10.1039/c7en00697g.

Razavi, B.S., Liu, S.B., Kuzyakov, Y., 2017. Hot experience for cold-adapted microorganisms: Temperature sensitivity of soil enzymes. Soil Biology and Biochemistry, 105: 236–243. Available from: https: doi.org/ 10.1016/j.soilbio.2016.11.026.

Razavi, B.S., Zarebanadkouki, M., Blagodatskaya, E., et al., 2016. Rhizosphere shape of lentil and maize: Spatial distribution of enzyme activities. Soil Biology and Biochemistry, 96: 229-237. Available from: https: doi.org/ 10.1016/j.soilbio.2016.02.020.

Reddy, M.V., Venkataiah, B., 1989. Influence of microarthropod abundance and climatic factors on weight loss and mineral nutrient contents of *Eucalyptus* leaf litter during decomposition. Biology and Fertility of Soils, 8(4): 319–324.

Reddy, P.V.L., Hernandez-Viezcas, J.A., Peralta-Videa, J.R., et al., 2016. Lessons learned: Are engineered nanomaterials toxic to terrestrial plants? Science of the Total Environment, 568: 470–479. Available from: https: doi.org/ 10.1016/j.scitotenv.2016.06.042.

Ren, R.F., Zhao, T.F., Wang, Q.Q., et al., 2015. Effect of Heavy Metal Pollution on Soil Animal Community Structure. Journal of Henan Agricultral Sciences,44(10): 90–94.

Rizwan, M., Ali, S., Qayyum, M.F., et al., 2017. Effect of metal and metal oxide nanoparticles on growth and physiology of globally important food crops: A critical review. Journal of Hazardous Materials, 322: 2–16. Available from: https: doi.org/ 10.1016/j.jhazmat.2016.05.061.

Romero-Freire, A., Sierra Aragón, M., Martínez Garzón, F.J., et al., 2016. Is soil basal respiration a good indicator of soil pollution? Geoderma, 263: 132–139. Available from: https: doi.org/ 10.1016/j. geoderma.2015.09.006.

Rousk, J., Bååth, E., Brookes, P.C., et al., 2010. Soil bacterial and fungal communities across a pH gradient in an arable soil. The ISME Journal, 4(10): 1340–1351. Available from: https: doi.org/ 10.1038/ismej.2010.58.

Rousk, J., Ackermann, K., Curling, S.F., et al., 2012. Comparative toxicity of nanoparticulate CuO and ZnO to soil bacterial communities. PLoS One, 7(3): e34197. Available from: https: doi.org/ 10.1371/journal.pone.0034197.

Scheckel, K.G., Luxton, T.P., El Badawy, A.M., et al., 2010. Synchrotron speciation of silver and zinc oxide nanoparticles aged in a Kaolin suspension. Environmental Science & Technology, 44(4): 1307–1312. Available from: https: doi.org/ 10.1021/es9032265.

Shah, K., Dubey, R.S., 1998. Cadmium elevates level of protein, amino acids and alters activity of proteolytic enzymes in germinating rice seeds. Acta Physiologiae Plantarum, 20(2): 189–196. Available from: https: doi.org/ 10.1007/s11738-998-0013-5.

Shahid, M., Shamshad, S., Rafiq, M., et al., 2017. Chromium speciation, bioavailability, uptake, toxicity and detoxification in soil-plant system: A review. Chemosphere, 178: 513–533. Available from: https: doi.org/ 10.1016/j.chemosphere.2017.03.074.

Shen, G.Q., Lu, Y.T., Zhou, Q.X., et al., 2005. Interaction of polycyclic aromatic hydrocarbons and heavy metals on soil enzyme. Chemosphere, 61(8): 1175–1182. Available from: https: doi.org/ 10.1016/j.chemosphere.2005.02.074.

Simon-Deckers, A., Loo, S., Mayne-L'Hermite, M., et al., 2009. Size-, composition- and shape-dependent toxicological impact of metal oxide nanoparticles and carbon nanotubes toward bacteria. Environmental Science & Technology, 43(21): 8423–8429. Available from: https: doi.org/ 10.1021/es9016975.

Singh, A., Singh, N.B., Afzal, S., et al., 2018. Zinc oxide nanoparticles: A review of their biological synthesis, antimicrobial activity, uptake, translocation and biotransformation in plants. Journal of Materials Science, 53(1): 185–201. Available from: https: doi.org/ 10.1007/s10853-017-1544-1.

Sinsabaugh, R.L., Lauber, C.L., Weintraub, M.N., et al., 2008. Stoichiometry of soil enzyme activity at global scale. Ecology Letters, 11(11): 1252–1264. Available from: https: doi.org/ 10.1111/j.1461-0248. 2008.01245.x.

Song, J.W., Shen, Q.L., Wang, L., et al., 2018. Effects of Cd, Cu, Zn and their combined action on microbial biomass and bacterial community structure. Environmental Pollution, 243: 510–518. Available from: https: doi.org/ 10.1016/j.envpol.2018.09.011.

Spisni, E., Seo, S., Joo, S.H., et al., 2016. Release and toxicity comparison between industrial- and sunscreen-derived nano-ZnO particles. International Journal of Environmental Science and Technology: IJEST, 13: 2485–2494. Available from: https: doi.org/ 10.1007/s13762-016-1077-1.

Spohn, M., Kuzyakov, Y., 2013. Distribution of microbial- and root-derived phosphatase activities in the rhizosphere depending on P availability and C allocation–Coupling soil zymography with 14C imaging. Soil Biology and Biochemistry, 67: 106–113. Available from: https: doi.org/ 10.1016/j.soilbio.2013.08.015.

Stark, S., Männistö, M.K., Eskelinen, A., 2014. Nutrient availability and pH jointly constrain microbial extracellular enzyme activities in nutrient-poor tundra soils. Plant and Soil, 383(1): 373-385. Available from: https: doi.org/ 10.1007/s11104-014-2181-y.

Stefanowicz, A.M., Kapusta, P., Zubek, S., et al., 2020. Soil organic matter prevails over heavy metal pollution and vegetation as a factor shaping soil microbial communities at historical Zn-Pb mining sites. Chemosphere, 240: 124922. Available from: https: doi.org/ 10.1016/j.chemosphere.2019.124922.

Stone, V., Nowack, B., Baun, A., et al., 2010. Nanomaterials for environmental studies: Classification, reference material issues, and strategies for physico-chemical characterisation. Science of the Total Environment, 408(7): 1745–1754. Available from: https: doi.org/ 10.1016/j.scitotenv.2009.10.035.

Suman, T.Y., Radhika Rajasree, S.R., Kirubagaran, R., 2015. Evaluation of zinc oxide nanoparticles toxicity on marine algae *Chlorella vulgaris* through flow cytometric, cytotoxicity and oxidative stress analysis. Ecotoxicology and Environmental Safety, 113: 23–30. Available from: https: doi.org/ 10.1016/j.ecoenv.2014.11.015.

Sun, L., Qiu, F.B., Zhang, X.X., et al., 2008. Endophytic bacterial diversity in rice (*Oryza sativa* L.) roots estimated by 16S rDNA sequence analysis. Microbial Ecology, 55(3): 415–424. Available from: https: doi.org/ 10.1007/s00248-007-9287-1.

Suresh, A.K., Pelletier, D.A., Wang, W., et al., 2010. Silver nanocrystallites: Biofabrication using *Shewanella oneidensis*, and an evaluation of their comparative toxicity on gram-negative and gram-positive bacteria. Environmental Science & Technology, 44(13): 5210–5215.

Tabelin, C.B., Igarashi, T., Villacorte-Tabelin, M., et al., 2018. Arsenic, selenium, boron, lead, cadmium, copper, and zinc in naturally contaminated rocks: A review of their sources, modes of enrichment, mechanisms of release, and mitigation strategies. Science of the Total Environment, 645: 1522–1553. Available from: https: doi.org/ 10.1016/j.scitotenv.2018.07.103.

Tourinho, P.S., van Gestel, C.A.M., Lofts, S., et al., 2012. Metal-based nanoparticles in soil: Fate, behavior, and effects on soil invertebrates. Environmental Toxicology and Chemistry, 31(8): 1679–1692. Available from: https: doi.org/ 10.1002/etc.1880.

Trujillo-Reyes, J., Majumdar, S., Botez, C.E., et al., 2014. Exposure studies of core–shell Fe/Fe_3O_4 and Cu/CuO NPs to lettuce (*Lactuca sativa*) plants: Are they a potential physiological and nutritional hazard? Journal of Hazardous Materials, 267: 255–263. Available from: https: doi.org/ 10.1016/j.jhazmat.2013.11.067.

Vance, E.D., Brookes, P.C., Jenkinson, D.S., 1987. An extraction method for measuring soil microbial biomass C. Soil Biology and Biochemistry, 19(6): 703–707.

Venkatachalam, P., Jayaraj, M., Manikandan, R., et al., 2017. Zinc oxide nanoparticles (ZnONPs) alleviate heavy metal-induced toxicity in *Leucaena leucocephala* seedlings: A physiochemical analysis. Plant Physiology and Biochemistry, 110: 59–69. Available from: https: doi.org/ 10.1016/j.plaphy.2016.08.022.

Waalewijn-Kool, P.L., Diez Ortiz, M., van Straalen, N.M., et al., 2013. Sorption, dissolution and pH determine the long-term equilibration and toxicity of coated and uncoated ZnO nanoparticles in soil. Environmental Pollution, 178: 59–64. Available from: https: doi.org/ 10.1016/j.envpol.2013.03.003.

Waalewijn-Kool, P.L., Rupp, S., Lofts, S., et al., 2014. Effect of soil organic matter content and pH on the toxicity of ZnO nanoparticles to *Folsomia candida*. Ecotoxicology and Environmental Safety, 108: 9–15. Available from: https: doi.org/ 10.1016/j.ecoenv.2014.06.031.

Wang, Z.Y., Xie, X.Y., Zhao, J., et al., 2012. Xylem- and phloem-based transport of CuO nanoparticles in maize (*Zea mays* L.). Environmental Science & Technology, 46(8): 4434–4441. Available from: https: doi.org/ 10.1021/es204212z.

Waring, B.G., Weintraub, S.R., Sinsabaugh, R.L., 2014. Ecoenzymatic stoichiometry of microbial nutrient acquisition in tropical soils. Biogeochemistry, 117(1): 101–113. Available from: https: doi.org/ 10.1007/s10533-013-9849-x.

Wei, L.L., Ding, J., Xue, M., et al., 2019. Adsorption mechanism of ZnO and CuO nanoparticles on two typical sludge EPS: Effect of nanoparticle diameter and fractional EPS polarity on binding. Chemosphere, 214: 210–219. Available from: https: doi.org/ 10.1016/j.chemosphere.2018.09.093.

Wu, J.J., Kim, H., 2017. Impacts of road salts on leaching behavior of lead contaminated soil. Journal of Hazardous Materials, 324: 291–297. Available from: https: doi.org/ 10.1016/j.jhazmat.2016.10.059.

Xiao, C., Fauci, M., Bezdicek, D.F., et al., 2006. Soil microbial responses to potassium-based black liquor from straw pulping. Soil Science Society of America Journal, 70(1): 72–77. Available from: https: doi.org/ 10.2136/sssaj2004.0339.

Xu, C., Chen, X.C., Duan, D.C., et al., 2015a. Effect of heavy-metal-resistant bacteria on enhanced metal uptake and translocation of the Cu-tolerant plant, *Elsholtzia splendens*. Environmental Science and Pollution Research International, 22(7): 5070–5081. vailable from: https: doi.org/ 10.1007/s11356-014-3931-3.

Xu, C., Peng, C., Sun, L.J., et al., 2015b. Distinctive effects of TiO$_2$ and CuO nanoparticles on soil microbes and their community structures in flooded paddy soil. Soil Biology and Biochemistry, 86: 24–33. Available from: https: doi.org/ 10.1016/j.soilbio.2015.03.011.

Xu, Z.J, Liu, G.S., 2005. Characteristics and Law of Lead Release from Red Soils Under the Influence of Simulated Acid Rain[J]. Journal of Agro-environment science,24(6): 1109–1113.

Yao, X.H., Min, H., Lü, Z.H., et al., 2006. Influence of acetamiprid on soil enzymatic activities and respiration. European Journal of Soil Biology, 42(2): 120–126.

Yin, X.Q., Meng, X.M., Zhang, Y., et al., 2018. Removal of V (V) and Pb (II) by nanosized TiO$_2$ and ZnO from aqueous solution. Ecotoxicology and Environmental Safety, 164: 510-519. Available from: https: doi.org/ 10.1016/j.ecoenv.2018.08.066.

Yruela, I., 2005. Copper in plants. Brazilian Journal of Plant Physiology, 17(1): 145–156 Available from: https: doi.org/ 10.1590/S1677-04202005000100012.

Zeng, L.S., Liao, M., Chen, C.L., et al., 2007. Effects of lead contamination on soil enzymatic activities, microbial biomass, and rice physiological indices in soil–lead–rice (*Oryza sativa* L.) system. Ecotoxicology and Environmental Safety, 67(1): 67–74.

Zhang, F.P., Li, C.F., Tong, L.G., et al., 2010. Response of microbial characteristics to heavy metal pollution of mining soils in central Tibet, China. Applied Soil Ecology, 45(3): 144-151. Available from: https: doi.org/ 10.1016/j.apsoil.2010.03.006.

Zhang, W., Long, J.H., Li, J., et al., 2019. Impact of ZnO nanoparticles on Cd toxicity and bioaccumulation in rice (*Oryza sativa* L.). Environmental Science and Pollution Research International, 26(22): 23119–23128. Available from: https: doi.org/ 10.1007/s11356-019-05551-x.

Zhang, W., Long, J.H., Li, J., et al., 2020. Effect of metal oxide nanoparticles on the chemical speciation of heavy metals and micronutrient bioavailability in paddy soil. International Journal of Environmental Research and Public Health, 17(7): 2482. Available from: https: doi.org/ 10.3390/ijerph17072482.

Zhang, X.C., Zheng, F.L., Wang, B., et al., 2013. Characteristics of Spatial Distribution of Soil Enzyme Activities in Sloping Farmland of Typical Area in Black Soil Region. Bulletin of Soil and Water Conservation, 33(2): 58–61. Available from: https: doi.org/ 10.13961/j.cnki.stbctb.2013.02.034 .

Zhao, L.J., Peralta-Videa, J.R., Ren, M.H., et al., 2012. Transport of Zn in a sandy loam soil treated with ZnO NPs and uptake by corn plants: Electron microprobe and confocal microscopy studies. Chemical Engineering Journal, 184: 1–8 Available from: https: doi.org/ 10.1016/j.cej.2012.01.041.

Zhao, X.Q., Huang, J., Lu, J., et al., 2019. Study on the influence of soil microbial community on the long-term heavy metal pollution of different land use types and depth layers in mine. Ecotoxicology and Environmental Safety, 170: 218–226. Available from: https: doi.org/ 10.1016/j.ecoenv.2018.11.136.

Zhu, H., Wu, C.F., Wang. J., et al., 2018. The effect of simulated acid rain on the stabilization of cadmium in contaminated agricultural soils treated with stabilizing agents. Environmental Science and Pollution Research International, 25(18): 17499–17508. Available from: https: doi.org/ 10.1007/s11356-018-1929-y.

Zuverza-Mena. N., Martínez-Fernández, D., Du, W.C., et al., 2017. Exposure of engineered nanomaterials to plants: Insights into the physiological and biochemical responses-a review. Plant Physiology and Biochemistry, 110: 236–264. Available from: https: doi.org/ 10.1016/j.plaphy.2016.05.037.

CHAPTER 5

Effect of biomass–based materials on enzyme activities in heavy metal-contaminated environment

5.1 Spatial-temporal distribution of enzyme activities in heavy metal-contaminated soil after application of organic fertilizers

5.1.1 Introduction

Soil contamination with chromium (Cr) has gained substantial attention worldwide because of the high risk of inducing hazardous consequences (Shahid et al., 2017). For instance, Cr contamination alters soil microbial community structure and activities and depresses plant growth and development (Mallick et al., 2010; Carpio et al., 2017; Ertani et al., 2017). The influence of Cr on plant growth is much stronger in the rhizosphere (e.g., a small volume of up to a few millimeters surrounding living roots) as compared to bulk soil (Hinsinger et al., 2009; Antoniadis et al., 2017a). The rhizosphere contains various rhizodeposits released by roots, which stimulate microbial activities and enzyme production and form one of the most dynamic and intense hotspots of enzyme activities in terrestrial ecosystems (Bais et al., 2006; Oburger et al., 2014). Through the rhizosphere, excessive Cr may inhibit plant growth by decreasing chlorophyll content, depressing plant growth and inhibiting germination (Rizvi and Khan, 2018).

Enzymes are the main biological drivers of soil carbon and nutrient cycling processes and an early indicator of soil microbial and root activities to reflect the impact of heavy metal contamination (Sinsabaugh et al., 2008). Sharp gradients are formed for rhizosphere enzyme activity and typically demonstrate a sigmoidal curve (Kuzyakov and Razavi, 2019). The distance from the root center to bulk soil where enzyme activities become stable has been generally defined as the rhizosphere extent of enzyme activity. Many previous studies have focused on visualizing the rhizosphere extent of enzyme activity using the in situ zymography technique (Razavi et al., 2016; Liu et al., 2017a, 2017b; Pu et al., 2019). For instance, the rice rhizosphere extent of enzyme activities in response to phosphorus and cellulose applications has been investigated (Wei et al., 2018). The total enzymatic hotspot area for the whole soil profile as affected by multiple heavy metals was also calculated (Duan et al., 2018).

Regarding the impact of heavy metals, the activities and dynamic behaviors of enzymes are mainly governed by the type, speciation and bioavailability of contaminants (Abdu et al., 2016, Shahid et al., 2017). For instance, Cr(VI) is more toxic to plants and microbes compared to Cr(III) because of its high reactivity with other elements. The potentially mobile fractions, for example, the exchangeable, reducible and oxidizable fractions, are considered to be bioavailable (Rinklebe and Shaheen, 2017). Many in situ approaches have been used to remediate heavy metal contamination, with the goal of weakening the migration of heavy metals and reducing their bioavailability (Ghosh et al., 2011). Increasing soil organic matter content has yielded promising outcomes with respect to amending Cr-contaminated soil and reducing stresses to microbial and root activities (Antoniadis et al., 2017b, 2018; Abbas et al., 2019). For instance, manure is the most commonly used organic fertilizer as it can stabilize heavy metals through various processes such as sorption, reduction, volatilization and rhizosphere modification (Park et al., 2011; Ghosh et al., 2011). In recent years, researchers have also proposed that manure-pyrolyzed biochar can not only enhance remediation of heavy metals but also promote plant growth (Abbas et al., 2019). Both manure and the biochar involved in remediating Cr-contaminated soils and improving microbial and root activities have their own specific mechanisms (Park et al., 2011, Li et al., 2017). First, although both manure and biochar can immobilize heavy metals by adsorption, promote the formation of soil aggregates and drive heavy metal reduction reactions, the capacities of manure and biochar are diverse and heavy metal-specific (Park et al., 2011; Li et al., 2017; Zhu et al., 2017). Second, the loading of indigenous enzymes and microbes from manure into soil will occur following their application (Criquet et al., 2007), which does not occur with manure-derived biochar. Third, manure and biochar provide organic compounds with various stabilities (Yanardağ et al., 2017). Manure generally maintains a high easily decomposed organic matter and nutrient content, while highly recalcitrant aromatic compounds are found in manure-derived biochar (Dinesh et al., 1998; Zornoza et al., 2016). This leads to a trade-off between more complete decomposition and mineralization and the incorporation of these compounds as soil organic matter. Though manure and its biochar have been widely recommended, there is a lack of mechanistic understanding regarding the enzymatic responses of plants and microorganisms to manure and biochar application in Cr-contaminated soil.

Here, we used direct soil zymography to visualize the spatial distribution of enzyme activities on the soil profile. Soil samples were collected from a chromium slag field. Maize (*Zea mays* L.) plants were grown in the rhizoboxes. We compared cow manure with its biochar application using three treatments. Manure: mixed manure throughout the soil; Biochar: mixed biochar throughout the soil; and Control: a control with only soil. The BCR sequential extraction procedure was used to determine the chemical speciation of Cr in soil. Considering the possible diverse impact of Cr for different enzymes, direct soil zymography was used to visualize the spatial and temporal distribution of the activity for the three enzymes: β-glucosidase, phosphomonoesterase and N-acetyl-glucosaminidase. β-glucosidase is respon-sible for catalyzing the hydrolysis of terminal 1,4-linked β-D-glucose residues from β-D-glucosides and is involved in the carbon (C) cycle (German et al., 2012). Phosphomonoesterase, which catalyzes the hydrolysis of organic phosphorus (P) compounds to inorganic P, is involved in the P cycle (Eivazi and Tabatabai, 1977; Malcolm, 1983). N-acetyl-glucosaminidase (chitinase), which catalyzes the decomposition of chitin to yield low molecular weight chitooligomers, is responsible for C- and nitrogen (N) acquisition (Huang, 2012). We hypothesized that: ①the application of manure and its biochar will reduce the bioavailability of Cr to varying degrees; ②the rhizosphere extent of enzyme activities will be wider than the bulk soil hotspots extent of enzyme activities because of the co-release of enzymes by roots and microbes within the rhizosphere (here the bulk soil hotspots were defined as the small volume of bulk soil where microorganisms gather and aggregate) (Ekschmitt et al., 2005); ③the extent will be narrower in manure-treated soil compared to biochar-treated soil due to high nutrient content and the integration of easily decomposed organic compounds.

5.1.2　Impact of organic amendments on soil acidification

Soil samples taken from the abandoned chemical plant were highly contaminated by chromium [$Cr_{(T)} = (2000 \pm 64)$mg·kg^{-1}] and the pH equaled 5.0. After incubation for 45 days, soil pH decreased for the Control, Manure and Biochar groups ($P < 0.05$, Figure 5.1A). Before the shoots were cut, soil pH in the Biochar group was higher than that in the Manure group ($P<0.05$). At day 45, soil pH in the Biochar group showed a sudden decrease and became lower compared to the pH in the Manure group. Soil pH in the

Control group (CK) also experienced a strong decrease within the first 15 days and was similar to the Biochar group, but strongly decreased after shoot cutting. Rhizosphere soil pH in the Manure group was lower than bulk soil ($P<0.05$), while the pH between the rhizosphere and bulk soil in the Control and Manure groups remained unchanged ($P>0.05$, Figure 5.2A).

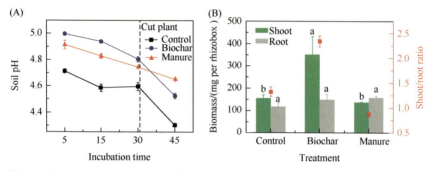

Figure 5.1 (A) response of soil pH to manure and its biochar application during the whole incubation period; (B) plant biomass and shoot/root ratio. The lower-case letters represent significant differences among treatments ($P<0.05$). Error bars represent standard error (SE). Control: a control soil without any addition. Biochar: soil homogeneously mixed with manure-derived biochar. Manure: soil homogenously mixed with cow manure. Credit: Reprinted with permission from Liu et al., 2020.

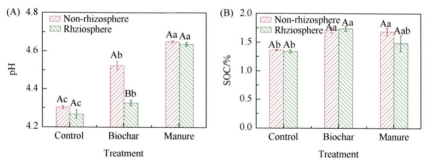

Figure 5.2 (A) pH of the rhizosphere soil and non-rhizosphere soil of each treatment group after culture; (B) Organic carbon content in rhizosphere soil and non-rhizosphere soil of each treatment group after culture. Error bars represent standard deviations ($\pm$SE). The lower-case letters above the bars indicate significant differences between treatments, separately for each soil ($P<0.05$). Capital letters indicate significant differences between rhizosphere soil and non-rhizosphere soil ($P<0.05$). Control: a control soil without any addition. Biochar: soil with homogeneously mixed manure-derived biochar. Manure: soil homogenously mixed with cow manure. Rhizosphere: soil from plant rhizosphere. Non-rhizosphere: soil from non-rhizosphere. Credit: Reprinted with permission from Liu et al., 2020.

Soil pH in the Control group strongly decreased by 0.4 units during the first 15 days of incubation and then became stable until day 30. In this process, the H^+ concentration increased by 2.5-fold, indicating that the soil became more acidic. Findings from previous studies have suggested that soil acidification could result from decomposition of organic matter, release of carbon dioxide or plant root exudation (Lu et al., 2005; Kabatapendias, 2010). Manure and its biochar application significantly increased soil pH (Figure 5.1A). The liming effect of biochar generally depends on its feedstock and the pyrolysis temperature, and more generally relates to the base concentration of the ash. In this study, biochar application induced a greater pH increase as compared to its feedstock before plant cutting, indicating that the liming effect of biochar was enhanced compared to its feedstock (Yuan et al., 2011). This enhancement was mainly attributed to the larger pH, ash content and BET specific surface area (Gomez-Eyles et al., 2013) in the manure-derived biochar (Table 5.1). A continuous reduction in soil pH was found in the three treatments after plant cutting at day 30. This could be because of the decomposition of plant root residues, which will release some organic functional groups, dissociate H^+ and accelerate soil acidification (Rukshana et al., 2009; Weil and Brady, 2016; Liu et al., 2018).

Table 5.1　Physicochemical properties of manure and biochar.

Properties	Manure	Biochar
Moisture/%	0.082 ± 0.005	0.043 ± 0.001
pH	8.60 ± 0.015	9.5 ± 0.033
Ash content/%	11.0 ± 0.066	27.0 ± 0.46
CEC/(mol·kg^{-1})	13.00 ± 0.67	44.0 ± 0.28
Cr/(mg·kg^{-1})	16.0 ± 1.6	57.0 ± 2.7
Total carbon/%	39.0 ± 0.30	50.0 ± 0.42
TN/(g·kg^{-1})	15.0 ± 0.34	26 ± 0.49
TP/(g·kg^{-1})	2.6 ± 0.12	5.6 ± 0.054
BET/(m^2·g^{-1})	0.95	7.0
Biochar yield	—	$30\% \pm 1.3\%$

Biochar application induced lower rhizosphere soil pH in contrast to the bulk soil ($P<0.05$, Figure 5.2A). Maize performance index (PI), as calculated in this study, was higher in the Biochar group as compared to the Manure group, indicating that maize growth is improved with biochar application (Kanchikerimath and Singh, 2001). This will induce greater release of root exudates in the Biochar group, which partly consist of organic acids, amino acids, fatty acids (Antoniadis et al., 2017a). The lower rhizosphere soil pH may be attributed to the residues of root exudates as compared to the Manure and CKs. Despite the nonsignificant difference between the pH values of rhizosphere and non-rhizosphere soil in the Manure and CKs, a slightly decreased pH in rhizosphere soil was demonstrated (Figure 5.2A). This result agrees well with a previous review study (Kuzyakov and Razavi, 2019), which stated that the gradient of soil pH between the root surface and nonrhizosphere soil decreases with decreasing soil pH. Even though the application of manure and its biochar attempted to neutralize soil acidity, the soil still remained acidic throughout the whole incubation period for the treatment duration. This will influence the bioavailability of chromium, microbial and enzyme activities, and plant growth.

5.1.3 Impact of organic amendments on plant characteristics

Shoot biomass was higher in the Biochar group compared to the Control and Manure groups ($P<0.05$, Figure 5.1B). The maize PI results calculated from shoot biomass showed that the Biochar group had better maize growth than the Manure group (PI_{Bio} = 2.27, PI_{Man} = 0.88). The ratio of shoot to root biomass was the highest in the Biochar group (~2.32), followed by the Control (~1.32) and Manure groups (~0.87).

A PI equal to or lower than 1.0 indicates that the amendment has no impact on improving plant growth. Therefore, the dramatically increased PI induced by manure-derived biochar, instead of manure, demonstrated that biochar application prompts plant growth in Cr-contaminated soil. A CK without Cr contamination is necessary to confirm Cr toxicity on plant growth and the mitigation of biochar application for Cr toxicity, but it is not possible to find soil uncontaminated by Cr from or close to the sampling field. Nevertheless, previous studies have stated that Cr application in soil strongly depressed maize growth (e.g., decreased plant biomass, length, grain protein) (Maiti, 2012; Rizvi and Khan, 2018). According to Fourier infrared (FT-IR) analysis results, the new peaks appearing in biochar also indicated the enhanced capacity of biochar to adsorb Cr and remediate hexavalent Cr toxicity (Figure 5.3A). Some environmental factors have also been proposed to modify plant biomass allocation patterns, such as nutrient levels, warming, precipitation gradients, etc. (Sun and Wang, 2016). Manure allocated many nutrients and easily available organic matter to soil, but a strong competition for nutrients between plants and microbes may limit nutrient uptake and depress plant growth (Liu et al., 2017a, 2017b). The lowest shoot to root ratio (i.e., shoot/root ratio = 0.87 < 1.0) in the Manure group demonstrated that manure application caused a wasteful competition between plant and microbes (Biedrzycki et al., 2010). In contrast, biochar application induced larger average enzyme activities on soil profile than control soil, indicating that organic matter decomposition in the Biochar group was stronger than the CK. This will result in more release of available nutrients and support plant growth (Flanagan and Van Cleve, 1983; Muscolo et al., 2007).

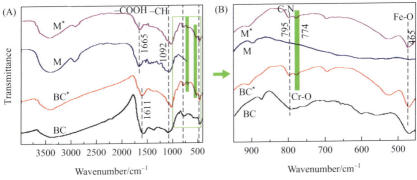

Figure 5.3 (A) The FTIR spectrum of manure and biochar before and after their application into Cr-contaminated soil; (B) magnified fragments spectra in the range of 1000–400cm^{-1} recorded before and after application. M: manure before the incubation. M*: manure after the incubation. BC: biochar before the incubation. BC*: biochar after the incubation. The green bars show the presence of Cr-O in manure and biochar after the incubation. The dashed vertical lines mark the positions of some functional groups such as -COOH, -CH, C-N and Fe-O. Credit: Reprinted with permission from Liu et al., 2020.

Our results also show that soil NO_3^--N, being more preferred by maize (Daryanto et al., 2019), and available P concentrations are higher in the Biochar group as compared to the CK (Table 5.2). With an abundant nutrient supply from soil, photosynthates will accumulate more in shoots instead of roots (Gan et al., 2002; Mašková and Herben, 2018). Our results also supported the "optimal partitioning" theory in biomass allocation of plants, that is, that plants respond to variable environmental conditions by allocating biomass among various organs to capture nutrients and light to maximize their growth rate (McConnaughay and Coleman, 1999; Shipley and Meziane, 2002).

Table 5.2 NO_3^- -N and available P in soils of three treatments after the incubation.

Treatments	NO_3^- -N	Available P
Control	0.97 ± 0.16 b	30 ± 7.58 c
Biochar	3.62 ± 1.46 a	58 ± 7.06 a
Manure	0.55 ± 0.03 c	54 ± 5.91 b

Note: a, b, c represent significant difference.

In contrast to the strongly increased shoot biomass, root biomass in the Biochar group had no significant difference between the CKs and Manure groups. This may be attributed to a specific characteristic of Cr, that is, that chromium retention inside the roots is a well-documented plant defense mechanism (Shahid et al., 2017). Soil pH in all treatments was less than 5, which is not favorable for maize root growth. Previous studies demonstrated that Cr application at a concentration of $50-204 mg \cdot kg^{-1}$ seriously depressed root growth and development, with a decrease in root biomass by 45%—61% (Mallick et al., 2010; Karthik et al., 2016; Rizvi and Khan, 2018). In this study, plants showed strong withering at day 30: leaves turned yellow and roots began apoptosis, which may be exacerbated by Cr in the soil (Figure 5.4).

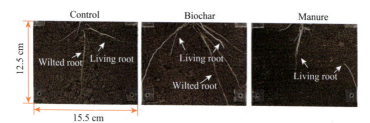

Figure 5.4 Plant root images in each treatment group at day 30. Control: a control soil without any addition. Biochar: soil with homogeneously mixed manure-derived biochar. Manure: soil homogenously mixed with cow manure. Credit: Reprinted with permission from Liu et al., 2020.

5.1.4 Cr concentration and speciation in soil

Manure and its biochar application induced significant chemical speciation dynamics of Cr (Figure 5.5 and Figure 5.6). Exceptions were the oxidizable and residual fractions, which were not influenced by manure and its biochar application during the whole incubation period. However, both fractions were the main speciation of Cr, which in total accounted for approximately 95% of Cr (T).

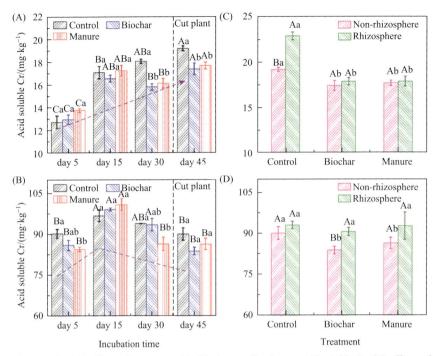

Figure 5.5 (A) effects of manure and its biochar application on acid soluble Cr; (B) effects of manure and its biochar application on reducible Cr; (C) acid soluble in rhizosphere soil and non-rhizosphere soil at the end of incubation; (D) reducible Cr in rhizosphere soil and non-rhizosphere soil at the end of incubation. Lower-case letters above the bars indicate significant differences among Control, Manure and Biochar ($P<0.05$). Capital letters in Figure 5.5C、D indicate significant differences between rhizosphere soil and non-rhizosphere soil ($P<0.05$). Error bars represent standard error ($\pm$SE). Control: a control soil without any addition. Biochar: soil homogeneously mixed with manure-derived biochar. Manure: soil homogenously mixed with cow manure. Rhizosphere: rhizosphere soil. Non-rhizosphere: non-rhizosphere soil. Credit: Reprinted with permission from Liu et al., 2020.

Acid soluble Cr increased over time, but the increment with time was smaller in the Manure and Biochar groups compared to the CK (Figure 5.5A). This resulted in a decline in acid soluble Cr by 10.66% and 12.38% in the Manure and Biochar groups, respectively, at day 30 as compared to the CK ($P<0.05$). A similar decline was also found at day 45 ($P<0.05$). A negative relationship between soil pH and acid soluble Cr was also confirmed, indicating the strong impact of soil pH on the dynamics of Cr chemical speciation (Table 5.3). Reducible Cr increased at day 15 and then decreased again ($P<0.05$).

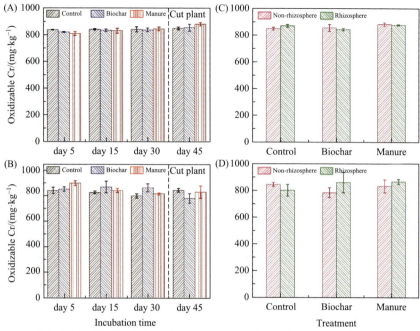

Figure 5.6 (A) Oxidizable Cr in each treatment group during the culture period. (B) residual Cr in each treatment group during the culture period. (C) oxidizable Cr in rhizosphere soil of each treatment group after culture. (D) residual Cr in rhizosphere soil of each treatment group after culture. There was no significant difference of the chromium in the oxidizable and residual state in the whole process($P<0.05$). Control: a control soil without any addition. Biochar: soil with homogeneously mixed manure-derived biochar. Manure: soil homogenously mixed with cow manure. Rhizosphere: soil from plant rhizosphere. Non-rhizosphere: soil from non-rhizosphere. Credit: Reprinted with permission from Liu et al., 2020.

Table 5.3 Correlation coefficients between enzyme activities of soil profile and chemical speciation of Cr based on Pearson correlation.

	pH	F1	F2	F3	F4	NAG	BG	PHOS
pH	1	−0.35*	0.24	−0.20	0.14	0.45**	0.38*	0.19
F1		1	0.25	0.48**	−0.23	0.12	−0.37*	−0.61**
F2			1	−0.23	0.09	0.51**	0.14	−0.13
F3				1	−0.40**	−0.27	−0.19	−0.29
F4					1	0.29	0.18	0.32
NAG						1	0.26	0.15
BG							1	0.64**
PHOS								1

Note: F1−F4 represents the acid soluble, reducible, oxidizable and residue fractions of heavy metals, respectively. NAG, BG, PHOS represents the N-acetyl-glucosaminidase, β-glucosidase and phosphomonoesterase enzyme activity, respectively. Significant correlations were shown in Bold. * Significant at $P<0.05$; ** Significant at $P<0.01$.

At day 45, acid soluble Cr in the rhizosphere soil of the CK was 19.25% higher compared to bulk soil (Figure 5.5C). Rhizosphere soil in the Manure and Biochar groups had higher reducible Cr versus bulk soil ($P<0.05$).

Acid soluble Cr content was negatively correlated with soil pH ($P<0.05$, Table 5.3). Although acid soluble Cr increased over time, the application of manure and biochar slowed down the increased rate by 47% and 55%, respectively, which supported our first hypothesis. Previous studies have indicated that soil pH played an important role in controlling the geochemical behavior of heavy metals in soil solid and liquid phases (Pietrzykowski et al., 2014). The geochemical behavior of Cr was greatly affected by soil pH and redox potential (Ashraf et al., 2017). According to the Eh-pH diagram of the chromium system ($C_{T, Cr} = 1\times10^{-3} mol\cdot L^{-1}$), when the soil is acidic and the redox potential is low, soil Cr mainly exists in the state of Cr(III) (Yuan et al., 2011). This was consistent with our results showing that the oxidizable Cr was one of the main fractions in the acidified soil and was positively correlated with acid soluble Cr ($P<0.01$). However, Cr(III) has low solubility only at pH<5.5 (Choppala et al., 2018), indicating that Cr(III) was almost precipitated, very stable and thus less toxic in severely acidified soil.

Manure and its biochar application not only mitigated soil acidification but also altered the chemical speciation of Cr through adsorption and complexation (Agegnehu et al., 2017). Animal manure contains substantial biodegradable organic matter which has a large number of oxygen-containing functional groups (Yang et al., 2018a). These functional groups can immobilize Cr ions and change their chemical fraction (Dai et al., 2017; Zhu et al., 2017). For instance, organic amendments caused Cr detoxification by reducing Cr(VI) to Cr(III) and subsequent precipitation as chromic hydroxide (Park et al., 2011). After pyrolysis, the amount and species of oxygen-containing functional groups in biochar increased, which enhanced the capacity to adsorb and immobilize inorganic and organic pollutants from contaminated soil (Zhou et al., 2019). In this study, compared with manure, the new peaks at $795 cm^{-1}$ and $465 cm^{-1}$ appearing in biochar corresponded to C-N and Fe-O (Figure 5.3A)(Yang et al., 2018b). After incubation, manure and its biochar were separated from soil and their FT-IR spectra were recorded. The new peaks at $774 cm^{-1}$ (Figure 5.3B) were intrinsic vibrations of the Cr-O bonds, respectively (Bhaumik et al., 2011). This was indicative of the adsorption of Cr onto manure and its biochar. Furthermore, plant roots secrete organic acids during growth and degradation processes, which can also affect the chemical speciation of Cr (Hinsinger et al., 2003).

5.1.5 Impact of organic amendments on enzyme activities and their extent

By applying manure and biochar at a dose of 5g $C \cdot kg^{-1}$ soil, the rhizosphere extent of enzyme activities varied among the three treatments and decreased gradually from day 5 to day 30 (β-glucosidase: reduction of ~1.1mm in the CK, 0.3mm in the Biochar group and 0.4mm in the Manure group; phosphomonoesterase: reduction of ~0.4mm in the CK, 0.5mm in the Biochar group and 0.2mm in the Manure group). Negative relationships between the extent of β-glucosidase and phosphomonoesterase activities and acid soluble Cr were found ($P<0.05$, Table 5.4). Manure and its biochar application reduced the rhizosphere extent compared to the CK ($P<0.05$; Figure 5.7—Figure 5.10). The rhizosphere extent of β-glucosidase and phosphomonoesterase activities was the narrowest in the Manure group, followed by the Biochar and CKs. Furthermore, the rhizosphere extent of enzyme activities was also wider than the bulk soil hotspots extent (Figure 5.8). The rhizosphere extent of β-glucosidase activities ranged from 1.2mm to 2.5mm, while the range of bulk soil hotspots extent ranged from 1.0mm to 1.8mm. Phosphomonoesterase activities in the rhizosphere extended from 1.3mm to 1.9mm, but its activities in bulk soil hotspots ranged from 1.1mm to 1.8mm.

Table 5.4 Correlation coefficients between rhizosphere or bulk soil hotspot extent of enzyme activities and soil pH or acid soluble Cr based on Pearson correlation.

	NAG		BG		PHOS	
	Rhizosphere	Bulk soil	Rhizosphere	Bulk soil	Rhizosphere	Bulk soil
pH	−0.09	0.064	−0.29	−0.11	−0.75**	0.10
F1	−0.09	−0.15	−0.62**	−0.60**	−0.47*	−0.56**

A positive relationship between average N-acetyl-glucosaminidase and β-glucosidase activities on soil profile and soil pH was found ($P<0.05$, Table 5.3), while phosphomonoesterase had no significant relation with soil pH. β-glucosidase and phosphomonoesterase were both negatively correlated with acid soluble Cr ($P<0.05$). Further, N-acetyl-glucosaminidase was positively correlated with the reducible fraction of Cr.

The rhizosphere and bulk soil hotspots extent of β-glucosidase and phosphomonoesterase showed a strong decrease with time for three treatments (Figure 5.8); the rate of such a decrement is reduced in the Biochar and Manure groups. The rhizosphere and bulk soil hotspots extent of enzyme activities generally indicates the capacity of roots and microorganisms to

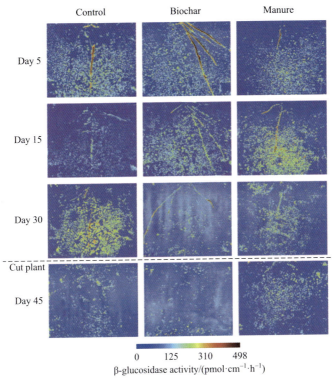

Figure 5.7 examples of zymograms for β-glucosidase activities. The actual images of soil profile at day 30 were shown in Figure 5.7. Three rows represent response of activities to three treatments: Control, Biochar and Manure. Figures from top to bottom are the measurements at days 5, 15, 30 and 45. The color bar corresponds to β-glucosidase activity (pmol·cm^{-2}·h^{-1}). Control: a control soil without any addition. Biochar: soil homogeneously mixed with manure-derived biochar. Manure: soil homogenously mixed with cow manure. Credit: Reprinted with permission from Liu et al., 2020.

immobilize nutrients and utilize carbon (Razavi et al., 2016). Therefore, the decreased extent demonstrated weakening root growth and microbial activities. Soil pH could be responsible for this weakening effect as our results illustrated that its extent decreased in parallel with decreasing soil pH and increasing acid soluble Cr (Figure 5.1, Table 5.4).

Soil enzyme activities were also strongly affected by pH, which is in accordance with a previous study (Turner, 2010). Both β-glucosidase and N-acetyl-glucosaminidase activities were inhibited with decreasing soil pH (Table 5.3). The exceptional nonsignificant relationship between soil pH and phosphomonoesterase activity may be induced by the coeffect of acid

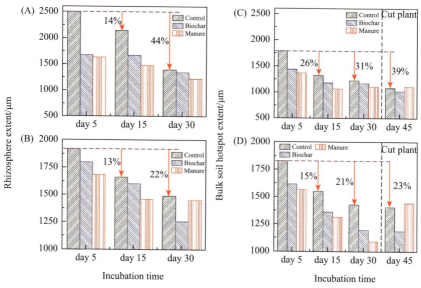

Figure 5.8 (A) rhizosphere extent of β-glucosidase activity. (B) rhizosphere extent of phosphomonoesterase activity. (C) bulk soil hotspot extent of β-glucosidase activity. (D) bulk soil hotspots extent of phosphomonoesterase activity. All the data were derived from logistic regression curve of enzyme activity as a function of distance from root center or bulk soil hotspot. The values beside the red line show the decreasing percentage of the extent of enzyme activities with time in Control. Control: a control soil without any addition. Biochar: soil homogeneously mixed with manure-derived biochar. Manure: soil homogenously mixed with cow manure. Credit: Reprinted with permission from Liu et al., 2020.

phosphatase and alkaline phosphatase because both enzymes coexist in soil and have varied optimum pH activities (Acostamartínez and Tabatabai, 2011). Further, negative relationships between enzyme activities and acidic soluble Cr also indicated that the enhanced bioavailability of Cr under acidic conditions may constrain enzyme activities. One exception is that we found a positive relationship between N-acetyl-glucosaminidase and the reducible fraction of Cr. This Cr fraction is generally bound to Fe/Mn oxides. Fe oxides may act as an electron acceptor and enhance nitrification process with a limited oxygen supply (Liptzin and Silver, 2009; Huang et al., 2016). This will cause decreased inorganic nitrogen (NH_4^+), which subsequently accelerates the production of N-cycling related enzymes (i.e., N-acetyl-glucosaminidase). The enhanced nitrification process could also contribute to the decreased soil pH because H^+ will be released to acidify soil (He et al., 2012).

Control   Biochar   Manure

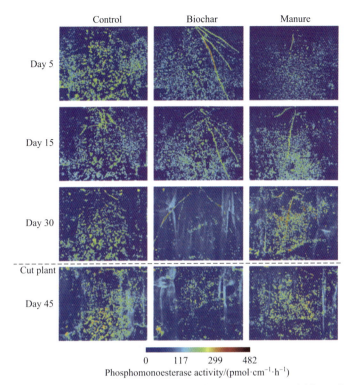

$$0 \quad 117 \quad 299 \quad 482$$
Phosphomonoesterase activity/(pmol·cm^{-1}·h^{-1})

Figure 5.9 Examples of zymograms for phosphomonoesterase activities. Three rows represent response of activities to three amendment strategies: Control, Biochar and Manure. Figures from top to bottom are the measurements at days 5, 15, 30 and 45. The color bar corresponds to phosphomonoesterase activity(pmol·cm^{-2}·h^{-1}). Control: a control soil without any addition. Biochar: soil with homogeneously mixed manure-derived biochar. Manure: soil homogenously mixed with cow manure. Credit: Reprinted with permission from Liu et al., 2020.

The rhizosphere extent of β-glucosidase and phosphomonoesterase activities was wider than the bulk soil hotspot extent of both enzymes. This result supported our second hypothesis. Plants roots not only exude various organic compounds but also release enzymes to decompose organic matter within their rhizospheres. Further, microorganisms in this area also become more active due to the impact of rhizodeposits. Therefore, a wider a rhizosphere extent was engendered. Among treatments, manure application resulted in the narrowest rhizosphere and bulk soil hotspot extent of β-glucosidase and phosphomonoesterase activities, followed by the Biochar and CKs. This was in accordance with our third hypothesis. The narrowest extent of this observation can be explained by the addition

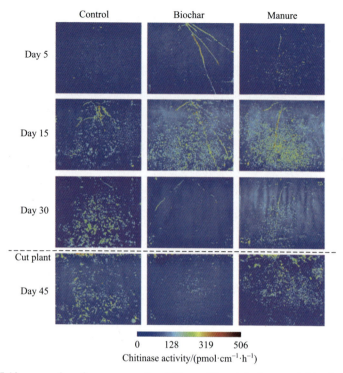

Chitinase activity/(pmol·cm^{-1}·h^{-1})

Figure 5.10 examples of zymograms for chitinase (N-acetyl-glucosaminidase) activities. Three rows represent response of activities to three amendment strategies: Control, Biochar and Manure. Figures from top to bottom are the measurements at days 5, 15, 30 and 45. The color bar corresponds to chitinase activity (pmol·cm^{-2}·h^{-1}). Control: a control soil without any addition. Biochar: soil with homogeneously mixed manure- derived biochar. Manure: soil homogenously mixed with cow manure. Credit: Reprinted with permission from Liu et al., 2020. Copyright 2020 Elsevier.

of labile organic compounds and nutrients following the manure application (Kuzyakov and Razavi, 2019). With abundant resources for growth and maintenance, microbes and plant roots do not necessarily extend enzymes very far.

5.1.6 Summary

For the first time, we identified and compared the impact of manure and its biochar application on the spatiotemporal distribution of soil enzyme activities under the stress of Cr contamination. Soil pH dramatically decreased during 45 days of incubation, which strongly increased the acid soluble fraction of Cr, decreased the activities of β-glucosidase and N-acetyl-glucosaminidase ($P<0.05$), and narrowed the rhizosphere extent of enzyme activities by 13%—44% compared to day 5. This indicated that the increased Cr bioavailability decreased soil microbial activities. Biochar application caused a larger liming impact compared to manure application because of the larger initial pH, ash content and BET-specific surface area in manure-derived biochar. Biochar group also had the highest soil NO_3^--N and available P concentration. In addition, the greatest shoot/root ratio was also found when biochar was applied, which reduced the wasteful competition between plants and microbes in Cr-contaminated soil. The rhizosphere extent of enzyme activities (~1.2—2.5mm) was wider in comparison with the bulk soil hotspot extent (~1.0—1.8mm), indicating higher demand for low-molecular-weight compounds and nutrients in the rhizosphere by roots and microbes. Further, manure application resulted in the narrowest extent of β-glucosidase and phosphomonoesterase activities as compared to the Biochar and CKs. This could be due to the addition of labile organic compounds and nutrients following manure application. Our study emphasizes the important role of pH on Cr bioavailability and enzyme activities and demonstrates that biochar application is more suited for remediating Cr-contaminated soil.

5.2 Passivation and stabilization mechanism of calcium-based magnetic biochar on soil Cr(VI) and its bioavailability

5.2.1 Introduction

For its advantages of small bulk density, large specific surface area, large content of functional groups, high stability, strong adsorption and cheap and easy affordability, biochar has been widely used in pollution remediation, agricultural environmental protection and other fields as a new environmental friendly functional substance. Although biochar has proved to be able to adsorb and degrade inorganic and organic pollutants in water, soil and other environmental media, it still has the problem of poor adsorption capacity. Therefore, a large number of studies on biochar modified materials have been carried out to enhance their adsorption capacity to target pollutants.

Biochar modification chiefly involves physical modification, chemical activation modification, metal salt or oxide modification, organic solvent modification and plasma modification. Among the numerous options, acid-base modification, combined modification of iron salts and iron oxides, combined modification of transition metals, and composite materials modification are the most widely used methods. The iron salts-iron oxides combined modification method has become one of the current research hotspots for its advantages of simple operation, controllable process, excellent modification effect and excellent application in a variety of environmental media. At present, iron-modified biochar can be prepared in various methods through different steps. Among them, the coprecipitation method is to load Fe onto biomass before pyrolysis through a chemical way, while the impregnation method is to impregnate biochar into Fe solution after pyrolysis, which is the most common loading method at present. In the meantime, there are various types of Fe that can be used for modification, such as Fe^0, Nano-Fe^0 and other elemental iron (Li et al., 2020a, 2020b; Mandal et al., 2017); $FeCl_2$, FeS, $FeCl_3$, $Fe(NO_3)_3$ and other iron-containing compounds (Han et al., 2016); hematite, magnetite and limonite in natural environment (Duan et al., 2017); Fe_3O_4, γ-Fe_2O_3, Fe_2O_3 and other iron oxides (Sun et al., 2015); complexes of Fe with other materials like Fe-Mn, Fe-Co, Fe-Mn-Ce, and Fe-Zn (Gao et al., 2020; Waa et al., 2021). Fe modification can largely improve the surface physical and chemical properties of biochar. Loading γ-Fe_2O_3 can add more adsorption sites on the surface of biochar and improve its adsorption capacity. Fe(III)-modified biochar features a specific surface area 3 times higher than that of the original biochar (Han et al., 2016). In addition, Fe(III) can also be complexed with biochar surface functional groups to increase the overall surface charge, thus facilitating the electrostatic adsorption. Meanwhile, some studies also found that Fe^0, Nano-Fe^0, Fe_3O_4, γ-Fe_2O_3, Fe-Mn and other modified biochar have strong adsorption capacity for Cd, Pb, As, Cr and other heavy metals (Zhang et al., 2021). However, the instability of Fe^0 and Fe(II) and the agglomeration of Fe^0 are the problems that must be encountered in the modification process.

Fe-modified biochar has been widely used in wastewater treatment due to its easy preparation, efficient adsorption and recyclability. Due to the complexity and variability of soil media, abatement of soil pollution by heavy metals has been troubled with problems such as high remediation cost and poor remediation effect. In recent years, the application of inexpensive and environmental friendly biochar-based materials in remediation of soil pollution by heavy metals has attracted extensive attention from scholars. To further improve the adsorption capacity of biochar-based materials, some studies have applied Fe-modified biochar to the remediation of soil pollution by heavy metals, and realized the effective fixation of a variety of heavy metals in soil.

Zhang et al. (2021) applied three kinds of exogenous Fe-modified biochar materials (BC), that is, Nano-Fe$_3$O$_4$ modified biochar, EDTA-Fe modified biochar and FeSO$_4$ modified biochar, and explored their effects on the root plaque formation and Cd enrichment of rice plants in Cd-contaminated paddy fields. Compared with EDTA-Fe and FeSO$_4$ modified biochar, Nano-Fe$_3$O$_4$ modified biochar increased the proportion of crystalline iron etching in rice root (31.8%—35.9%), with a Cd concentration of 7.64—13.0mg/kg. After the application of Nano-Fe$_3$O$_4$ modified biochar, the Cd transport factor of rice roots and that of rice stems and leaves decreased by 84.7% and 80.0%, respectively. Zhang et al. used Fe^0-modified Malaysian oil palm fiber biochar to restore Cd-As polluted paddy soil. Fe^0 and biochar could jointly promote the formation and dispersion of ferrous oxide minerals, and effectively improve the transformation of Cd and As in exchangeable/ water-soluble/adsorptive state bound to ferrous oxide minerals in soil into the Fe-Mn oxides, thus realizing the simultaneous fixation of As and Cd in soil. Moreover, the application of Fe^0-modified biochar could significantly lower the concentration of Cd and As in rice, by up to 50.2% and 35.6%, respectively. In the process of soil restoration, Fe^0 and biochar showed obvious synergistic passivation effect. Yao et al. (2021) applied the Fe$_3$O$_4$ modified biochar prepared by coprecipitation method to restore the rice field soil polluted by As. The application of Fe$_3$O$_4$-modified biochar significantly improved the cation exchange capacity of soil, reduced the bioavailability of As in soil, and increased the concentration of As in the crystalline part of hydration oxide in the soil. When the Fe$_3$O$_4$-modified biochar was added to 0.05%—1.6% (wt), the concentration of As in brown rice was reduced by 9.4%—47.3%, and the amount and thickness of iron spots on the root surface of rice were significantly increased. However, there have been few studies on the application of iron-modified biochar to Cr-contaminated soil. The study of Su et al. (2016) showed that for application of Nano-Fe^0 modified biochar in the remediation of Cr(VI) contaminated soil, in a 15-day remediation, a dosage of 8g·kg^{-1} could immobilize 91.94% of Cr(VI). In addition, Cr extracted from weak acid was almost all transformed into Fe-Mn oxides and organic binding states. Meanwhile, the application of Nano-Fe^0 modified biochar could effectively alleviate the Cr(VI) stress on Chinese cabbage seedlings and enhance growth of the plants. However, Nano-Fe^0 modified biochar itself could not maintain the stability of Fe^0 continuously, thus leading to the increase of the addition amount and the dissolution of Fe. However, some studies also showed that when the modified poultry manure biochar and modified sheep manure biochar synthesized with chitosan and Fe^0 were added, they featured a fixation rate of Cr in soil of 45% [adding amount of 5% (wt)] when the soil pH was 5.5. However, chitosan-Fe^0 modified biochar only slightly improved the fixation rate of Cr in soil, to 48% (Mandal et al., 2017).

Under the stress of soil pollution by heavy metals, the effects of biochar and iron-modified biochar on soil enzyme activity are highly variable, and this variability is related to the reaction between biochar and heavy metal ions (Bailey et al., 2011). On one hand, the adsorption of heavy metal ions in soil by biochar and iron-modified biochar is beneficial to the microbial activity and enzyme-catalyzed reaction in soil, and further improve the enzyme activity. On the other hand, biochar and Fe-modified biochar adsorb enzymes to protect the binding site of enzyme-catalyzed reaction, thus inhibiting the enzyme-catalyzed reaction (Cornelia, 2011). Meanwhile, the stress of soil heavy metal can also inhibit the microbial activity and enzyme activity. The study of Masto et al. (2013) found that the application of water hyacinth biochar could significantly increase the activity of (acid) alkaline phosphatase, catalase, dehydrogenase, fluorescein hydrolase and dehydrogenase in soil, and there was a "dose-activity" positive correlation between biochar and enzyme activity. Wen et al. (2020) collected the surface soil contaminated by As, Cd and Pb in a polluted paddy field in Shangyu County, China, and used rice husk biochar and Fe^{3+} modified biochar to repair it. Fe^{3+}-modified biochar significantly reduced the bioavailability of As and Pb in soil, but increased the concentration of Cd and Pb in straw and brown rice in pot experiment. Rice husk biochar increased the activity of soil catalase and urease, but Fe^{3+}-modified biochar decreased the activity of catalase and urease in soil. Fe^{3+}-modified biochar increased grain yield by 60% and 32%, respectively, under continuous inundated irrigation and alternating wet and dry irrigation. It has also been found that Fe-Mn-Ce modified biochar significantly promoted the activity of urease, catalase, alkaline phosphatase and peroxidase in soil, improved the tolerance of soil enzymes to heavy metals, and prevented the denaturation during the passivation and stabilization of As contaminated rice fields. Similarly, in As-contaminated paddy soil, the addition of Fe-Mn-La modified biochar only significantly increased the catalase activity in soil. When 2% (wt) of Fe-Mn-La modified biochar was added, the activity increased by 69.2%–268%. In Pb-Cd contaminated soil, Fe-modified coffee grounds biochar effectively enhanced the activity of urease, sucrose and catalase in soil, while Fe-modified sage biochar only enhanced the activity of urease among the three enzymes. Both of the two iron-modified BCs reduced the content of exchangeable Cd/Pb and reducible Cd/Pb in soil. Meanwhile, the contents of oxidizable and residual Cd/Pb were increased to achieve the fixation of heavy metals (Yu et al., 2020). The application of tea branch biochar and Fe-Mn modified biochar in Sb-Cd contaminated soil significantly improved the soil properties. Both of the two biochar remediation materials could significantly promote the catalase and urease activity in soil, among which Fe-Mn modified biochar had a more significant promoting effect. On the contrary, acid phosphatase activity in the soil was inhibited to a certain extent. Soil enzymes mark a standard biochemical index of soil that is highly sensitive to environmental stress. The response relationship between heavy metals and enzyme activity has aroused tremendous attention, but no firm conclusion has been reached at present that it is affected by a wide range of environmental factors, such as basic physical and chemical properties of soil, soil texture, soil enzyme types, heavy metal types, heavy metal concentrations, soil nutrients and plant species (Burns et al., 2013). Besides, they went up/down, or triggered no obvious responses.

In this study, rice husk biochar was taken as the substrate, and modified by coprecipitation method to prepare as magnetic biochar and calcium-based magnetic biochar composites. Besides, soil contaminated by different degrees of Cr(VI) was simulated in the laboratory by adding gradient concentrations of Cr(VI) exogenously. The three remediation materials prepared were used to passivate and stabilize the soil contaminated with gradient concentrations of Cr(VI), and corn was used as a model plant to investigate the passivation and stabilization effects of the three biochar composites on the soil contaminated with gradient concentrations of Cr(VI) at the time scale and concentration scale. Also, the stabilization and remediation mechanism of the soil contaminated with gradient concentrations of Cr(VI) was elucidated through the changes of forms of Cr(VI) in the soil, bioavailable state and soil properties. Also, the effects on the activity of enzymes involved in C, N and P cycles in soil, the microbial metabolic characteristics and the community structure during the passivation and stabilization of Cr(VI) contaminated soil were investigated. Combined with the changes of plant growth and other physiological indicators during the passivation and stabilization of contaminated soil, as well as the enrichment in plants, the bioavailability of different biochar composites applied in soil contaminated by gradient concentrations of Cr(VI) was evaluated, and the impact of passivation and stabilization process on soil ecosystem was further analyzed.

5.2.2 Characterization of calcium-based magnetic biochar materials

5.2.2.1 Scanning electron microscopy analysis of surface morphology of calcium-based magnetic biochar

In this study, Scanning electron microscopy (SEM)-EDS was used to characterize the surface morphology and element distribution of the BC/Fe-B/Ca-Fe-B material prepared, as shown in Figure 5.11 and Figure 5.12 below. According to the SEM patterns, all the three remediation materials form well-developed and dense pore structures. This is because the organic macromolecules inside the biomass are cracked into molecular gases such as CH_4, CO_2 and H_2O during the pyrolysis at high temperature, which form a large number of pore channels on biochar during the diffusion from inside to outside (Huong et al., 2016). Due to the well-developed pore structure of biochar, the specific surface area is greatly increased and more reactive sites are provided, thus facilitating the adsorption and complexation of pollutants. In addition, it is not difficult to observe from Figure 5.11B, C that there are loads on the surface of biochar, which are presumed to be part of the iron oxides and calcium carbonate added. The presumption is further confirmed by SEM Mapping test results.

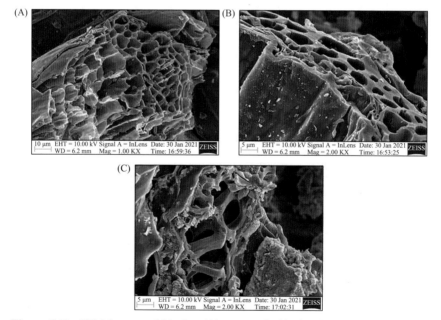

Figure 5.11 SEM images of biochar (A)、 magnetic biochar (B) and calcium-based magnetic biochar(C).

SEM mapping showed that the BC had the highest C and O content, which was consistent with its high aromatic hydrocarbon-rich structure. Fe was successfully loaded at the Fe-B surface, a proportion of 5.48%, and Fe and Ca were successfully loaded at the Ca-Fe-B surface, with a proportion of 7.46% and 5.93%, respectively. In addition, it can be seen from Figure 5.12D, F that Fe_3O_4 particles were evenly distributed on the surface of biochar without obvious agglomeration, which proved that Fe-B and Ca-Fe-B modified materials were successfully prepared.

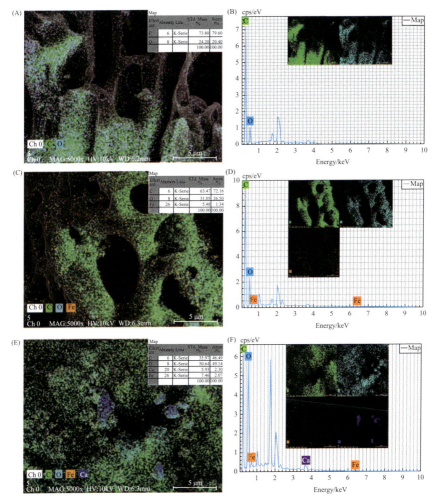

Figure 5.12 SEM mapping of biochar (A、B), magnetic biochar (C、D) and calcium-based magnetic biochar (E、F).

5.2.2.2 Fourier infrared FT-IR analysis of calcium-based magnetic biochar

Figure 5.13 shows the infrared spectra of BC, Fe-B and Ca-Fe-B materials.

In this study, the composition of BC/Fe-B/Ca-Fe-B surface elements and functional groups were analyzed by the FT-IR characterization. Among them, the typical —OH stretching vibration peak or —NH stretching vibration peak occurred at 3405cm^{-1} (Argun et al., 2009); 2922cm^{-1} was generated by CH$_2$s stretching vibration in alkyl groups; the broad peak at 2391cm^{-1} was generated by C≡C or —C≡N vibration of aromatic compounds; 1633cm^{-1} in BC and Fe-B IR spectrum and 1435cm^{-1} in Ca-Fe-B IR spectrum marked the stretching vibrations of aromatic compounds C=O and C=C; the strong absorption peak at 1078cm^{-1} was the antisymmetric stretching vibration peak of Si—O—Si bond or stretching vibration peak of phenolic compounds C—O; the absorption peak at 470cm^{-1} corresponded to the symmetric stretching vibration peak of Si—O bond (Khorram et al., 2017), which indicated that all the BC, Fe-B and Ca-Fe-B materials contained a large number of functional groups such as hydroxyl, carbon-oxygen double bond and carbon-carbon triple bond, and SiO$_2$ was also present in the structure. The characteristic peak at 803cm^{-1} in the infrared spectra of BC and Fe-B, and that at 875cm^{-1} in the infrared spectra of Ca-Fe-B represented the stretching vibration peaks of aromatic C—H. The C—O and C—H peaks were enhanced at 1078cm^{-1} and 803cm^{-1} for Fe-B and Ca-Fe-B as compared with BC, which was caused by the planar vibration of the C—H peak and the bending vibration of the C—O peak. However, a new characteristic peak appears at

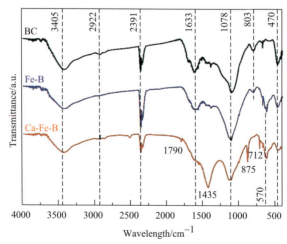

Figure 5.13 FT-IR diagram of biochar, magnetic biochar and calcium-based magnetic biochar.

$570cm^{-1}$ in the Fe-B and Ca-Fe-B mapping, which represents the characteristic peak of Fe—O stretching vibration in Fe^{3+} in the high spin state of Fe_3O_4. The results showed that Fe oxide was successfully loaded on the two modified BCs. In the Ca-Fe-B mapping, the characteristic peak at $1790cm^{-1}$ is generated by the vibration of CO_3^{2-} in calcite, and $712cm^{-1}$ represents the vibration absorption peak of the covalent bond of Ca—O, indicating that $CaCO_3$ is successfully loaded on Ca-Fe-B material. Some of it exists in the form of calcite, which is consistent with the characterization results of SEM and X-ray diffraction (XRD). FT-IR spectrogram showed that Fe-B and Ca-Fe-B composites were successfully prepared.

5.2.2.3 X-ray diffraction analysis of calcium-based magnetic biochar

Figure 5.14 illustrates the XRD pattern of BC/Fe-B/Ca-Fe-B materials. The standard magnetite powder (Fe_3O_4) diffraction peaks (PDF#28-0491) appear at 2θ=34.4°, 42.1°, 47.8° and 59.5° in the XRD pattern of Ca-Fe-B, corresponding to (−102), (−201), (210) and (−222) crystal planes, respectively.

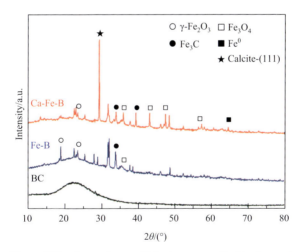

Figure 5.14 X-ray diffraction patterns of biochar, magnetic biochar and calcium-based magnetic biochar.

The diffraction peak (PDF#65-0390) representing maghemite γ-Fe_2O_3 appears at 2θ=23.8°, corresponding to (210) crystal plane. A wide baseline can be observed in the patterns, indicating the existence of poor crystallinity and amorphous morphology. However, the peak of calcite III diffraction peak (PDF#17-0763) appears at 2θ=29.5°, which proves that $CaCO_3$ is successfully loaded on Ca-Fe-B surface during modification, mainly in the form of calcite. In the XRD pattern of Fe-B, the diffraction peak (PDF#28-0491) of the standard magnet powder (Fe_3O_4) appears at $2\theta=34.4°$, corresponding to the crystal plane (−102). The diffraction peaks (PDF#65-0390) representing maghemite γ-Fe_2O_3 appear at $2\theta=18.42°$ and 23.8°, corresponding to (111) and (−201) crystal planes, respectively. It proves that there mainly are magnetite Fe_3O_4 and magnehemite γ-Fe_2O_3 on Ca-Fe-B. It indicates that Fe-B and Ca-Fe-B modified materials were successfully prepared, which is mutually verified with SEM and FT-IR characterization results. In addition, the diffraction peak at $2\theta=33.93°$ and 39.82° in the XRD pattern of Ca-Fe-B and that at $2\theta=33.93°$ in the XRD pattern of Fe-B represent Fe_3C (PDF#65-2411), corresponding to the crystal planes (102) and (002), respectively. In the XRD pattern of Ca-Fe-B, Fe^0 diffraction peak (PDF#65-4899) appears at $2\theta=65.01°$, corresponding to crystal plane (200), which indicates that some Fe exists in the material in the elemental and lower price states ($^{+2}$) and features the reductive property.

5.2.3 Passivation and stabilization of Cr(VI)-contaminated soil with calcium-based magnetic biochar

5.2.3.1 Effects of calcium-based magnetic biochar on soil pH

Soil pH is a comprehensive indicator of various biochemical properties of the soil, and the basis of almost all physical, chemical and biological reactions in soil. The environmental behavior of Cr in soil (adsorption, migration, transformation, etc.) is strongly influenced by the physical and chemical properties of the soil. Soil pH not only affects soil fertility, but also features a direct correlation with the occurrence form, availability, migration and transformation of Cr. Soil pH and organic matter content are also considered as key factors affecting the availability of heavy metals. It is generally argued that the closer the combination of heavy metals with soil, the lower their bioavailability, and that the increase of soil pH will further enhance the adsorption capacity of soil for heavy metals, thus reducing the content of available heavy metals in soil (Navarro-Pedreo et al., 2018).

As shown in Figure 5.15, the exogenous addition of Cr(VI) caused a decrease in soil pH to some extent. A possible reason is that on the premise of water as the extract, some Cr(VI) will produce competitive adsorption with H^+ to displace more H^+ into the soil, and consume OH- in the soil to generate precipitation in the conversion process to make soil more acidic.

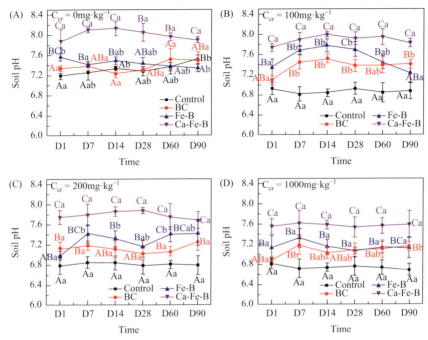

Figure 5.15 (A—D) Impacts of BC/Fe-BC/Ca-Fe-B on soil pH when Cr(VI) concentration in soil is 0mg·kg^{-1}, 100mg·kg^{-1}, 200mg·kg^{-1} and 1000mg·kg^{-1}. The uppercase letters represent the significant difference between different treatment groups at the same time ($P<0.05$), while the lowercase ones the significant difference between the same treatment group at different time points ($P<0.05$). The error line is the mean standard error of the three parallel samples ($\pm$SE).

On the whole, the application of BC, Fe-B and Ca-Fe-B significantly increased the soil pH under the pollution by Cr(VI) of gradient concentrations, which made alkaline soil environment. The increase of soil pH will also increase the surface charge of organic matters and clay minerals in soil, thus improving the adsorption capacity of soil for Cr(VI). Moreover, the obligate adsorption of a variety of heavy metals on the surface of oxides improves with the increase of soil pH, and most of the passively adsorbed heavy metals can be converted into obligate adsorption under a soil environment with higher pH. All above are conducive to the passivation and stabilization of heavy metals in soil by reducing their bioavailability. The study of Bandara et al. (2019)

also showed that the application of biochar-based materials can increase the soil pH and contribute a great deal to the remediation of heavy metal pollution. This is mainly because the large number of functional groups on the surface of biochar, the negative charge and developed pore structure can improve the soil pH, CEC, bulk density and other properties. However, with the increase of Cr(VI) pollution concentration, the improvement of soil pH by BC, Fe-B and Ca-Fe-B decreased gradually, and the remediation also showed a weakening trend.

When the Cr(VI) pollution concentration in soil was $0mg \cdot kg^{-1}$, Ca-Fe-B could significantly increase the soil pH by 0.4—1.1 units compared with BC and Fe-B. Fe-B had a slightly better effect on the soil pH improvement than BC, but there was no significant difference with the CK. When Cr(VI) pollution concentration was $100mg \cdot kg^{-1}$, BC, Fe-B and Ca-Fe-B could all significantly increase the soil pH, among which the effect of Ca-Fe-B was significantly better than that of BC and Fe-B, improving the soil by 0.8—1.2 units of pH, and the effect of Fe-B was significantly better than that of BC. After D28, the soil pH of the Fe-B treatment group showed a downward trend, which may be resulted from the release of more H^+ from exogenous Fe(II)/Fe(III) in the process of hydrolyzing and leaching from the surface of Fe-B materials (Yin et al., 2017).

When Cr(VI) pollution concentration was $200mg \cdot kg^{-1}$, Fe-B and Ca-Fe-B significantly increased the soil pH, and Ca-Fe-B outperformed Fe-B to increase the soil pH by 0.7—1.0 unit. BC could also increase the soil pH. Although its overall effect is not as good as Fe-B, the difference between the two is small. When the pollution concentration of Cr(VI) was $1000mg \cdot kg^{-1}$, Ca-Fe-B still could significantly increase the soil pH, showing the best effect among the three. But the increase effect was further reduced, by 0.5—0.7 unit only. BC and Fe-B also had a certain improving effect on soil, but there was almost no difference between the two after D14. By comparing the effects of BC, Fe-B and Ca-Fe-B of the same concentration on soil pH, it can be seen that Ca-Fe-B can significantly increase the pH of soil polluted by gradient concentrations of Cr(VI), and its effect is optimal among the three. This is because Ca-Fe-B is loaded with $CaCO_3$ on the surface, where the Ca^{2+} further increases the pH of Ca-Fe-B material itself, and the hydrolysis of soluble alkaline minerals (K, Ca and Na) also contributes to the increase of soil pH (Lu et al., 2014). The high ash content of biochar-based materials also significantly reflects its liming effect in soil (Yuan et al., 2011). The effect of Fe-B on soil pH was better than that of BC at low Cr(VI) concentrations (100mg·kg^{-1} and 200mg·kg^{-1}), but the difference between the two was small at high pollution concentrations, which could be attributed to the fact that when Cr(VI) pollution concentration exceeded the threshold, the impact of heavy metal on soil pollution was far greater than the remediation of materials, thus limiting the ability to repair.

As shown in Table 5.5, soil pH was negatively correlated with the exchangeable and oxidizable Cr in soil ($P<0.05$), and positively correlated with β-glucosidase and chitinase activity in soil ($P<0.05$).

Table 5.5 Analysis on soil pH based on pearson correlation coefficient.

	F1	F2	F3	F4	β-glu	NAG	Phos
pH	**−0.356***	−0.107	**−0.393***	0.13	**0.364***	**0.396***	−0.066

Note: β-glu, NAG and Phos represent the activity of β-glucosidase, chitinase and phosphatase, respectively.

F1 represents the exchangeable Cr, F2 the reducible Cr, F3 the oxidizable Cr, and F4 the residual Cr.

*$P<0.05$, ** $P<0.01$, bold fonts represent the significant correlation.

5.2.3.2 Effect of calcium-based magnetic biochar on morphological transformation of Cr

After entering the soil, heavy metals may take various chemical forms through dissolution, adsorption, complexation, precipitation, etc. Although the total amount of heavy metals in the soil can reflect the level of soil pollution to a certain extent, it cannot thoroughly evaluate the real potential ecological risks because different forms of heavy metals impose different impacts on the environmental ecology. Figure 5.16 illustrates the morphological transformation of Cr in soil polluted by Cr(VI) at concentrations of 100mg·kg^{-1}, 200mg·kg^{-1} and 1000mg·kg^{-1} of the Control treatment group, BC treatment group, Fe-B treatment group and Ca-Fe-B treatment group at D1, D7, D14, D28, D60 and D90.

As a whole, the exchangeable Cr is the dominant form in the CK, accounting for about 50% of the total. In the passivation and stabilization processes, the proportion of residual Cr under different Cr(VI) pollution concentrations in soil is basically in the order: Ca-Fe-B treatment group > Fe-B treatment group > BC treatment group > Control treatment group. When the Cr(VI) concentration in soil was 100mg·kg^{-1}, 200mg·kg^{-1} and 1000mg·kg^{-1}, at D90, Ca-Fe-B featured a fixation efficiency of 72.57%, 67.07% and 59.88%; Fe-B of 68.82%, 62.14% and 57.36%; BC of 67.18%, 60.02% and 54.20%, respectively. This shows that Ca-Fe-B exhibits a

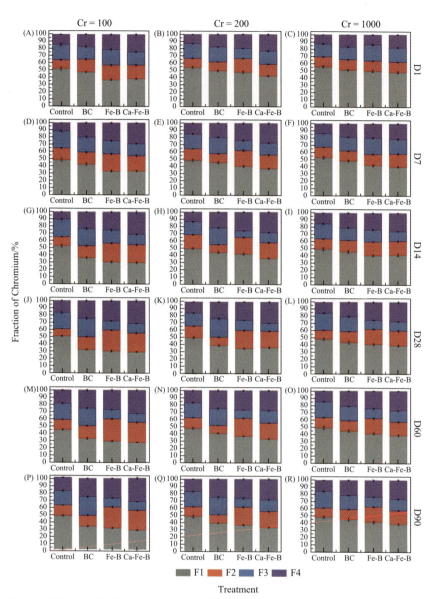

Figure 5.16 (A—R) illustrates the morphological transformation of Cr in soil polluted by Cr(VI) at concentrations of 100mg·kg^{-1}, 200mg·kg^{-1} and 1000mg·kg^{-1} of the control treatment group, BC treatment group, Fe-B treatment group and Ca-Fe-B treatment group on D1, D7, D14, D28, D60 and D90. F1: weak acid extraction state, F2: reducible state; F3: oxidizable state, F4: residual state. The error line is the mean standard error of the three parallel samples ($\pm$SE).

better Cr passivation effect than Fe-B and BC, because ①Ca-Fe-B can physically adsorb part of the Cr(VI) due to its developed pore structure and high specific surface area; ②a large number of active functional groups, such as —OH, —C=C—, —O—C=O and C=O on the material surface can also undergo displacement reaction or coordination complexation with Cr(VI). In the meantime, Fe(II) and Cr(VI) can form Fe(III)-Cr(III) complex precipitation obligate fixation. The loading of Ca^{2+} on the surface of Ca-Fe-B material resulted in a significant increase in soil pH, showing a strong correlation with the exchangeable heavy metals. There was a significant negative correlation between the soil pH and the exchangeable and oxidizable Cr in soil ($P<0.05$). In addition, when the soil pH is high, the acidic functional groups such as —OH and —COOH in the soil will undergo the deprotonation reaction, producing more soil particles with negative charge adsorption sites, and enhancing the electrostatic attraction of Cr(VI) by soil. Some Cr(VI) also converted into Cr(III) through soil REDOX reactions, while the alkaline soil is more likely to convert Cr(III) into precipitation or form coprecipitation with other substances so as to be adsorbed by soil minerals (Bolan et al., 2003). The synergistic effects of the three factors promoted the fixation of Cr in soil by Ca-Fe-B.

With the same restoration time, alongside the increase of Cr(VI) pollution concentration in soil, the difference in the proportion of exchangeable Cr in BC, Fe-B and Ca-Fe-B treatment groups gradually decreased as compared with the Control treatment group. When the Cr(VI) pollution concentration in soil was 100mg/kg, the content of exchangeable Cr in BC, Fe-B and Ca-Fe-B treatment groups decreased by 19.62%, 21.93% and 24.83%, respectively. When the Cr(VI) pollution concentration in soil was 200mg·kg^{-1}, the content of exchangeable Cr decreased by 14.86%, 17.76% and 20.95%, respectively. When the Cr(VI) pollution concentration in soil was 1000mg·kg^{-1}, the content of exchangeable Cr decreased by 10.04%, 14.1% and 16.62%, respectively. The application of BC, Fe-B and Ca-Fe-B increased the cation exchange capacity in soil, and adsorbed heavy metals and other cations on the colloid surface of the soil to reduce their migration. However, the limitations of remediation materials applied resulted in a limited increase in its ability to stabilize heavy metals. Therefore, BC/Fe-B/Ca-Fe-B could not effectively reduce the migration of Cr when the soil was polluted by too much Cr(VI), and the two showed a "dose-effect" relationship.

Different forms of Cr showed different rules of content in different treatment groups. When the concentration of Cr(VI) in soil was 100mg·kg^{-1}, the proportions of reducible Cr were 14.97%—17.66% in BC group, 20.2%—30.79% in Fe-B group, and 18.98%—27.86% in Ca-Fe-B group. When the Cr(VI) pollution concentration in soil was 200mg·kg^{-1}, the proportions of reducible Cr were about 10.76%—12.37% in BC group, 19.12%—25.06% in Fe-B group, and 16.15%—22.79% in Ca-Fe-B group. When the Cr(VI) pollution concentration in soil was 1000mg·kg^{-1}, the proportions of reducible Cr were 10.76%—12.37% in BC group, 19.12%—25.06% in Fe-B group, and 16.15%—22.79% in Ca-Fe-B group. When the concentration of Cr(VI) in soil was 100mg·kg^{-1}, the proportions of oxidizable Cr were 17.92%—26.06% in BC group, 12.62%—21.67% in Fe-B group, and 12.24%—18.21% in Ca-Fe-B group. When the Cr(VI) pollution concentration in soil was 200mg·kg^{-1}, the proportions of oxidizable Cr were about 20.77%—25.74% in BC group, 9.13%—14.37% in Fe-B group, and 11.06%—18.41% in Ca-Fe-B group. When the Cr(VI) pollution concentration in soil was 1000mg·kg^{-1}, the proportions of oxidizable Cr were 17.77%—20.79% in BC group, 12.59%—22.76% in Fe-B group, and 12.88%—20.53% in Ca-Fe-B group. When the concentration of Cr(VI) in soil was 100mg·kg^{-1}, the proportions of residual Cr were 16.83%—26.33% in BC group, 21.11%—28.09% in Fe-B group, and 24.34%—32.86% in Ca-Fe-B group. When the Cr(VI) pollution concentration in soil was 200mg·kg^{-1}, the proportions of residual Cr were about 17.49%—24.97% in BC group, 19.49%—26.87% in Fe-B group, and 23%—29.55% in Ca-Fe-B group. When the Cr(VI) pollution concentration in soil was 1000mg·kg^{-1}, the proportions of residual Cr were 16.68%—20.96% in BC group, 13.77%—24.79% in Fe-B group, and 17.83%—27% in Ca-Fe-B group. Under the same Cr(VI) pollution concentration, the content of reducible Cr basically followed the rule: Fe-B treatment group > Ca-Fe-B treatment group > BC treatment group; the proportion of oxidizable Cr was BC treatment group > Fe-B treatment group > Ca-Fe-B treatment group.

Fe-B treatment group featured the highest proportion of reduceable Cr, which can be attributed to the formation of a large number of Fe(III)-Cr(III) complexes on the surface of Fe-B materials during the passivation and stabilization, as shown in Figure 5.17A. BC treatment group had the highest proportion of oxidizable Cr, because a large number of functional groups on the surface of biochar, such as C=C, C—OH and O—C=O, participated in the adsorption of Cr(VI) in soil, and the application of biochar increased the content of organic matters in soil. Organic matters in soil can also adsorb the Cr(VI) in soil, and humic acid formed by the decomposition of humus can also be complexed with Cr(VI) in soil, thus increasing the concentration of metal-organic complex in soil (Antoniadis and Alloway, 2002). The overall increase of reducible Cr content in Fe-B and Ca-Fe-B treatment

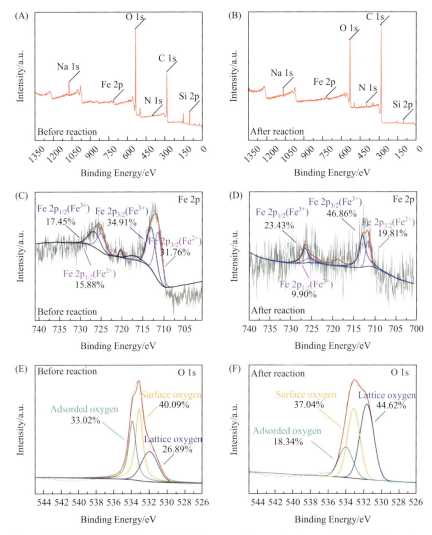

Figure 5.17 (A, B) Full X-ray photoelectron spectrogram before and after FE-B reaction, (C, D) Fe 2p spectrogram before and after Fe-B reaction, and (E, F) O 1s spectrogram before and after Fe-B reaction.

groups may be ascribed to the fact that Fe-B/Ca-Fe-B significantly increases the soil pH, while ferric and manganese oxide colloids in soil are amphoteric colloids. With the increase of soil pH, the solubility of organic matters in soil increases and the complexing ability of heavy metals improves.

5.2.3.3 Mechanism of calcium-based magnetic biochar's stabilization of Cr contamination of soil

1. X-ray photoelectron spectroscopy analysis of Fe-B and Ca-Fe-B materials

To further elucidate the passivation and stabilization mechanism of the Fe-B and Ca-Fe-B applied to Cr(VI)-contaminated soil, X-ray photoelectron spectroscopy was performed before and after the application of Fe-B and Ca-Fe-B materials to detect the changes of the valence of Fe and the morphology and quantity of Ca, C and O in Fe-B and Ca-Fe-B materials.

As can be seen from the comparison between Figure 5.17A and Figure 5.17B, the types of main elements in Fe-B materials stayed unchanged before and after the remediation, but the content of Fe, O, C and Si changed significantly. According to analysis of Figure 5.17C and Figure 5.17D, there are two sets of absorption peaks in Fe 2p spectrograms of Fe-B material before and after the restoration. Before restoration, the absorption peaks correspond to Fe $2p_{1/2}$ and Fe $2p_{3/2}$ orbits of Fe(II) at the electron binding energy of 725.12eV and 711.52eV. The absorption peaks at 726.77eV and 713.17eV correspond to Fe $2p_{1/2}$ and Fe $2p_{3/2}$ orbits of Fe(III). After repair, the absorption peaks of Fe $2p_{1/2}$ and Fe $2p_{3/2}$ orbits of Fe(II) and Fe(III) appear at 725.35eV, 711.749eV, 726.65eV and 713.05eV, respectively, and the peak deviation can be observed. The main peak energy distance of 13.6eV proves the existence of Fe_3O_4 in the material. Besides the main peak, the satellite peak at Fe $2p_{3/2}$ proves the existence of Fe_2O_3 in the material (Turu and Peter, 2008). This is consistent with the results of XRD characterization of Fe-B materials loaded with γ-Fe_2O_3 and Fe_3O_4. Figure 5.17C and Figure5.17D show that the relative content of Fe(II) and Fe(III) in Fe-B materials before restoration are 47.64% and 52.36% respectively, and those after restoration are 29.71% and 70.29% respectively. The results show that Fe(III)/Fe(II) is involved in the passivation and stabilization of soil Cr(VI) pollution, and acts as the electron donor during the reduction of Cr(VI) to Cr(III), oxidizing Fe(II) to Fe(III), and further forming the Fe(III)-Cr(III) complex on the surface of biochar.

According to Figure 5.17E and Figure 5.17F, the energy spectrogram of O 1s in Fe-B materials before and after restoration chiefly shows three peaks, which can be attributed to the lattice oxygen O^{2-}, OH compound type oxygen (surface oxygen) and adsorbed oxygen (Qin et al., 2018). Among them, lattice oxygen O^{2-} is considered to be a binding state between O and metal elements in Fe and Mn oxides, which conforms to the content of surface iron oxides indicated by XRD characterization of Fe-B materials. OH-type oxygen (surface oxygen) refers to functional groups such as —OH and —COOH on the surface of biochar, which is consistent with the results of functional groups on the surface of biochar shown by FT-IR characterization of Fe-B materials. Adsorbed oxygen may be in the adsorbed molecular water state (Barbero et al., 2006). The relative content of lattice oxygen O^{2-}, OH-type oxygen and adsorbed oxygen on the surface of Fe-B materials were 26.89%, 40.06% and 33.02% before restoration, and 44.62%, 37.04% and 18.34% after restoration, respectively. Before and after restoration, the content of lattice oxygen O^{2-} increased significantly, while that of the OH-type oxygen and adsorbed oxygen decreased materially, indicating that the Fe(III)/Fe(II) and the functional groups on the biochar surface, mainly —OH and —COOH, were involved in the Cr(VI) redox process in the soil, and that Fe(III)/Fe(II) played a major role in the passivation and stabilization of Cr(VI) in soil.

Figure 5.18 shows the full, C1s, O 1s and Ca 2p spectrograms of surface photoelectrons of Ca-Fe-B materials before and after the application. As can be seen from Figure 5.18B and Figure 5.18D, the C 1s spectrum can be divided into three sets of absorption peaks, due to the certain differences in the binding energy of C atoms in different functional groups. Absorption peaks of Ca-Fe-B materials at 284.6eV, 285.33eV and 287.01eV before remediation, those of Ca-Fe-B materials at 284.59eV, 285.26eV and 287.21eV after remediation correspond to carbon atoms in the functional groups —C=C—, C—OH and O—C=O at biochar surface, respectively. Before and after the passivation and stabilization process, certain changes in the relative content and binding energy of the three functional groups could be perceived, in which the relative content of —C=C— and C—OH decreased by 4.58% and 0.71% respectively, while the relative content of O—C=O increased by 5.29%. It indicated that all the three functional groups participated in the adsorption of Cr(VI) in soil during the passivation and stabilization of Cr(VI) pollution.

Figure 5.18E and Figure 5.18G show that Ca-Fe-B material had absorption peaks representing the lattice oxygen O^{2-}, surface oxygen and adsorbed oxygen at 531.37eV, 532.80eV and 533.55eV before remediation, and at 531.59eV, 532.81eV and 533.86eV after remediation, respectively. Compared with before remediation, the relative content of lattice oxygen

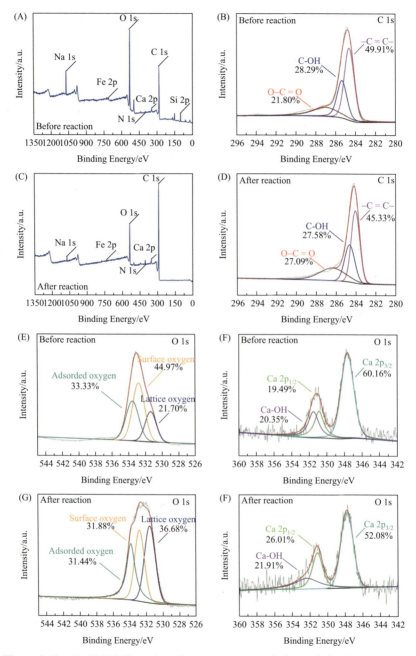

Figure 5.18 (A, C) Full X-ray photoelectron spectrogram before and after Ca-Fe-B reaction, (B, D) C 1s spectrogram before and after Ca-Fe-B reaction, (E, G) O 1s spectrogram before and after Ca-Fe-B reaction, and (F, H) Ca 2p spectrogram before and after Ca-Fe-B reaction.

O^{2-} on the surface of Ca-Fe-B after remediation increased by 14.98%, while the relative content of surface oxygen and adsorbed oxygen decreased by 13.09% and 1.89%, respectively. It suggests that Fe(III)/Fe(II) loaded on the surface of Ca-Fe-B material also participates in the reduction of Cr(VI) during the passivation and stabilization of soil, making it stabilize in the soil in the form of Cr_2O_3, $Cr(OH)_3$ and Fe(III)-Cr(III) complexes.

Figure 5.18F and Figure 5.18H show the sub-peaks of Ca 2p on the material surface. The absorption peaks of CA-Fe-B material at 347.70eV, 350.94eV and 351.57eV before remediation, and those at 347.81eV, 351.20eV and 352.41eV after remediation correspond to the Ca-O and Ca-OH bonds of Ca $2p_{3/2}$ and Ca $2p_{1/2}$ orbits, respectively (Wz et al., 2020). The change of its relative content and binding energy before and after the passivation and stabilization further proves that $CaCO_3$ is involved in the adsorption and stabilization of Cr(VI) in the soil. Given that Si still exists in the Ca-Fe-B material, hydroxy complexes of Ca will be formed on the surface of biochar under the overall alkaline soil environment, in which —O— and univalent hydroxyl Ca coordinate to form Si-O-Ca-OH bonds, making it one of the active sites on the surface of biochar. It can be adsorbed to anions to stabilize Cr(VI) in soil.

2. XRD analysis of BC/Fe-B/Ca-Fe-B materials after reaction

XRD analysis was carried out on the remedied BC/Fe-B/Ca-Fe-B material to determine the change of phase composition and crystallinity of the remedied material, further supporting the passivation and stabilization mechanism of BC/Fe-B/Ca-Fe-B material on the Cr(VI) pollution in soil. Figure 5.19 shows the XRD pattern of remedied BC/Fe-B/ Ca-Fe-B material.

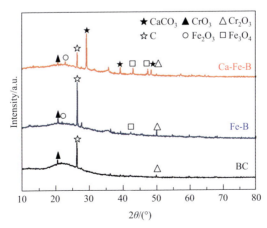

Figure 5.19 X-ray diffraction pattern of BC/Fe-B/Ca-Fe-B material after the reaction.

It can be seen from Figure 5.19 that the XRD patterns of BC/Fe-B/Ca-Fe-B show sharp peaks, indicating that they have a highly crystalline structure. The diffraction peaks of $CaCO_3$ appear at $2\theta=29.40°$, $39.41°$ and $48.51°$, which correspond to Planes (104), (113) and (116) of the calcite crystal structure (PDF#83-0577). Diffraction peaks appear at $2\theta=23.8°$, $43.05°$, and $47.80°$ of Ca-Fe-B material, and $2\theta=23.8°$ and $43.05°$ of Fe-B material, indicating that iron oxide crystals with cubic spinel crystal structure are loaded on the material, namely Fe_2O_3 and Fe_3O_4 characteristic diffraction peaks, corresponding to crystal planes (201), (400) and (210) of the crystal structure (PDF#65-0390; PDF# 99-00783; and PDF#28-0491). In the diffraction patterns of the three materials, the graphite-like microcrystal characteristic diffraction peak appears at $2\theta=26.6°$, which represents the chaotic graphite (111) plane (PDF#75-2078) as the amorphous structure characteristic peak of carbon. The characteristic diffraction peaks of CrO_3 and Cr_2O_3 also appear at $2\theta=21.13°$ and $50.2°$, representing the crystal planes (011) and (024) (PDF#07-0228; and PDF#06-0504). The results show that both Cr(VI) and Cr(III) exist on the surface of BC/Fe-B/Ca-Fe-B materials during the passivation and stabilization of Cr(VI) pollution in soil, and some ions exist in the form of CrO_3 and Cr_2O_3, which prove that BC/Fe-B/Ca-Fe-B has both physical adsorption and chemical reduction on Cr(VI).

3. Passivation and stabilization mechanism of BC/Fe-B/Ca-Fe-B on Cr pollution in soil

Coupled with the effect of BC/Fe-B/Ca-Fe-B on the transformation of soil pH and Cr form and the comparative analysis of material characterization before and after application in soil contaminated by Cr(VI) of gradient con-centrations, it can be seen that the Cr(VI) pollution in BC/Fe-B/Ca-Fe-B stabilized soil covers complex physical and chemical processes, with the main passivation and stabilization mechanisms shown in Figure 5.20.

(1) Due to its large specific surface area and porosity, biochar itself has certain adsorption characteristics, which can adsorb Cr(VI) in soil on its surface or diffuse it into the pore, and fix Cr(VI) through Van der Waals force and other effects, thus reducing the content of its available state.

(2) The application of BC/Fe-B/Ca-Fe-B to the soil contaminated by Cr(VI) of gradient concentrations can improve the soil pH to a certain extent, but the effect of pH improvement gradually decreases with the increase of Cr(VI) concentration. A high soil pH can not only promote the transformation of Cr(VI) morphology in soil, from exchangeable state to oxidizable state and

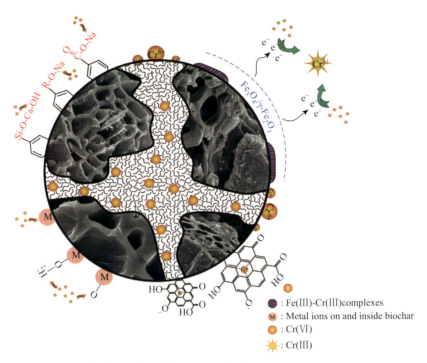

Figure 5.20 Passivation and stabilization mechanisms.

residue state, but also reduce its migration and increase the soil retention of Cr(VI). In addition, due to the decrease of H^+ content in the soil environment, and of the competitive adsorption with Cr(VI), it enhances the adsorption complex of Cr(VI) by BC/Fe-B/Ca-Fe-B, and promotes the conversion of Cr(VI) to Cr(III) so as to form Cr_2O_3, $Cr(OH)_3$ and other precipitates. In addition, the surface charge of BC/Fe-B/Ca-Fe-B can adsorb a certain amount of Cr(VI) through electrostatic action, and the increase of soil pH can increase the adsorption of Cr(VI) by the remediation material. The electrostatic adsorption will be enhanced to a certain extent with the increase of the initial concentration of Cr(VI). In the alkaline soil environment, Cr(III) is easy to be converted into precipitates or coprecipitates with other substances, and adsorbed by soil minerals.

(3) BC/Fe-B/Ca-Fe-B contains a large number of functional groups such as —OH, C=O, —NH, O—C=O, and C=C. These functional groups can improve the cation exchange capacity in soil and the stability of Cr(VI) in soil through ion exchange. In Fe-B/Ca-Fe-B, functional groups, dominated by —OH and —COOH, can also act as adsorption sites for coordination complexation with Cr(VI) to stabilize heavy metals. A large number of

oxygen-containing functional groups can store and transfer electrons, and participate in the stabilization process of Cr(VI) reduction as electron donors.

(4) A large number of Fe_3O_4 and $\gamma\text{-}Fe_2O_3$ are loaded on the surface of Fe-B/Ca-Fe-B, among which Fe(II) directly participates in the reduction of Cr(VI) in soil; Fe(II) forms an electron shuttle between Fe-B/Ca-Fe-B and Cr(VI), promotes the reduction of Cr(VI) to Cr(III), and produces Cr_2O_3, $Cr(OH)_3$ precipitates and $(Cr_xFe_{1-x})(OH)_3$ and $Cr_xFe_{1-x}OOH$ complex precipitates in alkaline soil (Reyhanitabar et al., 2012). The reaction process is shown as follows:

$$Cr_2O_7^{2-} + 6Fe^{2+} + 14H^+ \longrightarrow 2Cr^{3+} + 6Fe^{3+} + 7H_2O \tag{5.1}$$

$$xCr^{3+} + (1-x)Fe^{3+} + 3H_2O \longrightarrow (Cr_xFe_{1-x})(OH)_3 + 3H^+ \tag{5.2}$$

$$xCr^{3+} + (1-x)Fe^{3+} + 2H_2O \longrightarrow Cr_xFe_{1-x}OOH + 3H^+ \tag{5.3}$$

$$Cr^{3+} + 3OH^- \longrightarrow Cr(OH)_3 \tag{5.4}$$

Due to the formation of a large number of Fe(III)-Cr(III) complexes, the content of reducible Cr in soil increases.

(5) During the passivation and stabilization of Ca-Fe-B, the addition of $CaCO_3$ increases the pH of Ca-Fe-B, resulting in more negative charges on its surface to increase the adsorption of Cr(VI). Si-O-Ca-OH bond is formed on the surface of Ca-Fe-B, which is used as the active site to further adsorb Cr(VI).

(6) A large number of Na^+, Ca^{2+}, Mg^{2+} and Fe^{3+} cations on the surface of BC/Fe-B/Ca-Fe-B materials can be replaced with Cr(VI) in soil to achieve the fixation effect.

5.2.4 Effects of calcium-based magnetic biochar's passivation and stabilization on soil microbial activity

5.2.4.1 Effects of calcium-based magnetic biochar on enzyme activity

Soil enzyme is one of the most active bioactive substances in the process of energy flow and material circulation in the soil ecosystem, and a kind of macromolecule active protein. It can not only characterize the soil fertility and soil quality, but also fully reflect the response of soil microorganisms to environmental stress in the process of biogeochemical cycle. Therefore, soil enzymes are widely used as one of the main biochemical indexes to evaluate the quality of soil contaminated by heavy metals (Martinez et al., 2018). There are various soil enzymes. Affected by various environmental factors, such as soil pH and other physiochemical properties, nutrients and biological species, they would impose different impacts.

Among them, β-glucosidase, a cellulase, marks an important component of the cellulolytic enzyme system in soil. It can hydrolyze sugars such as fibrous disaccharides and cellulose oligosaccharides bound to the terminal nonreducing β-D-glucose bond, and release β-D-glucose and corresponding ligands. As an important carbon source of soil microorganisms and plants, glucose is further involved in soil C cycle. Chitinase (NAG), an extracellular hydrolytic enzyme, mainly participates in the reaction where chitin in soil is hydrolyzed to produce N-acetylglucosamine (Shimoi et al., 2020). As one of the main sources of mineralized nitrogen in soil, amino sugars account for about 10% of organic nitrogen in soil, and play a crucial role in nitrogen cycle in soil. Phosphatase, divided into alkaline phosphatase and acid phosphatase, mainly plays a role in the dephosphorylation of various substrates in the soil. It removes phosphate groups from different substrate molecules by hydrolyzing the phosphate esters, and generates hydroxyl and phosphate ions and other materials. However, element P plays an important role in the enzyme activation process in soil, cell synthesis of soil microorganisms, intercellular signaling and cell energy composition (Burns et al., 2013). Therefore, it is necessary to explore in detail the changes in the activity of β-glucosidase, chitinase and phosphatase in soil during the passivation and stabilization of soil contaminated by different concentrations of Cr(VI), by applying BC, Fe-B and Ca-Fe-B, so as to reveal the influence of soil C, N and P cycles.

Figure 5.21 shows the effects of BC, Fe-B and Ca-Fe-B on the activity of three soil enzymes at a Cr(VI) concentration of $100mg \cdot kg^{-1}$. As can be seen from the figure, the β-glucosidase activity in soil in the BC treatment group was significantly increased with the extension of the passivation and stabilization period. The β-glucosidase activity in the Ca-Fe-B treatment group was also improved as compared with the Control treatment group to a certain extent, an extent not as great as that in the BC treatment group. There was no significant difference in the changes of enzyme activity between Fe-B treatment group and CK. This can be attributed to the high aromatic hydrocarbon rich carbon structure of biochar-based materials. As an additional carbon source, they promoted the carbon cycle in soil to a large extent in the process of soil remediation, thus improving the enzyme activity related to the carbon cycle in soil (Khan et al., 2017). In addition, the presence of volatile compounds in biochar may also contribute to the high activity of this enzyme. The study by Li et al. found that the β-glucosidase activity was also significantly correlated with the microbial biomass carbon (MBC) in soil. In this study, there was a significant positive correlation between the two ($P<0.05$), which was consistent with previous research

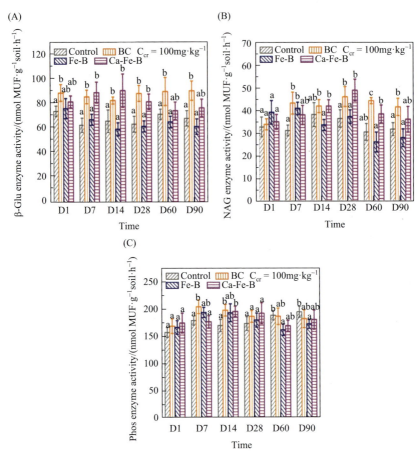

Figure 5.21 (A−C) The effects of BC/Fe-BC/Ca-Fe-B on the activity of β-glucosidase, chitinase and phosphatase in soil in the environment contaminated by 100mg·kg^{-1} of Cr(VI). Lower case letters represent the significant differences between different treatment groups at the same time ($P<0.05$). The error line is the mean standard error of the three parallel samples ($\pm$SE).

results. However, Fe-B material itself has certain biological toxicity, which can also affect the activity of β-glucosidase. Another possible reason is the large binding energy of β-glucosidase in iron-biochar complex, which facilitates the binding of enzyme and complex. Such interactions lead to the adsorption of β-glucosidase on the surface of iron-biochar complex, thus inhibiting the enzymes or enzyme substrates (Jain et al., 2016). Also, studies have shown that physiochemical properties of the soil can affect the enzyme activity. The study of Tang et al. found that soil pH was positively correlated with β-glucosidase activity (Tang et al., 2020). In this study, there was a significant positive correlation between the two ($P<0.05$). This can be attributed to the fact that the supply and quality of soil nutrients can affect the enzyme activity, especially the enzyme excretion, while low nutrient levels can inhibit the production of soil enzymes.

Compared with the CK, BC and Ca-Fe-B treatment groups significantly increased the activity of soil chitinase. Studies have shown that the soil nitrogen dynamics are affected by the application of biochar and the pH and texture of soil. Biochar and polymer can accelerate the transformation of nitrogen in soil and promote the activity of soil chitinase (Mahmoud et al., 2018). In this study, there was a significant positive correlation between soil pH and chitinase activity ($P<0.05$), which confirmed the rule of consistence. It may also be caused by the presence of nitrogen compounds in biochar, which act as substrates. Similarly, as the storehouse for carbon and nutrients, biochar boosted the microbial growth. Biochar is used to improve the chitinase activity in soil in an aim not only to enable more nitrogen conversion, but also to meet the higher microbial demand. Combined with the change of microbial community structure in soil, it was found that both the BC and Ca-Fe-B treatment groups promoted the relative abundance of Proteobacteria in soil. Some of the Proteobacteria can participate in the nitrogen cycle in soil and play an important role in the community structure and relative abundance of ammonia-oxidizing bacteria (Trivedi et al., 2013), suggesting that the degradation and stabilization of Cr(VI) contaminated soil is a biological-physical-chemical process. However, phosphatase activity in different treatment groups did not change regularly, and the response was not strong.

Figure 5.22 shows the effects of BC, Fe-B and Ca-Fe-B on the activity of the soil enzymes at a Cr(VI) concentration of $200mg \cdot kg^{-1}$. For β-glucosidase, the variation pattern among different treatment groups was basically consistent with the passivation and stabilization process in soil contaminated by Cr(VI) at a concentration of $100mg \cdot kg^{-1}$. BC and Ca-Fe-B treatment groups significantly increased the activity of β-glucosidase in soil. Among them, with the extension of the passivation and stabilization period, Ca-Fe-B treatment group had a more obvious improvement effect on β-glucosidase activity, perhaps because the passivation and stabilization effect of Ca-Fe-B material was obviously reflected with the increase of Cr(VI) pollution concentration in soil. The activity of soil chitinase in BC, Fe-B and Ca-Fe-B treatment groups first increased, then decreased, and tended to be stable with the extension of the passivation and stabilization period. BC and Ca-Fe-B treatment groups could still promote the activity of soil chitinase to a certain extent, but the promotion effect slowed down. This is perhaps because the addition of exogenous Cr(VI) reduced the soil pH, while the addition of remediation materials helped to alleviate the acidity conditions in soil and enable a high enzyme activity. Soil pH is crucial to the survival and function of enzymes (Martinez et al., 2018). Combined with the analysis of effects of BC, Fe-B and Ca-Fe-B on soil pH at this concentration, there was no significant difference between the soil pH of different treatment groups

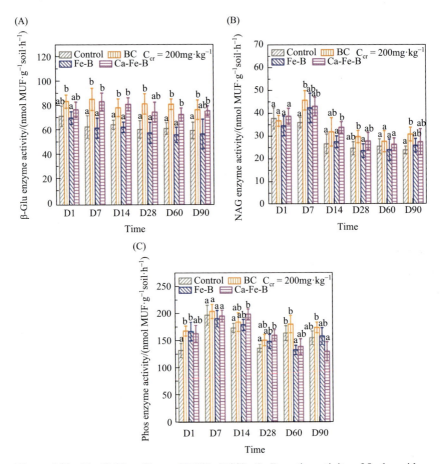

Figure 5.22 (A−C) The effects of BC/Fe-BC/Ca-Fe-B on the activity of β-glucosidase, chitinase and phosphatase in soil in the environment contaminated by 200mg·kg^{-1} of Cr(VI). Lower case letters represent the significant differences between different treatment groups at the same time ($P<0.05$). The error line is the mean standard error of the three parallel samples ($\pm$SE).

on D90, and the activity of chitinase also showed the same trend. With the increase of Cr(VI) pollution concentration, phosphatase activity in different treatment groups was also inhibited to different degrees, but there was no obvious regularity of the changes.

Figure 5.23 shows the effects of BC, Fe-B and Ca-Fe-B on the activity of the soil enzymes at a Cr(VI) concentration of 1000mg·kg^{-1}. As can be seen from the figure, under high pollution concentration, the effects of BC, Fe-B and Ca-Fe-B treatment groups on the activity of β-glucosidase,

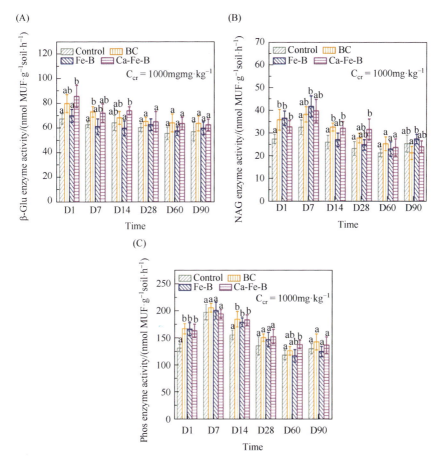

Figure 5.23 (A−C) The effects of BC/Fe-BC/Ca-Fe-B on the activity of β-glucosidase, chitinase and phosphatase in soil in the environment contaminated by 1000mg·kg^{-1} of Cr(VI). Lower case letters represent the significant differences between different treatment groups at the same time ($P<0.05$). The error line is the mean standard error of the three parallel samples ($\pm$SE).

chitinase and phosphatase was greatly limited. This can be attributed to the biotoxic effect of high concentrations of heavy metals in the soil. There are also cases where high concentrations of Cr(VI) in the soil can chelate some substrates or further react with enzyme-substrate complexes to inactivate soil enzymes (Mikanova, 2006). In the Ca-Fe-B treatment group, D1, D7 and D14 saw a certain promoting effect on enzyme activity, indicating that the remediation effect of Ca-Fe-B materials on soil Cr(VI) pollution was greater than the toxic effect of heavy metals in the short term. However, with extension of the remediation cycle, the toxic effect of heavy metals exceeded the remediation threshold, and activity of all the three soil enzymes showed a downward trend. The variation trend of enzyme activity under gradient concentrations also showed the feature that it "promoted at low concentration of heavy metals and inhibited at high concentrations."

In general, the activity of β-glucosidase, chitinase and phosphatase in different treatment groups decreased with the increase of Cr(VI) pollution concentration in soil. Among them, BC and Ca-Fe-B treatment groups could significantly promote the β-glucosidase and chitinase activity in the environment polluted by Cr(VI) at low concentration, while Fe-B treatment groups had lower promoting effects than BC and Ca-Fe-B treatment groups. The phosphatase activity in soil was less responsive to the application of BC, Fe-B and Ca-Fe-B materials. Due to complex environmental conditions and soil types, there are still many differences in the relationship between soil physiochemical parameters, biochar properties, and soil enzyme activity.

5.2.4.2 Effects of calcium-based magnetic biochar on soil MBC

As the most active part of soil phase, microorganisms participate in the whole metabolic cycle of soil. Although MBC accounts for a very small proportion in soil composition, it can reflect the soil fertility and microbial content to a large extent, as an important environmentally sensitive index to evaluate the microbial quantity in soil and soil fertility (Li and Chen, 2004). Therefore, through the change of MBC in soil, we can explore the impact of BC, Fe-B and Ca-Fe-B on soil microorganisms during the passivation and stabilization of soil contaminated by Cr(VI) at the time scale and concentration scale.

As shown in Figure 5.24, when there was no Cr(VI) pollution in the soil, the MBC concentration in soil of the CK and the BC treatment group showed a slow rising trend. Compared with the CK, the BC treatment group significantly boosted the growth of soil MBC, bringing its concentration to

298.35mg $C \cdot kg^{-1}$ in soil at the end of the culture period. Some studies also showed that the addition of biochar would lead to the increase of organic matter content in soil, thus significantly increasing the MBC in soil. However, Fe-B and Ca-Fe-B treatment groups significantly inhibited the MBC concentration in soil, which fluctuated slightly around 245mg $C \cdot kg^{-1}$ soil and 268mg $C \cdot kg^{-1}$ soil, respectively, during the whole remediation cycle. It also indicated that Fe-B and Ca-Fe-B materials had certain biological toxicity, which could inhibit soil microbial activity to a certain extent, and the biological toxicity of Fe-B materials was greater than that of Ca-Fe-B materials. When the Cr(VI) pollution concentration was 100mg $\cdot kg^{-1}$, comparison of the CK and BC treatment group could still significantly promote the increase of MBC concentration in soil and

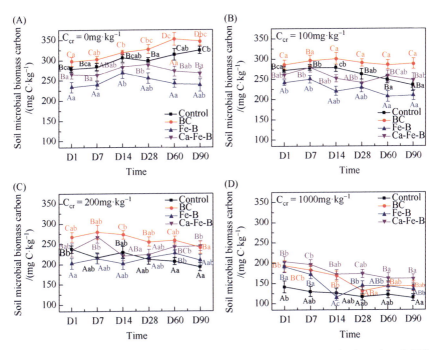

Figure 5.24 (A—D) Impacts of BC/Fe-BC/Ca-Fe-B on soil MBC respiration when Cr(VI) pollution concentration in soil is 0mg$\cdot$kg^{-1}, 100mg$\cdot$kg^{-1}, 200mg$\cdot$kg^{-1} and 1000mg$\cdot$kg^{-1}. The uppercase letters represent the significant difference between different treatment groups at the same time ($P<0.05$), while the lowercase ones the significant difference between the same treatment group at different time points ($P<0.05$). The error line is the mean standard error of the three parallel samples ($\pm$SE).

maintain a small fluctuation around 287mg $C \cdot kg^{-1}$ soil. The Ca-Fe-B treatment group showed a fluctuation in soil MBC concentration, but had no significant difference from the Control group. The overall trend of MBC decreased, while Fe-B treatment group still had a significant inhibitory effect on soil MBC, showing a continuous slow declining trend and a slight fluctuation around 217mg $C \cdot kg^{-1}$ soil.

When the soil Cr(VI) pollution concentration was 200mg·kg^{-1}, the BC and Ca-Fe-B treatment groups could significantly increase the soil MBC concentration compared with the CK, but the promotion effect of BC treatment group was better. After the end of the culture period, the MBC concentration of the two treatment groups remained around 247.83mg $C \cdot kg^{-1}$ soil. The results indicated that at this concentration, Ca-Fe-B's microbial stress relief effect on Cr(VI) was greater than its biotoxicity effect. Although the MBC concentration of the Fe-B treatment group fluctuated slightly, there was no significant difference with that of the CK. The four treatment groups showed a slow downward trend on the whole. When the soil Cr(VI) pollution concentration was 1000mg·kg^{-1}, the BC, Fe-B and Ca-Fe-B treatment groups significantly increased the soil MBC concentration, with the Ca-Fe-B treatment group behaving better than the other two. Upon end of the culture, the MBC concentration could be kept around 162.26mg $C \cdot kg^{-1}$ soil. There was no significant difference between the BC treatment group and Fe-B treatment group, both of which could increase the MBC concentration to 143.38mg $C \cdot kg^{-1}$ soil. All the treatment groups showed a decreasing trend on the whole.

In general, with the increase of soil Cr(VI) pollution concentration, the soil MBC concentration in all treatment groups showed a slow declining trend. In the environment under low concentration pollution (100mg·kg^{-1} and 200mg·kg^{-1}), biochar could be used as a carbon source to offer microorganisms nutrients to promote their growth and reproduction. In the BC treatment group, MBC concentration was always significantly increased, and the biological toxicity and pollution remediation effects of Fe-B and Ca-Fe-B materials were also reflected to some extent. However, in the environment of high concentration pollution (1000mg·kg^{-1}), the remediation effect of Ca-Fe-B materials on soil Cr(VI) pollution was more obvious. As shown in Table 5.6 below, soil MBC was negatively correlated with exchangeable Cr in soil ($P<0.01$), and positively correlated with β-glucosidase activity in soil ($P<0.05$).

Table 5.6 Analysis on correlation of all soil indicators based on pearson correlation coefficient.

	MBC	SBR	β-glu	NAG	Phos
pH	0.314	0.245	**0.364***	**0.396***	−0.066
F1	**−0.761****	−0.401	**−0.504***	−0.372	**−0.489***
MBC	/	**0.488***	**0.519***	0.228	0.115

Note: β-glu, NAG and Phos represent the activity of β-glucosidase, chitinase and phosphatase, respectively. MBC represents microbial biomass carbon, and SBR refers to basic CO_2 respiration rate of microbes. F1 stands for the interchangeable Cr. *$P<0.05$, ** $P<0.01$, bold fonts represent the significant correlation.

5.2.4.3 Effects of calcium-based magnetic biochar on CO_2 respiration in soil

As an indispensable component of soil, microorganisms are characterized by their large number, wide distribution and complex diversity. Participating in most of the biochemical functions in soil, such as the decomposition of organic matter, humus synthesis and nutrient conversion, they had a significant impact on the material cycle in soil, formation and development, and fertility evolution (Martinez et al., 2018). The basic CO_2 respiration in soil mainly comes from the metabolic activities of soil microorganisms, which is an important indicator of microbial biological activity and soil material cycling rate. Closely related to the environmental quality of soil, it can reflect the stress of soil pollution on the overall soil system to a certain extent.

As can be seen from Figure 5.25, when the soil is free from Cr(VI) pollution, the application of BC, Fe-B and Ca-Fe-B has no significant effects on soil respiration. When the soil Cr(VI) pollution concentration was 100mg·kg^{-1}, the soil CO_2 respiration in Ca-Fe-B treatment group was significantly higher than that in other treatment groups on D7, D14 and D90, and the difference between Ca-Fe-B treatment group and other treatment groups was reduced on D1, D28 and D60. When the soil Cr(VI) pollution concentration was 200mg·kg^{-1} and 1000mg·kg^{-1}, the soil respiration rate was not significantly affected by the application of BC, Fe-B and Ca-Fe-B. During the whole remediation period, despite the severity of soil Cr(VI) pollution, the soil respiration showed a decreasing tendency firstly and then an increasing one with progress of the remediation, and reached the lowest value on D14.

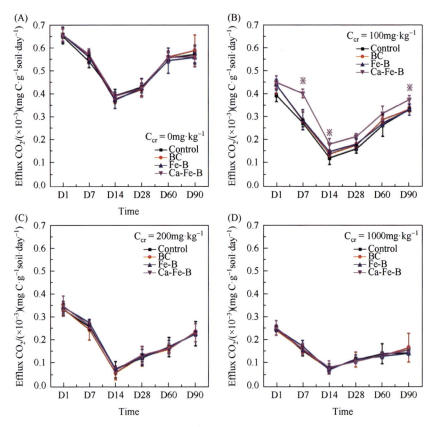

Figure 5.25 (A—D) Effects of BC/Fe-BC/Ca-Fe-B on soil CO_2 respiration when Cr(VI) pollution concentration in soil is $0mg\cdot kg^{-1}$, $100mg\cdot kg^{-1}$, $200mg\cdot kg^{-1}$ and $1000mg\cdot kg^{-1}$. ※ represents the significant differences between different treatment groups at the same time ($P<0.05$). The error line is the mean standard error of the three parallel samples ($\pm SE$).

When the soil was pollution-free, the three materials had no significant effect on the CO_2 release rate, indicating that biochar and its modified materials would not spur soil to produce higher CO_2 emissions. During the remediation, the CO_2 respiration in Cr(VI)-polluted soil showed a trend of decreasing before increasing, which could be attributed to the changes of the overall ecological environment caused by the passivation and stabilization in soil. The sensitivity and stress mechanism of soil microorganisms led to a temporary decrease in their activities. However, due to the toxic effect of Cr(VI) on the soil environment, the slow recovery of microbial activity is still limited with little chance to recover to the original level. While the soil

CO_2 respiration decreased, the soil MBC in this study showed an increasing trend, which also appeared in other studies (Zimmerman et al., 2011). This can be attributed to the increased carbon utilization efficiency, enabling the coexistence of microorganisms and carbon on the biochar surface, and reducing the enzyme production. The basic CO_2 respiration in the soil polluted by high concentration of Cr(VI) is lower than that in the soil polluted by low concentration of Cr(VI). The carbon dioxide emissions from the mineralization of soil organic matters vary to the Cr(VI) pollution degrees in soil, and there is a "dose-toxicity" relationship between the two, which decreases with the increase of Cr(VI) pollution concentration in soil. When the soil is polluted by Cr(VI), Cr(VI) combines with soil organic matters and reduces the available organic matters in soil. Meanwhile, the toxicity of heavy metal itself affects the quantity and activity of microorganisms, resulting in the instability of the soil structure. Ca-Fe-B had a certain remediation effect on the soil contaminated by Cr(VI) at a concentration of $100mg \cdot kg^{-1}$, and the soil respiration effect of this treatment group was significantly higher than that of the other three treatment groups. BC and Fe-B treatment groups showed no obvious effects on the soil respiration, possibly because the two materials have limited remediation effects on Cr(VI). This is also evidenced by the changes in basic CO_2 respiration in soil contaminated by high concentrations of Cr(VI) ($200mg \cdot kg^{-1}$ and $1000mg \cdot kg^{-1}$). There is a limit to the remediation capacity of the remediation material. When the concentration of Cr in soil environment reaches a certain threshold, the toxic effect of heavy metals on soil microorganisms is great. As shown in the correlation analysis in Table 5.6, there was a significant positive correlation between basic CO_2 respiration in soil and soil MBC ($P<0.05$).

5.2.4.4 *Effects of calcium-based magnetic biochar on microbial community structure in soil*

Bacteria, fungi and actinomycetes in soil can determine the distribution of microbial total and the decomposition and transformation of organic matters in soil.

The pollution, passivation and stabilization have certain effects on the microbial community structure and abundance of the soil. In the ecological study of microbial community structure, 16s rRNA amplica sequencing was used to detect the microbial population, quantity and abundance of the soil. The Chao1 and Shannon indexes were used to calculate the abundance and diversity of species. Chao1 was used to estimate the total number of species.

Shannon can characterize not only the biodiversity, but also the number and uniformity of species. Alpha diversity refers to the richness of sample species, and the sample Alpha diversity index reflects the evenness, diversity and abundance of microbial community.

As shown in Figure 5.26, in an environment where Cr(VI) pollution concentration was $200mg \cdot kg^{-1}$, the α-diversity of soil bacteria in the BC treatment group was significantly higher than that in the other treatment groups, indicating that biochar, as an exogenous carbon source, promoted the decomposition of organic matters in the soil and contributed to the survival and propagation of microorganisms in the soil. The addition of Fe-B and Ca-Fe-B deactivated the soil Cr(VI) to a certain extent and alleviated the heavy metal stress of microorganisms in soil.

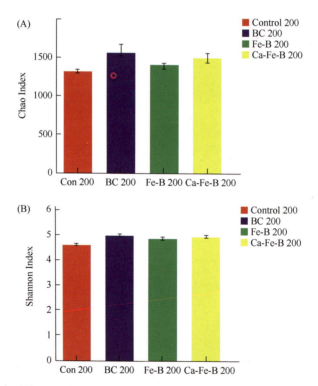

Figure 5.26 (A) Numerical values of chao1 index; (B) numerical values of Shannon index.

As shown in Figure 5.27, the abundance bar chart of phylum-level microbial species, in this study, Actinobacteriota, Firmicutes, Proteobacteria, Chloroflexi, Acidobacteriota and Patescibacteri are the main phylum-level microorganisms with a total relative abundance exceeding 90%. Among them, Actinobacteriota and Proteobacteria are the most abundant. In the Control treatment group, BC treatment group, Fe-B treatment group and Ca-Fe-B treatment group, Actinobacteriota account for 28.47%, 37.34%, 47.21% and 33.58% of the total bacterial sequence, respectively. Proteobacteria accounts for 21.14%, 32.97%, 17.85% and 31.02% of the total bacterial sequence, respectively. Proteobacteria is a type of bacteria with a fast growth rate in a nutrient-rich environment. It is mainly composed of β-*Proteobacteria*. It can participate in the nitrogen cycle in soil and has a great impact on the community structure and relative abundance of ammonia-oxidizing bacteria in soil (Trivedi et al., 2013). The relative abundance of bacteria in the BC treatment group and the Ca-Fe-B treatment group was significantly increased, indicating that BC and Ca-Fe-B promoted the transformation of nitrogen in soil, and further that they promoted the activity of chitinase related to the nitrogen cycle in soil. Studies have also shown that Proteobacteria have better tolerance to heavy metals, making them more easily to survive the Cr(VI) stress than other strains.

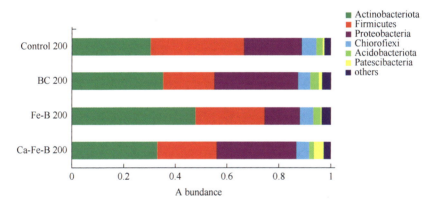

Figure 5.27　Abundance of Phylum-level Microbial Species [Cr(VI) Pollution Concentration in Soil of 200mg·kg^{-1}].

Actinobacteriota is a type of bacteria that can decompose organic matters in soil but grow slowly (Uksa et al., 2015). It usually grows well in soil improved by biochar. The thick peptidoglycan layer on the surface also makes it highly resistant to metal stress. It can be seen from the figure that the application of BC, Fe-B and Ca-Fe-B increases the relative abundance of Actinobacteriota in soil to a certain extent. In addition, studies have shown that the Actinobacteriota can produce indoleacetic acid (IAA) to promote the formation of stable compounds between ferric carriers and heavy metals. This is also confirmed by the significant increase in the relative abundance of this bacteria phylum in the Fe-B treatment group. Proteobacteria, Chloroflexi, Actinobacteriota and Gemmatimonadota have a dominant and symbiotic relationship, which is important for nitrogen cycle and decomposition of organic matters in soil. The genetic material of Firmicutes contains a variety of heavy metal resistance genes, enabling it to adapt to the metal-rich soil environment. Therefore, Firmicutes feature the largest relative abundance in the soil environment polluted only by Cr(VI), as high as 38.41%. As a phylum of Fe(III)-reducing bacteria, it also featured high relative abundances in Fe-B and Ca-Fe-B treatment groups, up to 28.23% and 24.79%, respectively.

On the whole, for their high tolerance, Actinobacteriota, Firmicutes, Proteobacteria and Chloroflexi become the dominant bacteria in Cr(VI) polluted soil.

5.2.5 Effects of passivation and stabilization of calcium-based magnetic biochar on plant growth

5.2.5.1 Effects of calcium-based magnetic biochar on bioavailable Cr

Due to the differences in physicochemical properties and biochemical reactions of the soil, heavy metals exist in soil in different forms and release in different rules. However, the total amount cannot completely determine their geochemical behavior and ecological effects. The content of available heavy metals in soil is better than the total content in measuring the ecotoxicity of heavy metals in soil. The available state of heavy metals in soil is the biologically available state, which usually refers to the element forms that can be actually absorbed and utilized by soil plants, including the acid-soluble state, water-soluble state, adsorbable state, chelate state of heavy metals, and some forms that can be directly absorbed and utilized by plants in a short time (such as the mineral state that is easy to be weathered, and the organic state that is easy to be decomposed). Determining the available heavy metals in soil with DTPA as a selective extraction agent has been a method widely used in the analysis of heavy metal morphology in soil (Liu et al., 2015).

In a soil environment polluted by gradient concentrations of Cr(VI), the application of BC, Fe-B and Ca-Fe-B all significantly reduced the bioavailable Cr(VI) in soil, showing a trend that the content of available Cr gradually decreased with the extension of culture time, which proved that the three materials could effectively stabilize the Cr(VI) pollution in soil (Figure 5.28). When Cr(VI) pollution concentration was 100mg·kg^{-1}, the

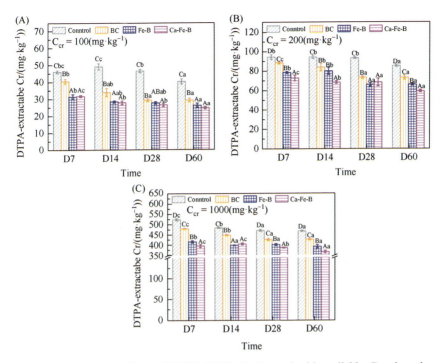

Figure 5.28 (A–C) Effects of BC/Fe-BC/Ca-Fe-B on the bioavailable Cr when the Cr(VI) pollution concentration in soil is 100mg·kg^{-1}, 200mg·kg^{-1} and 1000mg·kg^{-1}. The uppercase letters represent the significant difference between different treatment groups at the same time ($P<0.05$), while the lowercase ones respresent the significant difference between the same treatment group at different time points ($P<0.05$). The error line is the mean standard error of the three parallel samples ($\pm$SE).

content of available Cr in the Ca-Fe-B treatment group was significantly reduced, by the largest degree, during the culture period. Compared with the Control group, the content of available Cr in the treatment group was significantly reduced, by 38.66%, at D60. The content of available Cr in the BC and Fe-B treatment groups was also significantly reduced, by 27.89% and 34.57% respectively, at D60. Among them, Fe-B and Ca-Fe-B showed no significant difference in the passivation and stabilization of contaminated soil, but both were better than BC. When Cr(VI) pollution concentration was 200mg·kg^{-1}, all of the BC, Fe-B and Ca-Fe-B treatment groups significantly reduced the available Cr content in soil during the culture, and the passivation and stabilization effect was Ca-Fe-B > Fe-B > BC. The available Cr content in soil decreased by 14.33%, 22.71% and 30.62% respectively, on D60. When the Cr(VI) pollution concentration was

$1000mg \cdot kg^{-1}$, all the three materials could significantly reduce the available Cr content in soil, among which Ca-Fe-B demonstrated the best passivation and stabilization effect. The content of available Cr decreases by 8.91%, 16.13% and 21.63% respectively, on D60. The passivation and stabilization process reached a relatively stable state around D28. It can be seen that with the increase of Cr(VI) pollution concentration in soil, the passivation and stabilization effect of remediation materials on Cr(VI) pollution in soil showed a downward trend.

The available state of heavy metals is strongly affected by soil pH. The increase of soil pH, along with the deprotonation of acid functional groups containing (phenol) hydroxyl, carboxyl and amino on the surface of biochar, enables more negative charges on the surface of soil particles and remediation materials (Bolan et al., 2003). Therefore, a large number of negatively charged adsorption sites will be formed on the surface of BC, Fe-B and Ca-Fe-B to enhance the electrostatic attraction to Cr(VI) and reduce the content of available Cr. In addition, the $CaCO_3$ loaded on the surface of Ca-Fe-B further increases the pH of the material, enabling more negative charges on the surface of the material to enhance the adsorption of Cr(VI). In addition, soil pH also dictates the existential form of Cr(VI) in soil. With the increase of pH, Cr(VI) existing in the soil in the form of oxygen-containing anion is more easily converted into precipitated Cr_2O_3 and $Cr(OH)_3$, which greatly reduces its effectiveness. The application of BC, Fe-B and Ca-Fe-B has a significant effect on the pH of contaminated soil, which is also testified by their effects under different soil Cr(VI) pollution concentrations as Ca-Fe-B > Fe-B > BC. Meanwhile, it's found that the reduction rate of Cr(VI) in soil is seriously affected by such distribution characteristics as the soil pH, Fe(II), dissolved organic matters and soil particle sizes (Xiao et al., 2012). Ferric oxides on the surface of Fe-B and Ca-Fe-B materials, coupled with the Cr(VI), can form Fe(III)-Cr(III) complexes on the surface of Fe-B and Ca-Fe-B materials, thus facilitating the Cr(VI) adsorption. The Fe(II) on their surface also reduces Cr(VI) to $Cr(OH)_3$, $(Cr_xFe_{1-x})(OH)_3$, or $Cr_xFe_{1-x}OOH$ precipitates, thereby reducing their bioavailability.

It can be seen from Table 5.7 that the bioavailable Cr content in soil is closely related to the growth and enrichment of the corn, showing a very significant negative correlation with the biomass of aboveground and underground parts of the corn, and a very significant positive correlation with Cr enrichment inside the corn body ($P<0.01$).

Table 5.7 Analysis on correlation of all plant indicators based on pearson correlation coefficient.

	AGB	BGB	Cr in shoot	Cr in root
Available Cr	−0.911**	−0.887**	0.869**	0.938**
AGB	/	0.804*	−0.692	−0.661
BGB	/	/	−0.714	−0.773*
Cr in shoot	/	/	/	0.965**

Note: *$P<0.05$, ** $P<0.01$, bold fonts represent the significant correlation.

5.2.5.2 Effects of calcium-based magnetic biochar on plant biomass

Although the application of BC/Fe-B/Ca-Fe-B has a certain remediation effect on the soil contaminated by Cr(VI) of gradient concentrations, the impact of its passivation and stabilization on the soil plants cannot be ignored. The changes of plant biomass were closely related to the changes of available heavy metals in soil, which directly reflected the relief of plant stress in Cr(VI)-contaminated soil by the remediation material.

Figure 5.29 shows the effects of BC/Fe-B/Ca-Fe-B application on biomass of aboveground and underground parts of the corn in the soil contaminated by Cr(VI) in concentrations of 100mg·kg^{-1}, 200mg·kg^{-1} and 1000mg·kg^{-1}. On the whole, the biomass of aboveground and underground parts of corn in BC, Fe-B and Ca-Fe-B treatment groups increased significantly with the extension of time, as compared with the CK, under the low concentrations of Cr(VI) pollution (100mg·kg^{-1} and 200mg·kg^{-1}). However, the growth slowed down with the increase of pollution concentration. The growth in BC treatment group and Ca-Fe-B treatment group was significantly faster than that in Fe-B treatment group of both the aboveground and underground parts of the corn. This proved that the application of BC/Fe-B/Ca-Fe-B had a stable remediation effect on Cr(VI) pollution in soil in some sense, and the decrease of available Cr content in soil alleviated its stress effect on corn to a certain extent. The Ca-Fe-B treatment group showed a more significant Cr(VI)-stabilization effect in low-concentration polluted soil, and a more obvious alleviation of heavy metal stress on plants. The biomass of corn in the BC treatment group also increased significantly, because of the remediation effect of biochar on the one hand, and biochar's promotion of the corn growth as a carbon source on the other. However, Fe-B material itself showed certain biological toxicity, so there's a small increase in corn biomass in Fe-B treatment group. In the soil contaminated by high-concentration ($100\,mg\cdot kg^{-1}$) of Cr(VI), the biomass of aboveground and underground parts of corn in the three treatment groups increased with the extension of time, but there was little difference with the Control treatment group. The results showed that the toxic effect of Cr(VI) pollution on corn was much greater than the stable remediation of biochar remediation materials on heavy metals, resulting in the strong stress of Cr(VI) on corn.

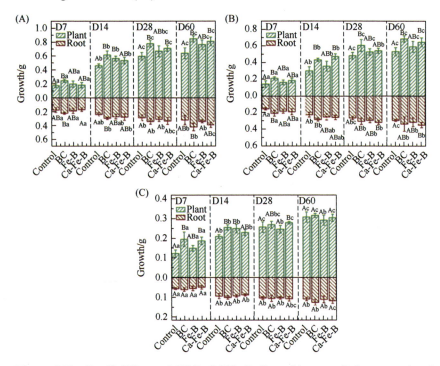

Figure 5.29 (A–C) Effects of BC/Fe-BC/Ca-Fe-B on biomass of aboveground and underground parts of corn when Cr(VI) pollution concentration in soil is $100\,mg\cdot kg^{-1}$, $200\,mg\cdot kg^{-1}$ and $1000\,mg\cdot kg^{-1}$. The uppercase letters represent the significant difference between different treatment groups at the same time ($P<0.05$), while the lowercase ones the significant difference between the same treatment group at different time points ($P<0.05$). The error line is the mean standard error of the three parallel samples ($\pm SE$).

5.2.5.3 *Effect of calcium-based magnetic biochar on Cr content in plants*

As a basic component of the biological chain, plants function as the main media through which soil pollution by heavy metals affects the environmental ecosystem and human beings. Therefore, it is imperative to study in detail the accumulation and enrichment of Cr(VI) in plants during BC/Fe-B/Ca-Fe-B's passivation and stabilization of soil contaminated by Cr(VI) of different concentrations.

Figure 5.30 shows the enrichment of Cr in the corn during the passivation and stabilization of aboveground and underground parts of corn. Regarding the Cr enrichment in the overground part, when the Cr(VI) pollution concentration in soil was $100mg \cdot kg^{-1}$ and $200mg \cdot kg^{-1}$, the addition of BC, Fe-B and Ca-Fe-B significantly inhibited the accumulation of Cr in the stems and leaves of corn plants, upon end of the culture period. Among them, the Ca-Fe-B treatment group had the best inhibition effect. Compared with the CK, Cr content in corn body decreased by 27.26% and 23.16%, respectively. When the Cr(VI) pollution concentration in soil was $100mg \cdot kg^{-1}$, the Fe-B treatment group featured less Cr content than the BC treatment group. When the Cr(VI) pollution concentration in soil was $200mg \cdot kg^{-1}$, there was no significant difference in the inhibitory effect on the Cr(VI) content in the corn between the two groups. When the Cr(VI) pollution concentration in soil was $1000mg \cdot kg^{-1}$, the application of BC, Fe-B and Ca-Fe-B had no obvious inhibition effect on the Cr(VI) content in corn during the entire remediation period, indicating that the pollution concentration had exceeded the threshold, the remediation effect of the remediation material was limited, and a large amount of available Cr accumulated in the corn. The internal Cr enrichment of aboveground parts of the corn during the passivation and stabilization was basically the same with the change of internal Cr content of the aboveground parts of the corn. When the Cr(VI) pollution concentration in soil was $100mg \cdot kg^{-1}$ and $200mg \cdot kg^{-1}$, the addition of BC, Fe-B and Ca-Fe-B significantly inhibited the accumulation of Cr(VI) in corn roots. The inhibitory effect was as follows: Ca-Fe-B treatment group > Fe-B treatment group > BC treatment group. Compared with the CK, Cr content in corn roots of Ca-Fe-B treatment group was reduced by 21.94% and 25.28%, respectively. When the Cr(VI) pollution concentration in soil was $1000mg \cdot kg^{-1}$, none of the three materials showed obvious remediation effect. It proves that Ca-Fe-B had the best remediation effect in soil contaminated by low concentration of Cr(VI), which effectively reduced the content of available Cr in soil to inhibit the Cr enrichment in the aboveground and underground parts of corn. In addition, by comparing the Cr enrichment in the aboveground and underground parts of corn, it was found that for the corn in polluted soil, Cr enrichment was concentrated in the roots of the plant, with only a small amount of Cr migrated to the stems and leaves. There were significant differences in Cr enrichment in the corn roots among different treatment groups, but little difference in the aboveground parts of corn. This echos the research results of some scholars. When Cr(VI) in soil is absorbed by root cells, it combines with the peptides, nucleotides, proteins and other compounds in protoplasm, and then enters the cell vacuoles for precipitation and storage. Thus, heavy metal ions were largely immobilized and inactivated in corn roots (Punz and Sieghardt, 1993).

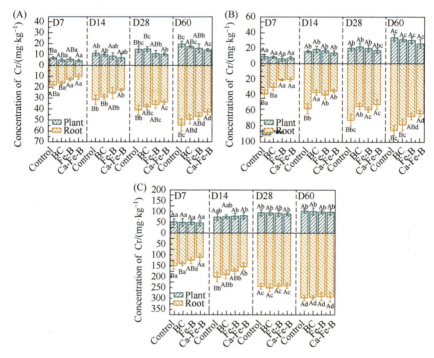

Figure 5.30 (A−C) Effects of BC/Fe-BC/Ca-Fe-B on Cr content inside the corn when Cr(VI) pollution concentration in soil is 100mg·kg^{-1}, 200mg·kg^{-1} and 1000mg·kg^{-1}. The uppercase letters represent the significant difference between different treatment groups at the same time ($P<0.05$), while the lowercase ones the significant difference between the same treatment group at different time points ($P<0.05$). The error line is the mean standard error of the three parallel samples ($\pm$SE).

5.2.6 Summary

In this study, calcium-based magnetic biochar was prepared and applied to the farmland soil contaminated with gradient concentrations of Cr(VI), to explore the impact of its passivation and stabilization mechanisms and restoration process on the bioavailability. The main conclusions are:

(1) In this experiment, rice husks were used as experimental materials to

prepare calcium-based magnetic biochar (Ca-Fe-B) and magnetic biochar (Fe-B) by the co-precipitation method. Magnetite, chiefly Fe_3O_4 and γ-Fe_2O_3 (with some Fe_3C and Fe^0), and $CaCO_3$ in the form of calcite were successfully loaded on the surface of Ca-Fe-B with a large number of functional groups such as —OH, —NH, —O—C=O, Si—O—Si, C≡C and C≡N. They provided favorable conditions for the remediation of soil pollution by heavy metals.

(2) The application of Ca-Fe-B could significantly increase the soil pH and make it generally alkaline. The degrees of increase were Ca-Fe-B> Fe-B>BC, which went down with the increase of Cr(VI) pollution concentration in soil. Ca-Fe-B treatment group featured the highest content of residual Cr under pollution of the same concentration of Cr(VI) on D90, which proved its better Cr passivation effect. Among the three, the Fe-B treatment group had the highest content of reducible Cr, while the BC treatment group the highest content of oxidizable Cr. Ca-Fe-B mainly promoted the conversion of exchangeable Cr to reducible and residual Cr in soil. The passivation and stabilization mechanism: Ca(II) increases the pH of the material itself, increases the surface negative charge, and strengthens the electrostatic attraction of Cr(VI) in the soil. Si-O-Ca-OH bond was formed on the surface, which could adsorb Cr(VI) as the active site. Ca(II) could also enhance the ion replacement of Cr(VI) as a harmless cation. As an electron donor, Fe(II) on the surface directly participated in the reduction of Cr(VI) to Cr(III), generating Cr_2O_3 and $Cr(OH)_3$ precipitates and $(Cr_xFe_{1-x})(OH)_3$ and $Cr_xFe_{1-x}OOH$ complex precipitates in the alkaline soil. A large number of oxygen-containing functional groups, chiefly —OH, —C=C— and O—C=O, adsorbed Cr(VI) in the complex soil.

(3) BC and Ca-Fe-B could significantly boost the activity of β-glucosidase and chitinase in soil, and the soil phosphatase gave no obvious response to the three materials. Enzyme activity showed a response characteristic of "promoting at low concentration and inhibiting at high concentration." Both Fe-B and Ca-Fe-B showed certain biological toxicity, imposing a certain inhibitory effect on soil MBC in the environment polluted by low concentration of Cr(VI). The basic CO_2 respiration in soil showed a trend of first decreasing and then increasing. All indexes of the soil microbial activity were significantly inhibited in soil polluted by high concentrations of Cr(VI). The application of Ca-Fe-B also significantly increased the relative abundance of Proteobacteria in the soil, while Fe-B significantly increased the relative abundance of Actinobacteriota in the soil. Due to their high tolerance, Actinobacteriota, Firmicutes, Proteobacteria and Chloroflexi became the

dominant bacteria in the Cr(VI)-contaminated soil.

(4) The application of BC/Fe-B/Ca-Fe-B significantly reduced the content of bioavailable Cr in soil, and Ca-Fe-B had better passivation and stabilization effects. Ca-Fe-B could significantly inhibit the accumulation of Cr in corn and promote the increase of biomass of aboveground and underground parts of corn in soil polluted by low concentrations of Cr(VI). The Cr enrichment in corn mainly concentrated in the roots.

Limited by the conditions, the research is far from perfect and further improvement will be done in the future. Here, the following suggestions are put forward:

(1) With only the case of single heavy metal contamination of soil considered, relevant studies on heavy metal-heavy metal and heavy metal-organic complex contamination can be carried out in detail to further improve the remediation process and related mechanisms.

(2) It can extend the experimental period of plants, and investigate in detail the heavy metal enrichment in fruits, seeds and different parts of plants in the whole growth stage.

(3) The effects of various proteases on the enzyme activity during the passivation and stabilization of heavy metals in plants can be delved deeper for joint analysis.

5.3 Immobilization and stabilization of lead-polluted soil by green tea biochar supported with nZVI

5.3.1 Introduction

With the progress of modern science and technology, there have been more and more frequent human activities in industry, manufacturing, agriculture, mining and other sectors. Improper disposal of wastes in the industry and manufacturing, excessive use of chemical fertilizers and pesticides in agricultural production, and excessive mining lead to pollutants cast to the soil environment. Although soil has a certain self-purification capacity, pollution will be caused when the pollutants generated by human activities exceed the maximum bearing capacity of soil environment. It's not rare that soil pollution causes problems to human health. In Bangladesh, 35—77 million people suffered health damage from arsenic poisoning (Shakoor et al., 2019). Each year, more than 600000 children worldwide experience mental retardation due to elevated blood lead (Keller et al., 2017). Many young children in countries such as Nigeria and Serbia have died from prolonged exposure to lead-contaminated soil or dust.

With the growing potential threat of soil pollution to human health, soil pollution remediation technology becomes more and more important. In situ remediation technologies, such as heat treatment, solidification and stabilization, soil leaching, phytoremediation and microorganisms, have been successively applied by researchers in the remediation of soil pollution by heavy metals and other related studies in an attempt to maximize the agricultural production and ensure food security. These soil remediation technologies have different working principles and pros and cons. There is a significant difference between the actual cost and the effect from laboratory to practical application (Khalid et al., 2017). Among these technologies, the soil modifier solidification and stabilization technology represented by biochar has gradually attracted the attention of researchers due to the simple accessibility of materials, low cost and environmental friendliness (He et al., 2019). Biochar, as a kind of solid carbon-based material, is the product of high temperature pyrolysis of plant or animal biomass. It can not only effectively reduce the mobility and bioavailability of heavy metals in heavy metal-contaminated soil but also improve soil characteristics and affect the soil redox process. Therefore, the application of biochar technology in the remediation of soil pollution by heavy metals has become the research focus of many scholars. Although many studies have underlined the advantages of biochar in soil applications, there's a lack of long-term studies on the risks of toxic substances (such as, polycyclic aromatic hydrocarbons, polychlorinated biphenyls and dioxins) that may be produced during biochar preparation and the secondary release of heavy metals after fixation. This blurs the potential risks of biochar to soil ecosystems and humans (Schimmel-pfennig and Glaser, 2012).

Engineering nanomaterials have received increasing attention in the field of environmental remediation due to their high reactivity and good mobility in porous media (Crane and Scot, 2012). In the past few decades, nano zero-valent iron (nZVI) has been widely used as an in situ remediation agent in soil and groundwater remediation. Existing studies and applications have proved that nZVI can play a role in the remediation of contaminated sites that contain pollutants such as nitrate, bromate, chlorate, pesticides, arsenic, lead and hexavalent chromium (Yirsaw et al., 2016). However, due to the strong magnetic properties and reactivity of nZVI, they would easily go through agglomeration and passivation, thus affecting the remediation effect. Therefore, researchers have attempted to load nZVI onto the surface of other materials or use polyelectrolytes (e.g., polyacrylic acid), surfactants (e.g., carboxymethyl cellulose and starch), doping of precious metal, vulcanization and other methods to improve the stability of nZVI. However, these methods may inhibit the reaction of zero-valent iron nanoparticles with pollutants, or adversely affect the application of zero-valent iron nanoparticles in environmental remediation (O'Conner et al., 2018; Rajajayavel and Ghoshal, 2015). Some studies have found that the composite made of carrier matrix and nZVI can retain the respective advantages of the two components and even show a synergistic effect for the removal of pollutants. The abundant functional groups, small pore size and good electrical conductivity on the surface of biochar can reduce the particle size of nZVI, delay the degree of corrosion, and improve the dispersion and electron transfer ability (Wang and Wang, 2019). Therefore, more and more studies have been conducted on the biochar-loaded nZVI.

In recent years, it has been reported that biochar-supported nZVI has obvious remediation effects on the organic and heavy-metal pollution in soil. Oleszczuk and Koltowski (2017) used biochar, nZVI, and biochar loaded zero-valent iron nanoparticles to remedy the PAH-contained soils at two pH values (4.3 and 6.6). The results showed that the remediation effect was limited when the nZVI was used alone; better remediation was achieved when the other two materials were adopted. Qiao et al. (2018) used biochar and biochar-loaded nano zero-valent Fe to remove the Cd^{2+} and As(V) from the contaminated rice fields, finding that biochar loaded with nano zero-valent Fe effectively reduced the concentration of pollutants in plants compared with biochar alone.

Although current studies have demonstrated the role of biochar loaded nZVI in the remediation of soil pollution by heavy metals, the potential toxicity of the material is also a major concern with the release of zero-valent iron nanoparticles on the material surface into the soil environment. At present, there are relatively few studies on the toxicity of biochar loaded nZVI, and relevant studies mainly deal with carbon-based materials, nano-metals and their metal oxides. Previous researchers have demonstrated while the toxicity of nZVI is uncertain, an excessive amount of iron, an essential element for the growth and development of almost all living things, is toxic (Ken and Sinha, 2020). Therefore, the toxicity of biochar loaded nZVI may be attributed to nZVI. As biochar loaded zero-valent iron nanoparticles enter the soil environment, the nZVI released by the material can be adsorbed on the cell membrane of microorganisms or enter the cells, to cause cell dysfunction (Auffan et al., 2008). Nanoparticles on the cell membrane can also block the cell ducts, cause structural changes in the cell membrane, and inhibit the nutrient intake, thereby causing death of microbes. Studies have shown that applying polymer coatings or surfactants on the surface of zero-valent iron nanoparticles can reduce their activity to microorganisms (Tang and Lo, 2013). Possible reasons include: ① such materials have a weak effect on the function of membrane of microbial cells; ② they reduce the stress response of microorganisms. The toxicity of zero-valent iron nanoparticles is also related to the varieties of microorganisms. Currently, the studies on the toxicity of zero-valent iron nanoparticles to microorganisms mainly focus on *Escherichia coli*, a variety highly sensitive to zero-valent iron nanoparticles. Studies have shown that the adverse effects of zero-valent iron nanoparticles on the microbial community will stop upon the termination of zero-valent iron oxidation, and the microbial community structure may return to the initial state. nZVI also has a positive effect on some microorganisms. Shin and Cha (2008) found that the presence of nZVI boosted the microbial reaction activity for nitrate reduction. Němeček et al. (2014) found that nZVI had no toxic effects on chillophilic bacteria and even stimulated the growth of gram-positive bacteria.

Today, nZVI is typically prepared through reduction with sodium borohydride and other chemical reagents. Although this method has a high yield and good quality of products, the use of too many chemical reagents will inevitably exert a certain negative impact on the environment. Especially, soil is a complex ecosystem, and any change in environmental factors will have a huge impact on flora and fauna in the soil. At present, most studies on the biochar loaded nZVI were done in the laboratory, shedding lights only on the immobilization efficiency of such materials on heavy metals in the soil, and rarely exploring the impact of biochar loaded nZVI on plants and microorganisms in the soil. As a result, it's difficult to systematically evaluate the risk of its use in the soil, and hinders the large-scale application of the biochar loaded nZVI and similar materials in the heavy metal contaminated sites. Therefore, relevant studies can explore the effect of biochar loaded nZVI on the flora, fauna and microorganisms in the soil after proving that such materials do well in immobilizing heavy metals.

Based on relevant studies on the biochar loaded nZVI remediation technology, this study explored the immobilization effect of the remediation technology on lead in the soil in the simulated lead-polluted soil environment, and the mechanism of solidification and stabilization through the transformation of lead forms. As nZVI is a type of nanoparticles with biological toxicity, this experiment scientifically evaluated the immobilization effect of biochar-loaded nZVI remediation technology on lead pollution and the ecological risk to soil microorganisms by studying the enzyme activity and the changes in microbial community structure and diversity. Eventually, it provided a feasible scheme for future soil remediation.

5.3.2 Preparation and characterization of green tea biochar-loaded nZVI

5.3.2.1 Particle size distribution and influencing factors of zero-valent iron nanoparticles reduced by green tea extract

The selection and preparation conditions for raw materials dictate the number of functional groups, the specific surface area, and the distribution of particle size on the material surface. These factors also determine the strength of material surface and the remediation ability of soil pollution by heavy metals. Therefore, it is very important to study the characteristics of the material itself. We studied the preparation parameters of green tea (GT) biochar-loaded nZVI in order to select the best reaction conditions. According to previous literature surveys, 2 hours was taken as the time for high temperature pyrolysis of biochar in this study (Weber and Quicker, 2018).

Polyphenols in GT, acting as both the reducer and wrapper, not only promotes the formation of nZVIs, but also affects its particle sizes (Khan et al., 2015). As can be seen from Figure 5.31A, when the mass fraction of GT is 2%, 4%, 6%, 8% and 10%, the size of corresponding zero-valent iron nanoparticles (GT-nZVI) is 115.6nm, 252.0nm, 310nm, 401.2nm and 710.0nm, respectively. The particle size goes heavily up with the increase of the mass fraction of GT. When the mass fraction of GT is 10%, a wide particle size distribution is shown in the figure. Only when the mass fraction of GT is 2%, a narrow particle size distribution can be observed. This is because with the increase of polyphenols in the GT extract, iron ions were reduced and aggregated, resulting in the growth of particle sizes and expansion of particle size distribution (Ding et al., 2011). The pH value and pyrolysis temperature also play a deciding role in the stability of zero-valent iron nanoparticles. Figure 5.31B shows that under different pHs, GT-nZVI particle size increases with the growth of GT mass fraction. As can be seen in Figure 5.31C, under the preset pyrolysis temperature in the experiment, the particle size of GT-nZVI prepared by 2% GT extract at 450℃ is the smallest.

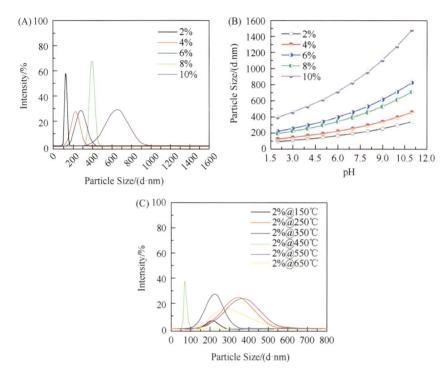

Figure 5.31 (A) Tea quality fraction; (B) pH; (C) influence of pyrolysis temperature on nZVI particle size.

5.3.2.2 Functional groups, pH and pH_{pzc} of green tea biochar-loaded nZVI

In this study, the difference of the number of surface functional groups and pH of GTBC-loaded nZVI was investigated. Figure 5.32A shows the concentrations of three major acidic groups (phenolic hydroxyl, lactone, and carboxyl) in GT-loaded nZVIs (nZVI@GTBC). As can be seen from the figure, the content of acid functional groups prepared at 450℃ (about 4381μmol·g^{-1}) is higher than that of those prepared at 250℃ (2571μmol·g^{-1}) and 350℃ (3335μmol·g^{-1}), and slightly higher than that of those prepared at 550℃ (4288μmol·g^{-1}) and 650℃ (4153μmol·g^{-1}). The phenolic functional groups of green tea biochar (GTBC) are mainly derived from the decomposition of lignin and tannin, accounting for 68% to 90% of the total acid functional groups. Abundant functional groups provide favorable conditions for nZVI@GTBC's immobilization of heavy metals (Stefaniuk et al., 2016).

Figure 5.32B shows that both the pH and pH_{pzc} of nZVI@GTBC prepared at 150℃, 250℃, 350℃ and 450℃ are less than 5.0. When the preparation temperature was 550℃ and 650℃, the two indicators were higher than 5.0, which may be caused by the decomposition of phenolic hydroxyl on the material surface during the high temperature pyrolysis process that has resulted in the reduction of the content, the oxidation of part of nZVI, the collapse of biochar pore structure, and the increase of ash content.

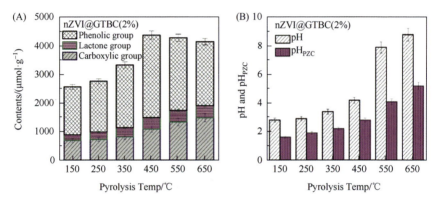

Figure 5.32 (A) Number of functional groups; (B) pH changes of green tea biochar loaded nZVI at different pyrolysis temperatures.

In conclusion, the mass fraction of GT added was determined as 2%, and the pyrolysis temperature of GTBC-loaded nZVI material was determined as 450℃ in this study.

5.3.2.3 *Characterization of physical and chemical properties of the material*
1. XRD and VSM characterization

XRD was used to investigate the crystallization properties of GTBC-biochar loaded nZVI (nZVI@GTBC) in the range of 15°−60°, and the GT, GTBC and (nZVI were compared. As shown in Figure 5.33A, the XRD pattern of GT shows a wide peak at 22°, which is associated with the high cellulose content in GT. After the pyrolysis at high temperature, GTBC showed a broad and slow peak at a large 2θ angle, which may be caused by the decomposition of loose and low-content organic components in the GT. The XRD pattern of nZVI shows an obvious peak at 44.5°, which is consistent with the diffraction pattern of FeO(110) crystal plane, indicating the high crystallinity of α-Fe_2O_3 and γ-Fe_2O_3, and a small peak at 34.9 corresponds to the oxidation of the outer layer of nZVI (Jiao et al., 2006). Zero-valent iron of nZVI@GTBC was also partially oxidized during the preparation, showing the peaks of FeO and Fe_3O_4, which was consistent with JCPDS 76-0956 (marked with # in the figure).

Figure 5.33B shows the hysteresis curves of nZVI and nZVI@GTBC, both of which present a typical S-shaped trend. The maximum saturation magnetic strength of nZVI is 0.32emu·mg^{-1}, showing superparamagnetism of high saturation magnetic strength. After binding with the GTBC, the material still showed superparamagnetism, with a maximum saturation magnetic strength of 0.24emu·mg^{-1}. This may be caused by the decrease of magnetic strength due to the coaction of tea polyphenol coating and biochar pores, but the difference between the two was not obvious. The stable magnetism of nZVI@GTBC helps the material maintain its stability in the remediation of soil pollution.

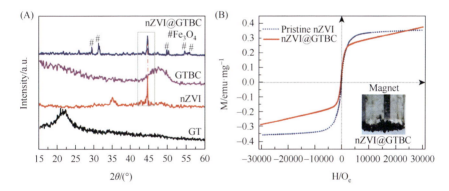

Figure 5.33 (A) XRD patterns of different materials; (B) nZVI and nZVI@GTBC hysteresis curves.

2. Structural characterization of functional groups

To identify the major functional groups in the GTBC-loaded nano-zero-valent iron (nZVI@GTBC), Fourier transform infrared spectroscopy (FTIR) was performed in the near infrared region (500–4000cm^{-1}) (Figure 5.34A). The wide peak at ~3367cm^{-1} in the figure corresponds to the tensile vibration peak of the carboxylic acid group (O—H). Tensile vibration peaks of C—C bond and carbohydrate (C—H) in aromatic rings were observed at ~1589cm^{-1} and ~1258cm^{-1} for nano zero-valent iron nZVI and nZVI@GTBC (Ma et al., 2017). The peaks of nZVI and nZVI@GTBC at ~591cm^{-1} and ~757cm^{-1} originated from Fe$_3$O$_4$ and Fe-O, but the peaks in nZVI@GTBC weakened upon the combination of nZVI and GTBC. This proves the hybridization between GTBC and nZVI (Neeli and Ramsurn, 2018).

In the results of UV-Vis scanning (Figure 5.34B), the GTBC showed no obvious absorption peak in the visible light region. Due to the presence of Fe^0, significant absorption peaks can be seen at 290nm for both nZVI and nZVI@GTBC (Singh et al., 2011). However, nZVI in nZVI@GTBC may form a complex with polyphenols in biochar, extending the absorption spectrum to 511nm.

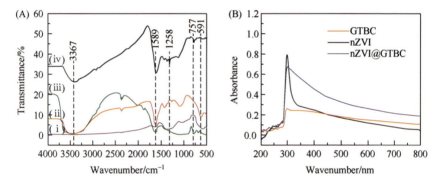

Figure 5.34 (A) Fourier transform infrared spectra of different materials (i: nZVI; ii: green tea; iii: green tea biochar; iv: green tea biochar nZVI); (B) UV-VIS spectra of different materials.

3. TGA-DSC characterization

The thermal stability of GT, nZVI, GTBC and nZVI@GTBC was studied at the temperature range from 25℃ to 800℃ (Figure 5.35). It can be observed that when the temperature rose to 400℃, the total mass loss of GT reached 46.91% (Figure 5.35A), which may be caused by the loss of water and volatile organic matters in GT with the increase of temperature. When the temperature exceeded 400℃, the mass loss of GT mainly came from the carbon oxidation. Due to the volatilization of surface water molecules, nZVI initially lost 7.72% of its mass (Figure 5.35B). A 2.32% increase in nZVI mass was observed after 400℃, possibly due to the oxidation of nZVI at high temperature. Mass losses of 1.74% and 4.35% can be observed for GTBC (Figure 5.35C) due to water evaporation at the initial stage of heating (100−200℃). The mass loss gradually increased after 200℃, and the trend of mass loss was more obvious at 500℃. A possible reason is that the high temperature caused the decomposition of cellulose in biochar, which produced a large amount of carbon and ash (Varhegyi et al., 2011). The final mass loss of GTBC was 44.78%, indicating that GTBC was unstable at high temperature. For nZVI@GTBC (Figure 5.35D), the mass loss came to about 2.53% at 250℃, and further to 23.12% at 450℃. However, compared with GTBC, nZVI@GTBC featured a smaller mass loss at high temperature, indicating a more stable material structure.

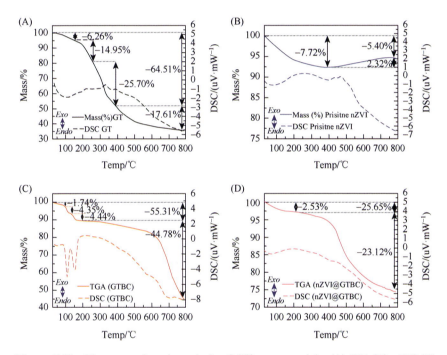

Figure 5.35 Thermogravimetric analysis of different materials: (A) GT; (B) nZVI;(C) GTBC; (D) nZVI@GTBC.

4. XPS characterization

The XPS characterization of nZVI and GTBC-loaded zero-valent iron nanoparticle (nZVI@GTBC) is shown in Figure 5.36.

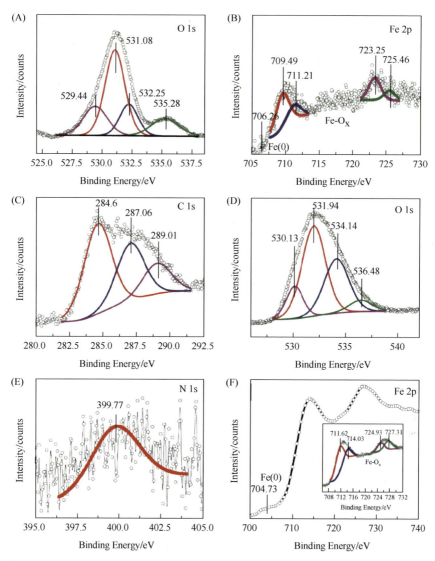

Figure 5.36 XPS characterization of nZVI: (A) O 1s spectrum; (B) Fe 2p spectrum; XPS characterization of nZVI@GTBC: (C) C 1s spectrum; (D) O 1s spectrum; (E) N 1s spectrum; (F) Fe 2p spectrum.

The O 1s spectrum of nZVI shows four characteristic peaks at ~529.44eV, ~531.08eV, ~532.25eV and ~535.28eV (Figure 5.36A), which may be attributed to the combination of iron in nZVI with oxygen to form Fe-O and Fe hydroxides (Shan et al., 2021). The peaks of nZVI's Fe 2p spectrum (Figure 5.36B) at ~709.49eV and ~723.25eV indicate that the Fe^{3+} spin-orbit splitting produces low energy band Fe $2p_{3/2}$ and high energy band Fe $2p_{1/2}$. The peaks of ~711.21eV and ~725.46eV indicate the formation of iron oxide (Fe-O_x) on the surface of nZVI. In addition, a small peak of nZVI at ~706.26eV indicates the presence of Fe^0 in the material. To sum up, there exists irons in nZVI, but the material surface is still partially oxidized and covered by an oxidized outer layer.

The binding energy at ~284.6eV, ~287.06eV and ~289.01eV of nZVI@ GTBC C 1s spectrum (Figure 5.36C) represents carbonyl (O—C=O), carbon-carbon (C-C) and carboxyl (-COOH), respectively (Ma et al., 2017). The binding energy of nZVI@GTBC O 1s spectrum (Figure 5.36D) at ~530.13eV, ~531.94eV, ~534.14eV and ~536.48eV is similar to that of nZVI. However, the increased binding energy of iron and oxygen on nZVI@GTBC surface indicates that the oxidation degree of the surface is lower than that of nZVI(Zhu et al., 2020). The peak of N 1s spectrum (Figure 5.36E) at ~399.77eV reflects the presence of amino group in nZVI@GTBC. The four peaks separated from the Fe 2p spectrum also indicate that the nZVI@GTBC surface is oxidized (Fe-O_x), but the intensity of Fe^0 (~704.73eV) is lower than that of nZVI.

5. Surface topography analysis

SEM and TEM were used to investigate the surface morphology and structure of the materials (Figure 5.37). According to the SEM image, under the joint action of nano-size effect and material magnetism, nZVI presents an aggregated spherical structure (Figure 5.37A). The high temperature pyrolysis causes the carbonization of GT, which releases the volatile components in GT, so that distinct micropore and vesicle structures can be observed on the GTBC surface (Figure 5.37B). In nZVI@GTBC, spherical nZVI particles are found to have formed on the biochar surface, and well adhered to and dispersed in the material surface and pores (Figure 5.37C). Further observation by TEM shows that nZVI is spherical and densely distributed, with an average particle size between 50nm and 250nm (Figure 5.37D). A disordered, dense and overlapping porous carbon layer with an average pore diameter above 500nm can be clearly seen on the surface of GTBC (Figure 5.37E). For nZVI@GTBC, spherical zero-valent iron particles (average size 50—200nm) can be seen loaded on the porous carbon layer (Figure 5.37F), which also proves that zero-valent iron nanoparticles are successfully loaded on the surface of GTBC.

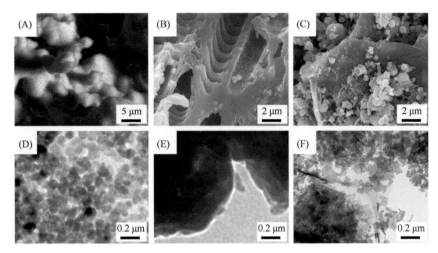

Figure 5.37 Material SEM diagram: (A) nZVI; (B) GTBC; (C) nZVI@GTBC; TEM image of materials: (D) nZVI; (E) GTBC; (F) nZVI@GTBC.

5.3.3 Study on lead-contaminated soil solidified and stabilized by green tea biochar-loaded nZVI

5.3.3.1 Stability of green tea biochar loaded nZVI material

In this study, the stability of nZVI and nZVI@GTBC was compared under different pHs (pH=4±0.5, pH=6.8±0.5 and pH=8.0±0.5) (Figure 5.38). It can be seen that the absorbance of both the two materials decreases at 508nm under different pHs as time goes by. At pH=4.0±0.5, pH=6.8±0.5 and pH=8.0±0.5, the absorbance of nZVI is reduced by 5.41%, 4.34% and 6.95%, respectively, within 20 minutes; by 56.67%, 17.39% and 92.26%, respectively, at 90 minutes (Figure 5.38A). For nZVI@GTBC, under the same pH, the absorbance decreases by 9.97%, 2.29% and 15.63%, respectively, at 20 minutes; by 42.34%, 8.76% and 55.59%, respectively, when the reaction time reaches 90 minutes (Figure 5.38B). Although the absorbance of both nZVI and nZVI@GTBC decreases at 508nm as time goes by, the relatively strong magnetism and agglomeration of nZVI result in a faster settlement rate and a greater decline in absorbance. When nZVI was loaded onto the surface of GTBC, the magnetism and aggregation of nZVI were significantly reduced, and the stability of nZVI was gradually improved, especially at pH=6.8±0.5. The deposition rate of nZVI@GTBC increases at a high pH (pH=8.0±0.5), possibly because under the alkaline conditions, the oxygen-containing functional groups and hydroxyl groups on the surface of biocha oxidize the nZVI in the pores of biochar to iron

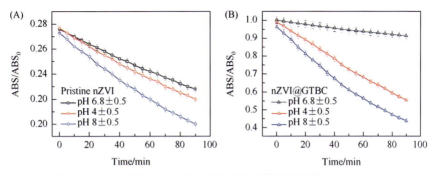

Figure 5.38 Material stability test: (A) nZVI; (B) nZVI@GTBC.

oxides, leading to an increase in particle density and eventually accelerating the material settling (2016a-Su). In a word, nZVI@GTBC has good stability.

5.3.3.2 Immobilization effect of green tea biochar-loaded nZVI on lead-contaminated soil

The immobilizing and stabilizing ability of nZVI@GTBC on lead pollution under different pHs (pH=4.0±0.5, pH=6.8±0.5, and pH=8.0±0.5) was investigated, and the two materials of nZVI and GTBC were compared during a 30-day experiment. In Figure 5.39, E1 represents the lead-contaminated soil with no materials added, while E2 the lead-contaminated soil with nZVI added, E3 the lead-contaminated soil with GTBC added, and E4 the lead-contaminated soil with nZVI@GTBC added. On the first day of the experiment, the immobilization efficiency of E2 and E4 for lead in the soil (pH=6.8±0.5) was 55.41% and 73.3%, respectively. On the 15th and 30th days, the immobilization efficiency of E2 and E4 for lead was 61.66% and 87.91%, and 79.58% and 98.33%, respectively. The immobilization efficiency increased with passage of the experiment time. For nZVI@GTBC, the increasing trend was also observed in the soil (pH=4.0±0.5), but the immobilization efficiency of Pb was lower than that of the treatment group (pH=6.8±0.5). Although nZVI@GTBC also showed a good immobilization efficiency in soil (pH=8.0±0.5) on day 30, it was greatly affected during the experiment. A possible reason may be that the high soil pH increased the surface charge of the material, reduced the mobility and activity between biochar and nZVI, increased the electrostatic repulsion between them, and eventually caused nZVI to be oxidized continuously. Therefore, under the conditions of this experiment, the pH of the soil itself (pH=8.0±0.5) is a more suitable condition for the use of GTBC-loaded nZVI materials. For the detection of bioavailability of lead contamination in the experiment through the physiology-based extraction experiment (PBET) (Mandal et al., 2020a, 2020b), as shown in Figure 5.39D—Figure 5.39F, on day 1, the proportion of bioavailable lead in E1, E2, E3 and E4 treatment groups was 99.37%, 44.58%, 77.29% and 26.66%; on day 30, the proportion of bioavailable lead in the four treatment groups became 94.16%, 20.41%, 57.08% and 1.66%. Table 5.8 also lists the lead immobilization efficiency of such materials, all lower than that of nZVI@GTBC prepared in this experiment.

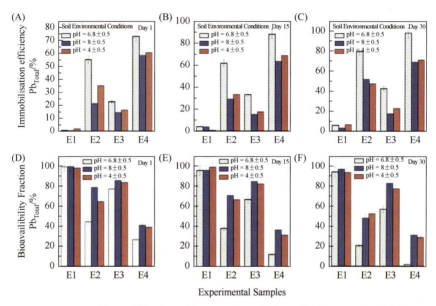

Figure 5.39 Lead immobilization effect (A—C) and bioavailability (D—F) of different materials under different pHs E1: blank control; E2: with nZVI added; E3: with GTBC added; E4: with nZVI@GTBC added.

Table 5.8 Lead Immobilization Efficiency of Different Materials.

Materials	Pollutants	Dosage	Immobilization efficiency	Reference
Magnetic wheatgrass biochar		7.5%	42.2%	Gong et al., 2020
Corn stalk biochar	Pb	60mg/kg	64.35%	Xu et al., 2020
Walnut leaf biochar		2%	34.9%	Kabiri et al., 2019
Rice husk biochar		2%	76%	Derakhshan-Nejad and Jung, 2019

In summary, nZVI@GTBC can effectively immobilize the heavy metal lead and reduce its bioavailability.

5.3.3.3 *Effect of green tea biochar loaded nZVI on form transformation of lead*

The continuous extraction method (SEP) was used to analyze the morphological changes of lead in soil of different treatment groups at different sampling time. Figure 5.40 shows the proportions of different forms of lead in each treatment group under different pHs (pH=4.0±0.5, pH=6.8±0.5, and pH=8.0±0.5) on the 1st, 15th and 30th days of the experiment. F1 represents the exchangeable state (EX), while F2 the carbonate bound state (CB), F3 the iron-manganese oxidation state (OX), F4 the organic state (OM) and F5 the residual state (RS).

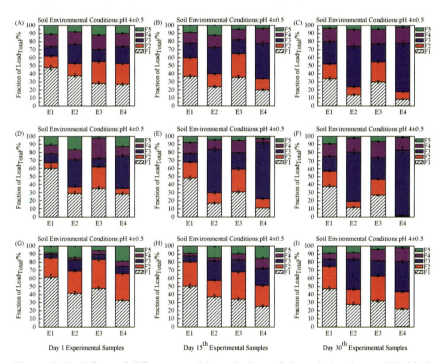

Figure 5.40 Effects of different materials on lead morphology during the test. E1: blank control; E2: with nZVI; E3: with GTBC; E4: with nZVI@GTBC. F1: exchangeable state; F2: carbonate binding state; F3: iron manganese oxidation state; F4: organic state; F5: residual state.

On the first day of the experiment, in the soil environment (pH=6.8±0.5), the proportion of F1 in E2 (lead-contaminated soil with nZVI added) and E4 (lead-contaminated soil with nZVI@GTBC added) was significantly smaller than that in E1 (lead-contaminated soil with no materials added), and most of them was converted into F3 with the proportion increased by 22.30% and 29.85%, respectively. The proportion of F4 in E3 (lead-contaminated soil with GTBC added) was higher than that in other treatment groups. The conversion trend of E2 and E4 may be caused by the strong reaction between nZVI and lead in the material that results in the formation of $Pb\text{-}Fe_2O_4$, $Pb\text{-}Fe_2O_3$ and other substances. In E3, GTBC reduces the content of exchangeable lead by forming organic-binding compounds with lead. During the whole experiment period, the proportion of lead in the exchangeable state gradually decreased, and on day 15, the F1 and F2 components in E2 and E4 were largely converted to F3. On day 30, almost all the F1 and F2 components in E4 were converted into F3 (> 80%).

The experimental results show that the material can effectively lower the availability of lead. But soil pH can affect this result. As can be seen in Figure 5.40, when soil pH=4.0 ± 0.5, the effect of nZVI is slightly weakened; when soil pH=8.0 ± 0.5, the immobilization effect of E2 and E4 is greatly weakened, and the proportion of available F1 and F2 increases significantly. The possible reason is that electrostatic repulsion is incurred between nZVI and lead under alkaline conditions, while other available metals in the soil participate in the reaction to form a competitive relationship with lead, thus reducing the conversion efficiency of available lead (Mandal et al., 2020a, 2020b). In addition, the high pH reduces the mobility of nZVI and oxidizes its surface, thus reducing its reactivity. Alkaline conditions will also lead to slowed microbial growth, which may also affect the interaction between materials and lead.

5.3.3.4 Study on the mechanism of lead immobilization and stabilization in green tea biochar-loaded nZVI

In this study, the properties of soil after immobilization experiment were also explored, with the changes of soil chemical components before and after immobilization mainly considered. Figure 5.41 The following figure shows the presence of carbon (C), oxygen (O), iron (Fe) and lead (Pb) in soil with nZVI@GTBC added, respectively. The peaks of the C-1s spectrum at ~284.6eV, ~286.2eV and 287.8eV correspond to carbonyl ($O\text{—}C\text{=}O$), carbocarbon ($C\text{—}C$) and carboxyl ($\text{—}COOH$) (Figure 5.41A), respectively. The peak at ~289.1eV may be the $PbCO_3$ formed by carbonate and lead in soil. The peaks in the O 1s spectrum (Figure 5.41B) at ~529.6eV, ~531.9eV and~534.3eV represent the functional groups of —C—O and —OH and O—C=O, respectively (Shan et al., 2021). The peaks in the O 1s spectrum are very wide but display less structures. We speculate that the lead with low binding energy in the spectrum of Pb 4f (Figure 5.41D) may be oxidized into PbO. As the experiment progressed, more lead was converted into oxides and hydroxides, resulting in lower peaks at ~529.6eV and ~534.3eV. The Fe 2p spectrum (Figure 5.41C) shows the change in the Fe state and its interaction with lead in the soil. The peak at ~708.5eV corresponds to nZVI, and that at ~712.4eV (Fe $2p_{1/2}$) to the oxidation state of Fe (III). Similarly, the peaks at ~716.0eV, ~719.08eV (Fe $2p_{3/2}$), ~722.6eV (Fe $2p_{1/2}$) and ~726.2eV (Fe $2p_{1/2}$) correspond to the oxidation state of Fe (II), respectively. The higher peaks of Fe (III) also indicate that Fe (II) is further oxidized to Fe (III) after the reaction of nZVI with lead oxidizes. The peaks at ~712.4eV and ~719.08eV may be the offset of the combination of iron and lead in the soil. The peaks of Pb

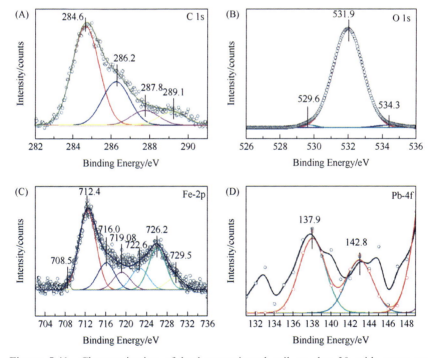

Figure 5.41 Characterization of lead-contaminated soil on day 30 with green tea biochar loaded nZVI added.

4f spectrum (Figure 5.41D) near ~137.9eV and ~142.8eV indicate that lead is attached to the material surface to form substances with PB-O/PB-O-Fe/ PB-O structure. The said analysis of Fe $2p_{3/2}$ is only for reference, because Pb-O-Fe cannot be determined by Fe $2p_{3/2}$ (Figure 5.41C) for its low content, and the overlap of spectra causes difficulties in the fitting analysis process of various iron-containing substances (Biesinger et al., 2011). It can be seen from the above results that the interaction between nZVI@GTBC and lead gradually forms complex compounds in the soil, and complexation occurs over time (Ho et al., 2017).

Since the soil environment is complex, the reaction between the materials and the lead is dictated by many conditions. The GTBC contains a large number of polyphenols and has distinctive aromatic properties. After its interaction with the soil, the surface functional groups are lost in the form of H_2, CO_2, NH_4^+, PO_4^{3-} and H_2O. When the nZVI is combined with the GTBC, the charges of the surface polyphenol groups will be transferred (to zero-valent iron), thus temporarily maintaining their stability and forming a complex (Stefaniuk et al., 2016). Polyphenol groups can also form highly stable metal complexes with metals through chelation, in which metal ions bind in a dicoordinate form. The change in soil pH and charge density on the material surface will affect the surface functional groups of nZVI@GTBC. When the pH is below 3, the carboxyl and phenolic hydroxyl groups would decompose, and the surface compounds would carry no charges. When the pH is above 3, the surface compounds would carry negative charges (Lei et al., 2018; Xiong et al., 2009).

The common reaction mechanisms of biochar and its composites can be summarized as follows: ion exchange, surface complexation, co-precipitation, electrostatic adsorption and redox. Based on SEP, XPS and other characterization and experimental results, nZVI@GTBC's immobilization of lead covers the following steps: ①the pore structure and oxygen-containing functional groups (C—O, C=O, —OH, etc.) on the biochar surface immobilize Pb^{2+} to the material surface through electrostatic adsorption or ion exchange; ②nZVI on the material surface reduces Pb^{2+} to Fe-Pb compounds or lead oxides through redox and fixes on the material surface, and Fe itself converts to Fe (III); ③as time goes by, the lead oxide on the material surface changes to a more stable state under the action of complexation (Figure 5.42).

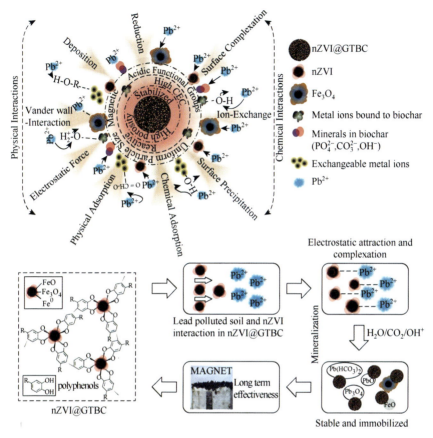

Figure 5.42 Reaction Mechanism Diagram.

5.3.4 Effects of green tea biochar loaded nZVI on plants and soil microorganisms

5.3.4.1 Change of soil pH

As can be seen from Figure 5.43, all treatment groups show a general trend of increase before decrease, while the planting of corn has little impact on the change of pH and the change trend of each treatment group. Among them, the soil pH of the four treatment groups, S1, S2, S5 and S6, increased significantly from the beginning of the experiment and reached the maximum as revealed by the sampling on day 15. The soil pH slowly decreased and tended to be stable after day 15 of the experiment, and was still higher than the initial value on day 60 of the experiment. However, the soil pHs of S2 and S6 (with GTBC added) were slightly higher than those of the CK S1 and S5. The possible reason may be that after the addition of GTBC in soil, negatively charged functional groups, such as phenolic groups, hydroxyl groups and carboxyl groups on the biochar surface, combined with H^+ in soil (Gul et al., 2015); or the biochar gradually released alkali or alkaline earth metals.

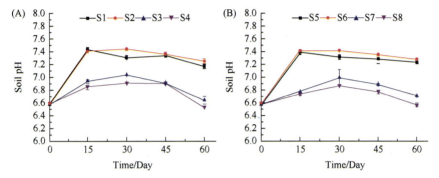

Figure 5.43 Changes of soil pH in different treatments. With the planting of corn: S1: blank control group; S2: GTBC; S3: GT-nZVI; S4: nZVI@GTBC. Without the planting of corn: S5: blank control group; S6: GTBC; S7: GT-nZVI; S8: nZVI@GTBC.

Studies have shown that the addition of zero-valent iron nanoparticles to soil generally increases the soil pH, but the oxidation process of zero-valent iron nanoparticles sometimes reduces the soil pH (Pei et al., 2020) or imposes no obvious effect, depending on the type of zero-valent nano iron, the dosage used and the type of soil (Sun et al., 2020). In this experiment, it can be observed that the soil pH of the four treatment groups, S3, S4, S7 and S8, all increased slowly from the beginning of the experiment, and the maximum was measured in the sampling on day 30 of the experiment. This is because the surface of the zero-valent irons nanoparticles, even though prepared under high hypoxia conditions, would always be oxidized to a certain extent, thus forming nanoparticles with the Fe^0 core and oxide layer. Once they enter the soil environment, the surface oxide layer is rapidly thickened due to the oxidation of zero-valent iron (Crane and Scott, 2012). In this process, the nanoparticles will release ferrous (Fe^{2+}) and ferric (Fe^{3+}) ions on the surface and gradually oxidize them to form Fe (II) and Fe (III) oxides, and OH^- to raise the soil pH. The process lasts until the Fe^0 is completely oxidized. It was also observed that during the whole experiment period, the pHs of S3 and S7 (with GT-nZVI added) were slightly higher than those of S7 and S8 (with nZVI@GTBC added), which might be caused by the fact that biochar delayed the release of nZVI in nZVI@GTBC to finally affect the oxidation process (Fajardo et al., 2020).

5.3.4.2 Length of shoot and root and their iron content

To explore the toxicity of materials to plants, the lengths of corn roots and stems were measured on day 15, 30, 45 and 60 in this experiment, with the results shown in Figure 5.44A In the CK S1, the plant stem length increased with the extension of experiment time. Compared with S1, S2 promoted the growth of corn to a certain extent due to the addition of GTBC. From day 30 of the experiment, the stem length increased by 13.4%, 19.8% and 13.1%, respectively, compared with S1. The addition of GT-nZVI into the soil

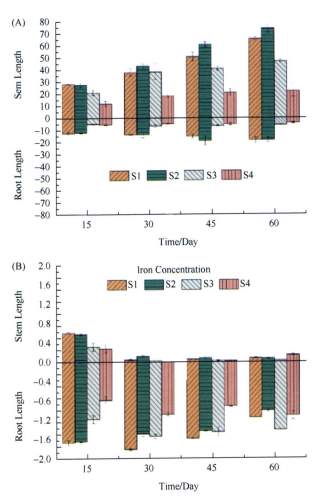

Figure 5.44 Length of shoot and root and their iron content. With the planting of corn: S1: blank CK; S2: GTBC; S3: GT-nZVI; S4: nZVI@GTBC.

inhibited the growth of corn in S3. The plant height on day 15 was 28.4% lower than that in the CK S1, and the length of corn stem did not increase significantly from day 30. Compared with S3, S4 had a more obvious inhibitory effect on the stem growth due to the addition of nZVI@GTBC. On day 15, the stem length of corn in S4 was shortened by 65.9% compared with S1, and the corn stem length of S4 showed no obvious growth after day 30, which only increased by 5.7% as compared with that on day 60. For the root length of each treatment group, S2 featured no obvious difference with the CK S1 during the whole experimental cycle, indicating that GTBC had no obvious promoting effect on the growth of corn root. Both S3 and S4 had significant inhibitory effects on the growth of corn roots, as compared with the root length of CK S1. On day 15, 30, 45 and 60, the root length of S3 and S4 was only 39.1% and 43.4%, 48.4% and 34.1%, and 32.6% and 23.3% of that of S1, respectively. In summary, both the GT-nZVI and nZVI@GTBC used in this experiment had inhibitory effects on the growth of corn rhizomes, while that of nZVI@GTBC was more obvious.

To further explore the reasons for the inhibition of corn growth, the corn rhizomes of each treatment group were dissolved, respectively, with the iron concentration of each treatment group shown in Figure 5.44B. It can be seen that in all treatment groups, the total iron concentration in corn roots is significantly higher than that in corn stems, indicating that corn roots are the preferred organs for storing iron (Zand et al., 2020), and the iron in the soil itself is sufficient to maintain the normal life activities of plants. In the figure, it can be observed that the iron concentration of roots in S3 and S4 is smaller than that in S1 and S2 30 days before the experiment. This is perhaps because the soil of both the two treatment groups contains nZVIs, and under the reduction condition, ferrous ions (Fe^{2+}), the by-product of zero-valent iron oxidation, may be further oxidized into trivalent iron ions (Fe^{3+}) with low solubility by the oxidants released from plant roots; meanwhile, insoluble iron compounds formulated on surface of the plant roots may block the surface structure of cell membranes or deposition in cell walls and cells, and obviously decrease the efficiency of water and nutrient uptake by plant roots (Ma et al., 2013). This phenomenon also affects the absorption of irons by corn roots, resulting in the reduction of iron concentration in corn roots, and in turn leading to the reduction of iron content in corn stems. Iron is an essential nutrient for plant growth and development because it is involved in many important enzyme-catalyzed reactions such as nitrogen fixation, DNA synthesis (ribonucleotide reductase), and hormone synthesis (lipid oxidase and ACC oxidase). After 30 days, the iron concentration in corn roots in S2 gradually decreased, while that in S3 and S4 increased bit by bit, and the difference between the four treatment groups decreased stepwise. This is perhaps because the GTBC in S2 adsorbed part of the iron in the soil, and reduced the amount absorbed by the plants. However, for S3 and S4, zero-valent irons gradually form an iron oxide film on the surface of corn roots, resulting in the increase of iron content in the root (Wu et al., 2020). Therefore, we speculate that nZVI@GTBC damaged the structure of corn root by releasing nZVIs to inhibit the absorption, transportation and utilization of nutrients required for normal life activities of the corn, thus affecting its growth.

Although adding nZVI@GTBC of certain amount required by the experimental design to the soil inhibited the growth of corn, studies have shown that the influence of nanomaterials on plant growth depends on not only the type of plant planted, but also the type and concentration of soil nanomaterials used (Yoon et al., 2019). Therefore, it is necessary to further explore whether the material can have a positive effect on plants at a lower dosage.

5.3.4.3 *Microbial community in soils*

To further explore the influence of the materials on the structure of microbial community, 16r-s-RNA high-throughput sequencing was conducted on the last batch of soil samples on day 60, and the indexes and data representing the abundance and diversity of microbial community, including Ace, Chao 1, Shannon and Simpson, were obtained to study the differences of microbial community structure and diversity in different treatment groups. The meanings of different indexes are as follows:

Chao1: It is commonly used in ecology to estimate the total number of species, and was firstly proposed by Chao.

Ace: It is often used in ecology to estimate the total number of species, and was firstly proposed by Chao using different algorithms.

Shannon: It is typically used to indicate the community diversity. The larger the Shannon index, the higher the community diversity.

Simpson: It is typically used to indicate the community diversity. The larger the Simpson index, the lower the community diversity.

All the indexes are listed in Table 5.9. The results show that the planting of corn hardly affected the microbial abundance and diversity (S1 and S5, S2 and S6, S3 and S7, and S4 and S8) in the treatment groups with the addition of same materials. Overall, compared with the CKs (S1 and S5), those with GTBC added (S2 and S6) showed a slight increase in soil microbial abundance and almost the same microbial diversity. The microbial abundance and diversity of the treatment groups with GT-nZVI (S3 and S7) and nZVI@GTBC (S4 and S8) added were lower than those of the CK, those of the former slightly higher than those of the latter (Figure 5.45).

Table 5.9 Abundance and diversity index of soil microbial community.

	Chao1	Ace	Shannon	Simpson
S1	2448.04	2439.09	6.28	0.05×10^{-1}
S2	2666.23	2656.41	6.28	0.05×10^{-1}
S3	2175.30	2211.53	5.33	0.18×10^{-1}
S4	2286.48	2295.56	5.72	0.11×10^{-1}
S5	2518.28	2511.59	6.21	0.06×10^{-1}
S6	2555.45	2558.81	6.33	0.05×10^{-1}
S7	2073.94	2124.04	5.18	0.17×10^{-1}
S8	2112.00	2070.72	5.18	0.24×10^{-1}

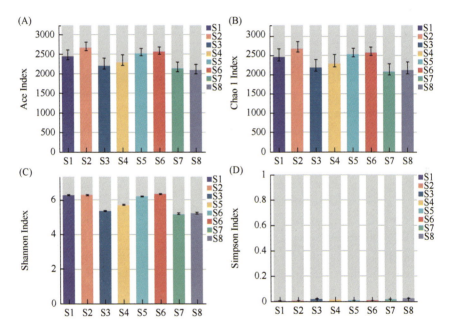

Figure 5.45 Abundance and diversity index of soil microbial community. With the planting of corn: S1: blank CK; S2: GTBC; S3: GT-nZVI; S4: nZVI@GTBC. Without the planting of corn: S5: blank CK; S6: GTBC; S7: GT-nZVI; S8: nZVI@GTBC.

Although the surface containing nZVI was coated with surfactants (tea polyphenols), which could provide nutrients for microorganisms while enhancing the stability and dispersion of nZVI, the nZVI in the material was released into the soil to incur a series of physical, chemical and biological actions to destroy the cell membrane integrity, degrade proteins on the cell membrane, interfere with the respiration and destroy the cell DNA (Auffan et al., 2008), because its physical and chemical properties can simply adhere to the microbial surface through Brownian motion, electrostatic action, hydrogen bonding and other actions. The said reactions can trigger rupture of the cell membrane or interfere with the bioenergy exchange and information transmission in the cell membrane, thus causing microbial death (Lyon et al., 2008) to reduce the abundance and diversity of soil microorganisms. The said experimental results showed that GTBC had a positive effect on soil microorganisms, while GT-nZVI and nZVI@GTBC had a negative effect. There was little difference between the two materials in the adverse effect on the microbial community (Figure 5.45).

nZVI features a certain selectivity to the microbial community. While inhibiting the growth of some microbial communities, it will promote the growth of those resistant to nZVI (Ken and Sinha, 2020). To further explore the effects of materials on the soil microbial community, the soil microorganisms in all treatment groups were analyzed on day 60 at the phylum level. It can be seen from the figure that there are 9 main phyla in all treatment groups, with Proteobacteria, Actinobacteria, Firmicute and Acidobacteria as the four dominators, accounting for more than 80% of the total OTUs in the community. See Figure 5.46 and Table 5.10 for the specific data.

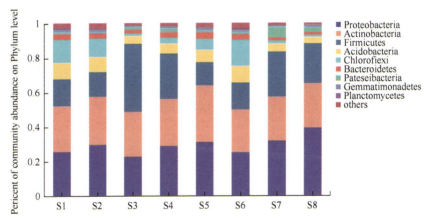

Figure 5.46 Proportion of soil microbial community. With the planting of corn: S1: blank CK; S2: GTBC; S3: GT-nZVI; S4: nZVI@GTBC. Without the planting of corn: S5: blank CK; S6: GTBC; S7: GT-nZVI; S8: nZVI@GTBC.

Table 5.10　Proportion of microorganisms of different phyla in soil (%).

	S1	S2	S3	S4	S5	S6	S7	S8
Proteobacteria	25.5	29.6	22.6	28.8	31.2	25.1	31.7	39.1
Actinobacteria	26.3	27.8	25.8	27.1	32.5	24.8	25.4	25.8
Firmicutes	15.8	14.3	39.5	26.2	13.5	15.5	26	23.1
Acidobacteria	13.2	10.5	1.6	3.4	6.3	15.1	1.8	1.1

Proteobacteria, a relatively dominant bacterial colony, account for more than 20% in all treatment groups. This is because Proteobacteria, being widely found in all kinds of soils, can adapt to different environment with a complex lifecycle, and can grow quickly when granted access to nutrients (Fierer et al., 2012). In the treatment group with planting of corn, the proportion of Proteobacteria in S2 (29.6%) was the largest among the four treatment groups, indicating that GTBC could promote the growth of Proteobacteria. The proportion in S4 (28.8%) was higher than that in S3 (22.6%), and that in the CK S1 (25.5%), indicating that nZVI@GTBC had a certain promotion effect on Proteus. In the treatment group without planting of corn, the proportion of Proteobacteria in S7 and S8 was higher than that in the other two treatment groups, and that in S8 (39.1%) was higher than that in S7 (31.7%), indicating that the presence of plants had a certain influence on the growth of Proteobacteria, and nZVI@GTBC featured a better selectivity for Proteobacteria. Actinobacteria, a slow-growing microorganism, can decompose organic matters in the soil. Regarded as a consumer of carbon-rich and hard-to-degrade materials, it usually grows well in the soil modified by biochar. The thick peptidoglycan layer on the actinomycete surface also makes them highly resistant to metals (Mackie et al., 2015). The proportions of actinomycetes in all treatment groups were above 20%, yet there was little difference among different treatment groups, indicating that the material added had little influence on the actinomycetes. But the proportion of Actinobacteria in S2 (26.3%), S3 (25.8%) and S4 (27.1%) was slightly higher than that in S6 (24.8%), S7 (25.4%) and S8 (25.8%) with the same material added, a difference possibly caused by the symbiotic relationship between actinomycetes and plants (Ventura et al., 2007). The proportion of Firmicutes in soil microorganisms was 39.5%, 26.2%, 26.0% and 23.1% in S3, S4, S7 and S8, respectively, which was much higher than that in other treatment groups. This is perhaps because there are a variety of heavy metal-resistant genes in Firmicutes' genetic material, which enables the bacteria of this phylum to adapt to the metal-rich soil environment. In the experiment, the proportion of Firmicutes in S3 and S7 was higher than that in S4 and S8. This could be a result of the interaction between the functional groups on biochar surface in nZVI@GTBC and nZVI, which delayed the release of nano iron and weakened the activity of Fe (III) reducing bacteria of Firmicutes to trigger a relative decrease in the proportion. Acidobacteria are taken as the main microorganisms in soils with a low mineralization rate, which contribute to the TOC accumulation and DOC reduction in soils. Although the proportion of Acidobacteria in all treatment groups was below 15%, that in S1, S2, S5 and S6 was higher than that in other treatment groups, indicating that GT-nZVI and nZVI@GTBC would affect the accumulation of organic carbon in soils.

5.3.4.4 *Enzyme activity*

In the treatment groups with the planting of corn (S1–S4), at the early stage of the experiment, the release of nZVI from S3 and S4 into the environment was delayed due to the action of tea polyphenols and biochar, and iron promoted the activity of β-glucosidase secreting microorganisms at low concentrations. Therefore, the enzyme activity in S3 and S4 was higher than that in the CK S1. With the increase of nZVI's toxicity on microorganisms in soil environment, it was observed that the enzyme activity in S3 and S4 was weaker than that in S1 on day 30 (Figure 5.47). However, on day 45, β-glucosidase activity in S3 and S4 was slightly increased again, perhaps because β-glucosidase in S3 and S4 was involved

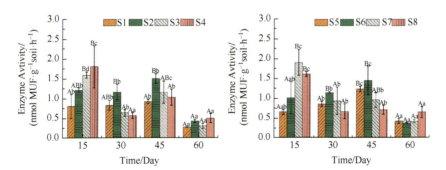

Figure 5.47 Changes of soil β-glucosidase activity in different treatments. With the planting of corn: S1: blank CK; S2: GTBC; S3: GT-nZVI; S4: nZVI@GTBC. Without the planting of corn: S5: blank CK; S6: GTBC; S7: GT-nZVI; S8: nZVI@GTBC.

in the hydrolysis and biodegradation of various β-glucosides in the remains of corn root; this is consistent with the growth of corn root and stem in the two treatment groups after 30 days (Adetunji et al., 2017). As GTBC was added to treatment group S2, cellulose and other substances in the material could provide carbon and more sources for microorganisms, the β-glucosidase activity in treatment group S2 gradually increased to a level higher than that in other treatment groups 45 days before the experiment (Merino et al., 2016). Apart from the materials, no nutrients was added to the soil during the whole plant culture cycle in this experiment, which may be the reason for the low β-glucosidase activity in the soil of all treatment groups on day 60.

The variation tendency of β-glucosidase activity in the treatment groups without the planting of corn (S5—S8) was similar to that in those with the planting of corn. At the initial stage of the experiment, the zero-valent iron nanoparticles in S7 and S8 released less iron ions into the environment, which promoted the activity of β-glucosidase secreting microorganisms at low concentrations. However, with the gradual enhancement of the toxicity of iron nanoparticles, the β-glucosidase activity was inhibited from day 30. Because cellulose and other substances in the GTBC provided carbon sources for microorganisms in S6, the β-glucosidase activity was promoted and gradually increased 45 days before the experiment. Still, no nutrients was added to the soil during the whole plant culture cycle in this experiment, which may be the reason for the low β-glucosidase activity in the soil of all treatment groups on day 60.

β-glucosidase is an extracellular enzyme involved in the carbon cycling process in soil, subject to influence of the organic carbon in soils (Nigam et al., 2019). According to the above enzyme activity results, the addition of GT-nZVI (S3 and S7) and nZVI@GTBC (S4 and S8) into the soil showed an overall inhibitory effect on the β-glucosidase activity, and (S2 and S6) a promoting effect of GTBC on the β-glucosidase activity. This is consistent with the proportion of Acidobacteria in the soil microbial community.

5.3.5 Summary

In this study, GT was used as the biomass material to produce GTBC loaded nZVI material (nZVI@GTBC) by the chemical impregnation method. The basic physiochemical properties of the material, the immobilizing and stabilizing efficiency and action mechanism of lead pollution in soil, as well as the effects on plants, microorganisms and enzyme activity were studied to get the following conclusions:

(1) The synergistic effect between biochar and zero-valent iron nanoparticles also improves the dispersion and stability of zero-valent iron nanoparticles in materials. Although Fe^0 exists in the material and part of it may be oxidized, the rich functional group structure in the material also provides a favorable prerequisite for the remediation of soil pollution by heavy metals.

(2) The study on lead immobilization and stabilization shows that the material demonstrates the best immobilization and stabilization effect in the soil of a pH=6.8±0.5. The lead immobilization efficiency of nZVI@GTBC reached 98.33% during the 30-day experiment. The morphological changes of lead indicate that nZVI@GTBC mainly converts the available lead into the iron and manganese oxidation state.

(3) The soil XPS results show that lead oxides and hydroxides were formed during the immobilization and stabilization process. Based on the SEP and XPS results, the lead immobilization mechanism of nZVI@GTBC is as follows: the pore structure and oxygen-containing functional groups (C—O, C=O, —OH, etc.) on the biochar surface first immobilize Pb^{2+} to the biochar surface through electrostatic adsorption or ion exchange. Then nZVI on the material surface reduces Pb^{2+} to Fe-Pb compounds or lead oxides by redox, and fixes them on the material surface. Fe itself also transforms to Fe (III). As time goes by, the lead oxides on the material surface change into a more stable state under the action of complexation.

(4) In the 60-day experiment with the planting of corn, the nZVI@GTBC inhibited the growth of corn rhizomes, and the lengths of corn rhizomes on day 60 were only 23.3% and 34.1% of that in the CK. The small concentration of iron in nZVI@GTBC roots may be a result of the nZVI in nZVI@GTBC that interferes with the plant roots' absorption of water and nutrients through adsorption, precipitation and redox on the surface of plant root cells, and affects the plant growth. The coating of tea polyphenols delayed the release of zero-valent iron nanoparticles. Therefore, nZVI@GTBC promoted the β-glucosidase activity to a certain extent at the initial stage of the experiment, but gradually turned to inhibit it with the toxicity of iron nanoparticles. The nZVI@GTBC can reduce the microbial abundance in the soil and increase the proportion of heavy metal-resistant microorganisms, such as Firmicute, in the whole microbial community. The decreasing proportion of Acidobacteria in the microbial community indicates that nZVI@GTBC may have a certain negative effect on the accumulation of TOC in soil.

Limited by the experimental conditions, the research is far from perfect and further improvement will be done in the future. Here, the following suggestions are put forward.

(1) A short experimental period is set up for the study on the immobilizing and stabilizing efficiency of lead pollution in soil, which could not fully demonstrate whether the lead immobilized on the material could be released and re-enter the soil environment after 30 days of the experiment. In future studies, the experimental time can be extended appropriately, and the remediation effect of the material on the composite pollution can be studied.

(2) With the dosage set up in this experiment, nZVI@GTBC had an inhibitory effect on the plant growth, and the plant experiment was not considered to be carried out in contaminated soils. In future experiments, treatment groups of different nZVI@GTBC dosages can be arranged to study whether lower material dosage promotes the plant growth. Besides, plant experiments can also be conducted in contaminated soils.

5.4 Effects of biochar slow-release nitrogen fertilizer on microbial community and plant growth in copper-contaminated soil

5.4.1 Introduction

Pesticides, herbicides, bacteriostatic agents and other agricultural chemicals, such as the commonly used Bordeaux mixture, are often sprayed into the soil to increase the crop yields. Copper-based agricultural chemicals can increase crop yields and control plant pests, but play as a major source of excess copper in cultivated soil as well (Husak, 2015). The excessive copper in soils may damage the DNA and cell membrane integrity, and affect the respiration, photosynthesis and changes of enzyme activity to inhibit the plant growth or even threaten the plant survival (Chaffai et al., 2007). Copper in the soil can also accumulate in plants and harm the animal and human health. As a chemical remediation method for copper pollution in soil, passivation/stabilization has been widely used. Currently, some materials have been developed for remediation of copper pollution in soil, but problems such as high cost and secondary pollution still persist. Research on low-cost and environmental friendly green remediation agents has become a hot topic today. Meanwhile, to improve the crop yield, soil fertility is often improved through fertilization (with urea, etc.). However, when urea is applied into the field, 20%–70% of nitrogen is lost to the environment through nitrogen leaching and NH_3 volatilization due to the low utilization efficiency of the element, resulting in a serious waste of fertilizer. Unabsorbed fertilizers may also pollute the environment, leading to soil degradation and eutrophication of rivers and lakes (Tilman et al., 2002). Therefore, it is very important for agricultural production to reduce the nitrogen loss and improve the nitrogen utilization efficiency.

Biochar is the product of pyrolysis of biomass (such as plant straw, animal feces, and sludge) under the hypoxia or oxygen restriction. Its diversified characteristics, such as high specific surface area and abundant pores, make it an effective antidote of the soil pollution by heavy metals and slow release urea. The biochar adsorbs heavy metal ions in the soil through ion exchange, electrostatic attraction, precipitation, redox and complexation, changes the form of heavy metals, and reduces the toxicity of heavy metals, thus realizing the passivation of heavy metals in soil. Meanwhile, biochar can bind nutrients through such mechanisms as electrostatic attraction, pore adsorption and functional group complexation. Some scholars have developed the slow-release urea materials based on biochar and demonstrated their slow-release performance, proving that the strong adsorption of BCs limits the release of nitrogen and reduces the loss of nitrogen in soil (Wen et al., 2017). Adding biochar to the soil can also improve the fertility, cation exchange capacity and microbial activity of the soil. Therefore, by recycling biological wastes in farmland, a low-cost and environmentally friendly biochar slow-release nitrogen fertilizer is developed to not only remedy the soil pollution by heavy metals to a certain extent, but also reduce nitrogen loss and improve nitrogen use efficiency, which is helpful for improving the soil environment and providing a theoretical basis for the development of environmentally friendly slow-release fertilizer. This is of great significance for rational utilization of resources.

Biochar can absorb and slow release nutrients because its surface is rich in active functional groups, such as hydroxyl, carboxyl, and carbonyl, which can chemically react with fertilizers to load certain nutrients. In addition, part of the chemical functional groups on the biochar surface can ionize and generate charges, enabling it to have a high ion exchange ability to absorb nutrients through ion exchange. The abundant pore structure of biochar gives it a large specific surface area and thus a strong adsorption capacity, allowing its surface to absorb and hold certain nutrients. Biochar can also hold nitrogen, phosphorus and potassium nutrients in fertilizer through loading, adsorption and absorption, and reduce the dissociation of nutrient ions from the surface of soil particles, thus decreasing the nutrient runoff/leaching and volatilization losses.

The effects of biochar slow-release fertilizer on crop production were mainly achieved through its impact on the water, fertilizer, air and heat in the soil, water and fertilizer absorption channels of the crops, with the emphasis laid on the control of water and fertilizer, two key factors for crop growth. The key to the regulatory effect of biochar slow release fertilizer is the slow release performance of carbon-based carrier and the regulation characteristics of biochar. A large number of pot and field experiments have confirmed that the application of biochar slow release fertilizer can promote the growth and development of main crops in field, and increase their dry matter accumulation and economic yield. Biochar slow-release fertilizer increased the yield of not only staple crops such as wheat, corn and rice, but also grain and economic crops such as peanuts, potatoes and cotton. The application of biochar slow-release nitrogen fertilizer can better meet the demand of crop growth and reduce the leaching loss of soil nitrogen, than the common fertilizers. In addition, the release time of slow-release fertilizer is obviously postponed, which can provide nutrients to crops for more than 120 days with the validity period extended for at least 80 days. The nitrogen utilization efficiency is 38% higher than that of ordinary fertilizer, which is conducive to the balanced absorption of nutrients by plants (Liang and He, 2002).

At present, the research on biochar is relatively scattered, and that on the performance of biochar is relatively fragmented. For example, there's very limited research on whether biochar slow-release fertilizer can remedy the soil pollution by heavy metals while slowing the release of nutrients. After it is applied into the soil, there are still some uncertainties and unresolved problems about the impact on the soil and plant growth: ①The effects of biochar slow-release nitrogen fertilizer on soil characteristics, such as pH, enzyme activity and microbial community structure, are still unclear. Will the said effects change under the soil pollution by heavy metals? What impact will biochar bring to the soil pollution by heavy metals? These are questions that need to be answered. ②Existing studies on the effects of biochar slow-release fertilizer on nitrogen release and microbial activity in the soil have not gone deeper, with scarce attention paid to the key action mechanisms. For example, in what way biochar binds with the fertilizer, what effect biochar slow-release fertilizer has on the soil nitrogen cycling-related microorganisms, and how it affects the nitrogen utilization efficiency remain unanswered. ③The research on the effects of biochar slow-release fertilizer on the nitrogen utilization efficiency and the remediation of soil pollution by heavy metals mainly focused on its effects on the plant growth by changing the soil properties. Combined with the effects on the plant growth, it could come to a more comprehensive understanding of the influencing mechanism.

Based on the previous findings, this study prepared the biochar slow-release nitrogen fertilizer, using corn straw biochar and urea as the materials and sepiolite and bentonite as the mineral binders, and applied it to the soil contaminated or uncontaminated by copper. Through soil and plant tests, the following contents are intended to be discussed: ①The effects of biochar slow-release nitrogen fertilizer on the dynamic variation characteristics of different nitrogen forms (NH_4^+-N, and NO_3^--N) in the soil and the ability to supply nitrogen nutrients to the plant growth; ②effects of biochar slow-release nitrogen fertilizer on soil microorganisms (such as structure and abundance of soil microbial community, and enzyme activity); ③effects of biochar slow-release nitrogen fertilizer on plant growth (such as plant height, root length, leaf SPAD, biomass, and rhizosphere enzyme activity); ④effects of biochar slow-release nitrogen fertilizer on passivation efficiency and bioavailability of copper in soil.

5.4.2 Preparation and characterization of biochar slow-release nitrogen fertilizer

5.4.2.1 Fourier transform infrared spectroscopy

The BC and biochar slow-release nitrogen fertilizer (BN) were tested with FT-IR in the wavenumber range of 500–4000cm^{-1} (Figure 5.48), which was used to identify the important functional groups in the material.

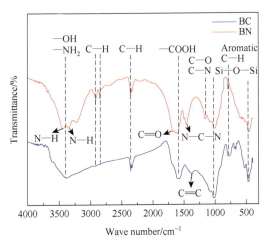

Figure 5.48 Infrared spectra of biochar (BC) and biochar slow-release nitrogen fertilizer (BN).

The absorption peaks of BC and BN are mainly located around $3400cm^{-1}$, $1630cm^{-1}$, $1320cm^{-1}$, $1100cm^{-1}$, $800cm^{-1}$ and $500cm^{-1}$. The broad absorption peak at wavenumber $\sim3400cm^{-1}$ may be the O—H tensile vibration of alcohol or N—H tensile vibration of secondary amine; the band at wavenumber $\sim2382cm^{-1}$ may be caused by the weak stretching of alkyl (C—H); the absorption peak at the wavenumber $\sim1630cm^{-1}$ may be the tensile vibration of C=O, indicating the existence of functional groups such as —COOH; the absorption peak at wavenumber $\sim1320cm^{-1}$ is the bending vibration of O—H; the characteristic absorption peak at wavenumber $\sim1100cm^{-1}$ may be the simultaneous occurrence of C=O tensile vibration and C—N. The absorption peak at the wavenumber $\sim834cm^{-1}$ may be the tensile vibration of —CH (Xu et al., 2013), while the tensile vibration of Si—O—Si may produce the absorption peak at the wavenumber $\sim500cm^{-1}$. Oxygen-containing functional groups, such as hydroxyl and carboxyl groups, allow biochar a higher polarity, which greatly improves its nutrient retention ability, especially the polar ionic compounds such as NO_3^-. The active functional groups on the biochar surface have strong adsorption on heavy metal ions, and certain adsorption on some non-metallic nutrient elements (such as N and P). For example, under acidic conditions, functional groups, such as hydroxyl and carboxyl, combine with H^+ to form positively charged groups, thus gaining the ability to absorb NO_3^-, PO_4^{3-} and other anions. Amine groups and hydroxyl groups can also form hydrogen bonds to bind nitrogen-containing groups, thus making the binding of nitrogenous nutrients and biochar more stable (Tan et al., 2015).

In the infrared spectrum of BN, a new absorption peak appears, and BN has a double peak at $3200-3500cm^{-1}$. According to the interpretation of the peak shape, it can be inferred that there should be two different substances here. Among them, one is the self-associative —OH and the other is nitrogen in the form of adipose amine. The absorption peak at $2800-3000cm^{-1}$ is attributed to the tensile vibration of C—H bond; the absorption peak at the wavenumber $\sim1682cm^{-1}$ is due to the tensile vibration of C=O bond in urea molecule; the absorption peak at the wavenumber $\sim1615cm^{-1}$ may be attributed to the bending vibration of —NH_2; the absorption peak at wavenumber $\sim1465cm^{-1}$ can be attributed to the shear vibration of N—C—N bond; the absorption peak at wavenumber $\sim1154cm^{-1}$ is due to the tensile vibration of C—N bond. The presence of such groups and chemical bonds demonstrated biochar's successful adsorption of urea.

5.4.2.2 Scanning electron microscope

SEM is mainly used to characterize the surface morphology of BC and BN, which can detect the pores and channels in biochar and the incorporation of urea.

In Figure 5.49A and Figure 5.49B refer to SEM images of the surface and cross section of the BC, respectively. Biochar has a smooth and dense porous structure. During the pyrolysis, the organic macromolecules inside the biochar are cracked into small gas molecules including H_2O, CO_2, CO and CH_4. When they are diffused from the inside to the outside, a channel with a large aperture is formed in this process (Huong et al., 2016). This makes it capable of adsorbing heavy metals and microorganisms in the soil, storing and retaining nutrients, and storing water during irrigation or rainfall. In the process of biomass pyrolysis, the specific surface area of biochar increases to promote the adhesion between the surface functional groups of urea and biochar, and increase the possibility of chemical reactions.

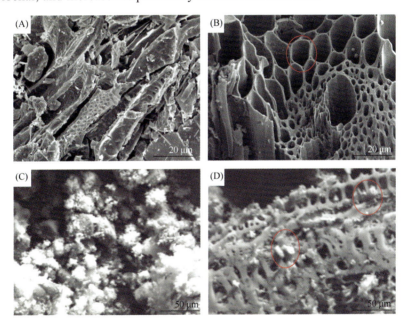

Figure 5.49 SEM Images of Biochar (BC) and Biochar Slow-release Nitrogen Fertilizer (BN).

In Figure 5.49C and Figure 5.49D refer to the SEM images of the surface and cross section of the BN, respectively. It can be seen from Figure 5.49C that BN presents a rough and undulated surface, which is attributed to the grafting polymerization of urea with biochar and bentonite, and the filling of pores and channels of biochar by urea and bentonite. This polymer network structure controls the rate at which water penetrates the porous biochar, and nutrients are released into the soil. Some urea particles filled the pores of biochar, resulting in uneven crystallization on the biochar surface, cracks and collapsed positions (Figure 5.49D). When biochar is in a humid environment, there is a high water potential gradient between the inside and outside of the pore channel that make available the release of nutrients (Liu et al., 2019).

5.4.2.3 X-ray photoelectron spectroscopy

The chemical composition and oxidation state of BC and biochar slow-release nitrogen fertilizer (BN) are characterized with XPS (Figure 5.50).

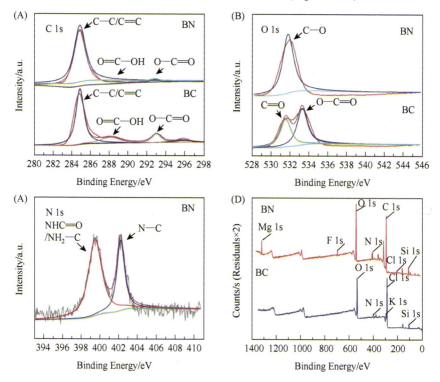

Figure 5.50 XPS Spectra of Biochar (BC) and Biochar Slow-release Nitrogen Fertilizer (BN).

Figure 5.50A shows the contour of the comparison peak of the C 1s spectrum. For BC and C 1s spectra, two strong peaks were observed at 284.4eV and 293.2eV, corresponding to C—C/C=C and O—C=O respectively, indicating the existence of aromatic and aliphoid groups (Tiwari et al., 2013). The peaks appearing at 288.8–289.1eV corresponded to the C=O—OH, referring to the carbon in carboxyl group or ester group. For BN, the peaks at 284.4eV and 288.8–289.1eV were significantly weaker than that at BC, indicating that the carbon in biochar formed other kinds of chemical bonds, possibly the result of the chemical bond reaction between urea and biochar. Figure 5.50B shows the contour of the comparison peak of the O 1s spectrum. For BC, there are two strong peaks in the O 1s spectrum. The peak at 531.0–531.9eV corresponds to the C=O in carbonyl oxygen; that at 533.1–533.8eV, corresponding to O—C=O, are the non-carbonyl (ether) oxygen atoms in esters and anhydrides (Lakshminarayanan et al., 2004); for BN, 532.3–532.8eV peaks, corresponding to C—O, are carbonyl oxygen atoms in esters and anhydrides and oxygen atoms in hydroxyl groups.

Figure 5.50C shows the contour of the comparison peak of N 1s spectra of BN material. The peaks near 400eV were attributed to N 1s, which included two with different binding energy values (the one at 398.3–399.8eV corresponded to NHC=O/NH$_2$—C) indicating that urea reacted with some organic compounds in biochar (Shi et al., 2020); the other at 402eV corresponded to N—C, assigning the bond in this form to the nitrogen atom (Yang and Jiang, 2014). Figure 5.50D shows the general distribution of elements in BC and BN materials. Both BC and BN contain nitrogen, and BN contains an apparently higher content, proving the incorporation of urea. Due to the incorporation of urea, the content of C and O in BN increased. The high content of Si in BN came from silicates in clay minerals, with a small part from the parent biological carbon.

5.4.2.4 X-ray diffraction

The position of the diffraction peak in the XRD pattern of biological slow-release nitrogen fertilizer (BN) was similar to that of biochar. The diffraction peaks of the two samples appeared in the 2θ ranges of 14°–26°, 20°–30° and 30°–40°, respectively. The diffraction peaks in the 2θ range of 14°–26° may be mainly attributed to the organic substances. Corresponding to the diffraction peak of graphite (2θ between 20° and 30°), there was a diffused diffraction peak, namely 002 peak, indicating that the amorphous aromatic structure was formed in biochar after heat treatment. This structure, prone to having electrophilic substitution reactions with heavy metal ions and urea molecules, would lead to the high delocalization of π electron. Diffraction peaks with the 2θ ranging from 30° to 40° were attributed to the Mg in biochar (Tang et al., 2015)(Figure 5.51).

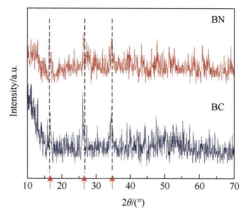

Figure 5.51 XRD images of biochar (BC) and biochar slow-release nitrogen fertilizer (BN).

5.4.3 Effects of biochar slow-release nitrogen fertilizer on microbial communities in copper-contaminated soil

5.4.3.1 Changes in soil pH

In the soil uncontaminated by copper, the pH of the soil treated with biochar (BC), urea (UR) and biochar slow release nitrogen fertilizer (BN) increased by 0.32(D60)–0.49(D14), 0(D60)–0.53(D7) and 0.12(D60)–0.55(D7) compared with the CK, respectively (Figure 5.52A). Due to the alkaline nature of biochar, the soil pH increased upon its application without significant changes brought about in the later period. Upon application of nitrogen in the soil, the following reactions will be triggered.

Hydrolysis reaction:

$$CO(NH_2)_2 \xrightarrow{H^+} NH_4^+ \qquad (5.5)$$

$$NH_4^+ \longrightarrow NH_3(aq) + H^+ \qquad (5.6)$$

Nitrification reaction:

$$NH_4^+ \longrightarrow NO_3^- + H^+ \qquad (5.7)$$

It can be seen that when urea was applied to the soil, hydrolysis reaction would occur soon to deplete the H^+ in the soil, thus increasing the soil pH; the NH_4^+ produced by hydrolysis was very unstable in the soil, which quickly transformed into NH_3 to volatilize and lose to the atmosphere, and NO_3^- and H^+ were generated through nitrification reaction, which reduced the soil pH. Therefore, the soil pH of the UR and BN treatment groups increased first and decreased then, while that of the BN treatment group was always higher than that of the UR treatment group, indicating that the hydrolysis reaction of urea was slowed down due to the adsorption of biochar on urea. In the soil contaminated by copper, the addition of Cu^{2+} consumed the OH^- in the soil, and the soil pH of all treatment groups decreased to a certain extent compared with the cases uncontaminated by copper. The soil pH of BC(Cu200), UR(Cu200) and BN(Cu200) treatment groups were 0.42(D60) −0.55 (D7), 0.2(D60) −0.89 (D7) and 0.19(D60) −0.83 (D7) higher than that of CK(Cu200) treatment group, respectively (Figure 5.52B). The increase extent of soil pH in the UR(Cu200) and BN(Cu200) treatment groups was significantly higher than that in the UR and BN treatment groups uncontaminated by Cu, which may be caused by the overall decrease of soil pH and the increase of H^+ in the soil upon the addition of Cu, making the urea hydrolysis accelerated and subsequent nitrification reaction inhibited to a certain extent.

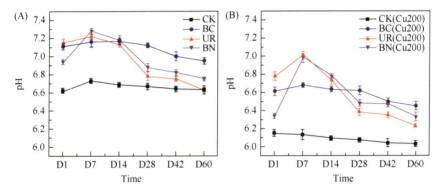

Figure 5.52 Changes in Soil pH in Each Treatment Group during the Culture Period. Uncontaminated by copper (A), contaminated by copper (B). The error line is the mean standard error of three parallel samples.

5.4.3.2 *Changes in different Cu forms in soil*

The BCR four-step leaching method was used to analyze the different forms of copper in the soil: exchangeable copper (Figure 5.53A), oxidizable copper (Figure 5.53B), reducible copper (Figure 5.53C) and residual copper (Figure 5.53D). Exchangeable copper is a form that is easily absorbed and utilized by plants and microorganisms in soil, and a form of heavy metal pollution to soil. At the early stage of culture, the content of exchangeable copper in all treatment groups was the highest among all the forms, accounting for about 50% of the total. It can be seen from Figure 5.53A that the content of exchangeable copper in the soil of BC(Cu200) and BN(Cu200) treatment groups gradually decreased; that in the soil of

UR(Cu200) treatment group decreased slightly after 7 days of culture, and then tended to be stable. After 60 days of culture, that in soils treated with BC(Cu200), UR(Cu200) and BN(Cu200) decreased by 26.53%, 4.08% and 18.37%, respectively. It can be seen that biochar-contained BC and BN have a great influence on the change of exchangeable copper content. The possible reason is that biochar features a porous structure and contains rich oxygen-containing functional groups, such as C—O, C=O and —OH, which helps with the interaction with copper, thus binding copper to biochar (Mandal et al., 2020a, 2020b). However, the effect of BC is better than that of BN, probably because some pores and functional groups in BN are occupied by incompletely released urea, and there are less binding sites with heavy metals.

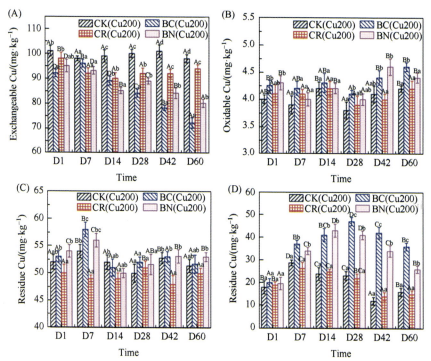

Figure 5.53 Changes in the content of exchangeable copper (A),oxidizable copper (B), reducible copper (C) and residual copper (D) in the soil of each treatment group during the culture period. The uppercase and lowercase letters represent the significant differences between one treatment group at different sampling times and between different treatment groups at the same sampling time, respectively ($P<0.05$). The error line is the mean standard error of three parallel samples.

The oxidizable copper in the soil is a metal that combines with the organic matter and sulfide in soil. When the soil environment becomes oxidized, this metal will be released into the environment. As can be seen from Figure 5.53B, after 60 days of culture, the oxidizable copper in the soil of the UR(Cu200) treatment group almost stayed unchanged, while that in the soil of the BC(Cu200) and BN(Cu200) treatment groups increased by 9.52% and 4.76%, respectively.

The reducible copper in soil is generally formed by adsorption or coprecipitation of cations, such as iron and manganese oxide, whose content is affected by the changes in redox conditions and soil pH. The higher REDOX potential and soil pH are conducive to the formation of reducible copper. It can be seen from Figure 5.53C that on the day 7 of culture, the reducible copper content in the soil of BC(Cu200) and BN(Cu200) treatment groups increased by 11.54% and 7.69%, which may be because the soil pH increased to a certain extent upon the addition of biochar. Cu^{2+} diffused at the same time can chelate the functional groups (phenol/carboxyl groups) on biochar (Shen et al., 2018). After 60 days of culture, there was no significant difference in the reducible copper content between the UR(Cu200), BC(Cu200) and BN(Cu200) treatment groups and CK(Cu200) treatment group. It may be because the soil pH decreased slightly in the later stage of the long incubation period, and the reducible copper formed was reduced in the soil.

The residual copper in the soil, relatively stable in the soil, exists in the sediment for a long time. It is not easy to be activated and released, thus difficult to be absorbed and utilized by plants. It can be seen from Figure 5.53D that during the whole culture period, the residual copper in the soil of the UR(Cu200) treatment group always maintained the same level as that of the CK(Cu200) treatment group, indicating that urea had no remedying effect on the copper pollution in the soil. However, the residual copper content in the soil of BC(Cu200) and BN(Cu200) treatment groups increased, to the maximum level around day 28, and then gradually decreased, probably due to the desorption of part of the copper from biochar in the later period (Liu et al., 2017a, 2017b). After 60 days of culture, the residual copper content in the soil of BC(Cu200) and BN(Cu200) treatment groups increased by 115% and 67.4%, respectively. The result shows that by changing the form of copper, biochar converts the form easily absorbed and utilized by plants into one not so, and reduces the toxicity of copper, so as to passivate the heavy metal copper in soil and remedy the copper pollution in soil. However, the remediation effect of BC(Cu200) treatment group was better than that of BN(Cu200) treatment group, probably because some of the pores and functional groups of biochar in BN were occupied by residual urea. The changes of different copper forms in soil will affect the changes of other indicators in soil.

5.4.3.3 *Dynamic variation characteristics of soil nitrogen*

1. Dynamic variation characteristics of NH_4^+-N in soil

According to Figure 5.54, the content of NH_4^+-N in the soil treated with urea (UR) and biochar slow-release nitrogen fertilizer (BN) showed a trend of first increasing and then decreasing with/without copper pollution. In the soil without copper pollution, the content of NH_4^+-N in the BN treatment group went significantly higher than that of the UR treatment group (Figure 5.54A) from day 14. The possible reason may be that upon application to the soil, the simple nitrogen would soon have the hydrolysis reaction, resulting in the formation of NH_4^+, followed by the rapid formation of NH_3 and NO_3^-. Therefore, the content of NH_4^+-N in the UR treatment group began to decrease from day 7, while that in the BN group began to decrease from day 14. The decrease rate was slower than that of UR group, indicating that biochar had the effect of slow release of urea. In the soil contaminated with copper, the decrease of pH accelerated the hydrolysis of urea and inhibited the nitrification reaction to a certain extent, thus resulting in a rapid increase of NH_4^+-N content in the soil of the UR(Cu200) and BN(Cu200) treatment groups in the early stage, which was significantly higher than that of the UR and BN groups uncontaminated by copper (Figure 5.54B). In the later stage, the content of NH_4^+-N decreased rapidly because most of the urea hydrolysis was completed and the nitrification reaction was dominant in the soil.

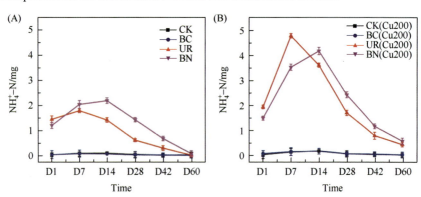

Figure 5.54 Changes in soil NH_4^+-N content in each treatment group during the culture period. Uncontaminated by copper (A), contaminated by copper (B). The error line is the mean standard error of three replicates.

2. Dynamic variation characteristics of NO_3^--N in soil

Due to the nitrification, NO_3^- gradually accumulated in the soil. Figure 5.55A shows that in the soil without copper pollution, the NO_3^--N content in the soil of the urea (UR) treatment group reaches the maximum on day 14, and then decreases, because the NH_4^+ produced by urea hydrolysis is very unstable, and only part of NH_4^+ is converted to NO_3^- through loss in volatilization. The NO_3^--N content in the soil of the biochar slow-release nitrogen fertilizer (BN) treatment group continued to increase, indicating that due to the slow-release effect of biochar on urea, urea was released into the soil continuously instead of at one time. Thus, NH_4^+ was continuously generated and converted into NO_3^-, which accumulated in the soil. On day 60, the content of NO_3^--N in BN treatment group was 121.37% higher than that in UR treatment group. Upon involvement of copper pollution, the NO_3^--N content in the soil of UR(Cu200) and BN(Cu200) treatment groups was lower than that of the soil without copper pollution (Figure 5.55B), which may be a result of the decrease of pH inhibiting the nitrification to a certain extent and lessening the NO_3^- generated in the soil. Similarly, the content of NO_3^--N in the BN(Cu200) treatment group was higher than that in the UR(Cu200) treatment group at the later stage, and the content of NO_3^--N in the BN(Cu200) treatment group was 69.98% higher than that in the UR(Cu200) treatment group on day 60, which also indicated the slow-release efficiency of biochar for urea.

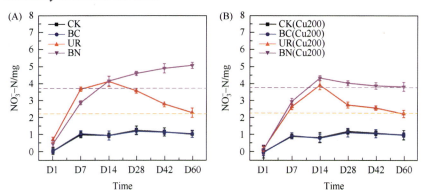

Figure 5.55 Changes in Soil NO_3^--N Content in Each Treatment Group during Culture. Uncontaminated by copper (A), contaminated by copper (B). The error line is the mean standard error of three replicates.

5.4.3.4 *Changes in enzyme activity*

The biological and biochemical processes that occur in soil are fundamental to the functioning of terrestrial ecosystems, which persist due to the presence of soil enzymes. As one of the major driving forces of soil ecosystem metabolism, enzyme has a specific catalytic effect that enables all biological and biochemical processes in the soil. The enzymes in soil come from different sources and feature a wide range of varieties. To date, about 60 kinds of soil enzymes have been identified as active in soil. Closely related to some physiochemical properties of the soil, enzyme activity can reflect the quality, fertility, and microbial activity of the soil. A high enzyme activity can promote the decomposition and mineralization of organic matters in the soil, thus promoting the nutrient circulation. Enzyme catalysis plays an important role in the circulation and migration of C, N, P and other elements in soil. Enzyme activity has been recommended as one of the biological indicators for the assessment of soil pollution by heavy metals (Merino et al., 2016).

1. Chitinase

Chitinase, an extracellular hydrolytic enzyme involved in the conversion of hydrolyzed chitinase into aminosaccharides, serves as the main source of mineralized nitrogen in soil. Therefore, the activity of the chitinase decomposing the chitin is the key to control the effect of soil nitrogen cycle, and plays an important role in the nitrogen cycle. As can be seen from Figure 5.56A, without copper pollution, the soil chitinase in the three treatment groups of biochar (BC), urea (UR) and biochar slow-release nitrogen fertilizer (BN) was increased to varying degrees compared with the CK. This is because carbon is not only an important part of soil enzymes, but also an energy available for soil microorganisms (Merino et al., 2016). Therefore, since biochar provides a considerable amount of unstable carbon, the soil chitinase activity in BC treatment group increased significantly in the early stage, reaching 54.62%. This echos the research results of Bailey et al. (2011). In the later stage, the extent of improvement decreased, possibly due to the adsorption or blocking of enzymes or substrates, meanwhile the soil enzymes got adapted to the carbon-rich environment. On day 60, the soil chitinase in the BC treatment group was increased by 30.58%. As the addition of urea-nitrogen source promoted the cycling and migration of nitrogen in soil, the UR treatment group had the best promotion of soil chitinase activity in the first 14 days, with a maximum rate of 80.77%. Then, the extent of improvement gradually decreased because of the instantaneous release of simple urea. Given the loss in later volatilization and nitrification, the soil nitrogen content decreased, and the soil chitinase activity increased by 45.21% in UR treatment group on day 60. In the first 14 days, the activity of chitinase in BN treatment group increased less than that in UR treatment group, with a maximum rate of

61.88%. In the later stage, due to the slow release of urea by biochar, nitrogen was continuously released into the soil. On day 60, the soil chitinase in BN treatment group was increased by 69.7%. Upon copper contamination, the chitinase activity was inhibited on the whole (Figure 5.56B). In the first 14 days, the effect of copper on chitinase activity was not obvious, which even slightly increased. This is perhaps because copper is a trace element required for microbial growth, and an appropriate number of Cu ions can promote the expression of *hydA*, a functional gene of typical cellulose degrading bacteria, and enhance its activity and content in soil, thus promoting the cell growth and reproduction (Xu et al., 2015). With the aggravation of heavy metal pollution after day 14, the content of available copper in soil exceeded the threshold of microbial endurance, destroying the enzyme system, reducing the synthesis and secretion of enzyme, and leading to the decline of enzyme activity. After 60 days of culture, the chitinase activity in soil of UR(Cu200) treatment group was not significantly different from that in the CK, while that in BC(Cu200) and BN(Cu200) treatment groups was increased by 31.61% and 41.21%, respectively, indicating that BC and BN could alleviate copper's inhibition of soil chitinase activity to a certain extent.

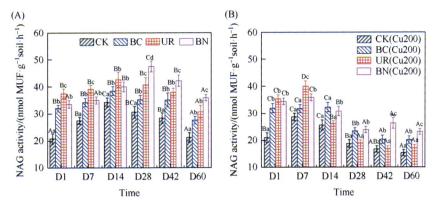

Figure 5.56 Changes of chitinase activity in each treatment group during the incubation. Uncontaminated by copper (A), contaminated by copper (B). The uppercase and lowercase letters represent the significant differences between one treatment group at different sampling times and between different treatment groups at the same sampling time, respectively (*P*<0.05). The error line is the mean standard error of three parallel samples.

2. β-glucosidase

The glucose produced by the decomposition of cellulose residues by β-glucosidase is an important carbon source for soil microorganisms (Merino et al., 2016). Closely related to the organic matters, microbial activity and carbon cycle in the soil, it can reflect the change of organic carbon in the soil. As can be seen from Figure 5.57A, without copper pollution, since biochar provides a considerable amount of unstable carbon, it is easily used by soil microorganisms as the energy source in a relatively short time and decomposed by β-glucosidase in soil (Smith et al., 2010). Biochar (BC) and biochar slow-release nitrogen fertilizer (BN) treatment significantly increased the β-glucosidase activity in soil as a whole. In the early stage, β-glucosidase activity in the BC treatment group was better improved than that in the BN treatment group; however, there was no significant difference between the two in the middle stage; in the late stage, the BN treatment group outperformed the BC group. On day 60, the β-glucosidase activity of soil in BC and BN groups was increased by 21.16% and 38.65%, respectively. The addition of nutrient elements not only balances the soil nutrients, but also improves the enzyme activity. Since urea provided nutrient elements, the urea (UR) treatment group gave the β-glucosidase activity in the soil certain improvement in the first 14 days of culture. Afterwards, there was no significant difference between the UR treatment group and the control group (CK) in the β-glucosidase activity, because the instantaneous release of urea was lost from the soil through various ways, and the nutrient elements were reduced. When the soil was put under copper pollution, the β-glucosidase activity was also boosted in the first 7 days. Later with the deterioration of heavy metal pollution, the β-glucosidase activity gradually decreased (Figure 5.57B). It can be seen that the β-glucosidase

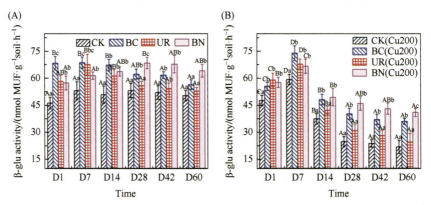

Figure 5.57 Changes of β-glucosidase activity in soil of each treatment during the incubation. Uncontaminated by copper (A), contaminated by copper (B). The uppercase and lowercase letters represent the significant differences between one treatment group at different sampling times and between different treatment groups at the same sampling time, respectively ($P<0.05$). The error line is the mean standard error of three parallel samples.

activity in BC(Cu200) and BN(Cu200) treatment groups still enhanced to a certain extent. The BN(Cu200) treatment group outperformed the BC(Cu200) treatment group, possibly because BN could provide both unstable carbon source and nutrient nitrogen.

3. Phosphatase

Soil phosphatase, a hydrolytic enzyme of organophosphorus compounds, can release inorganic phosphorus to be absorbed and utilized by plants through promoting the hydrolysis of organophosphorus compounds. The activity of phosphatase directly affects the decomposition of soil organophosphorus and its biological availability, playing an important role in phosphorus cycle in the soil. As can be seen from Figure 5.58A, without copper pollution, all the three treatment groups of biochar (BC), urea (UR) and biochar slow release nitrogen fertilizer (BN) improved the activity of soil phosphatase in the first 14 days, showing no significant difference among them. On the 28th day, there was no significant difference between the BC treatment group and the CK in soil phosphatase activity. After 28 days, the soil phosphatase activity in the BC and UR treatment groups was not significantly different from that in the CK, while that in the BN treatment group was still improved, because soil phosphatase activity was related to soil pH to a certain extent. The biochar in BN was alkaline, which could increase the soil pH and improve the acidic soil. When the soil pH was increased, nutrients required by microorganisms were released, thus improving the soil phosphatase activity. Soil pH in the UR treatment group also increased to a

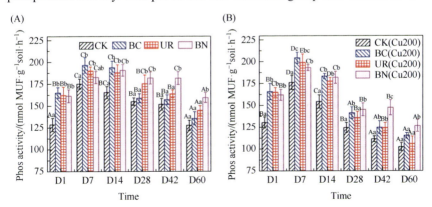

Figure 5.58 Changes of soil phosphatase activity in each treatment group during the incubation. Uncontaminated by copper (A), contaminated by copper (B). The uppercase and lowercase letters represent the significant differences between one treatment group at different sampling times and between different treatment groups at the same sampling time, respectively ($P<0.05$). The error line is the mean standard error of three parallel samples.

certain extent in the early stage. In the first 14 days after the involvement of copper pollution, the soil phosphatase activity was increased in all treatment groups due to the stimulation of copper, showing no significant difference among BC(Cu200), UR(Cu200) and BN(Cu200) treatment groups. In the later period, soil phosphatase activity was significantly decreased due to the aggravation of copper pollution. Nevertheless, BC(Cu200) and BN(Cu200) treatment groups still showed some improvement (Figure 5.58B) as compared with CK(Cu200) treatment group. This is because the overall decrease of soil pH upon involvement of copper pollution leads to the decrease of soil phosphatase activity, while BC and BN have certain remedying effects on the soil copper pollution, thus slowing down copper's inhibition of soil phosphatase activity to a certain extent (Table 5.11).

Table 5.11 Analysis on correlation of all soil indicators based on pearson correlation coefficient.

	pH	NH_4^+-N	NO_3^--N	Exchangeable Cu	NAG	β-glu	Phos
pH	1	0.913[*]	−0.741	−0.873[*]	0.811	0.880[*]	0.833
NH_4^+-N		1	−0.309	0.329	0.862[*]	0.681	0.792
NO_3^--N			1	−0.239	0.544	−0.317	0.495
Exchangeable Cu				1	−0.360	−0.199	−0.367
NAG					1	0.866[*]	0.874[*]
β-glu						1	0.830[*]
Phos							1

Note: NAG, β-glu and Phos represent activity of the chitinase, β-glucosidase and phosphatase, respectively. Bold fonts indicate significant correlations. * $P<0.05$; ** $P<0.01$.

5.4.3.5 Changes in structure and abundance of microbial community

Due to the high sensitivity of soil microorganisms to changes in the soil environment, the structure and abundance of soil microbial communities have been widely used as the indicators of soil quality changes (Chu et al., 2007). It has been found that adding biochar to the soil may affect the community structure and richness of soil microorganisms. When the organic matters are changed, the biomass and community composition of soil microorganisms also change. Such changes are related to the carbon content in the soil, because the mineralization of organic matters boosts the activity of soil microorganisms. The effects of fertilizers on soil microorganisms are variable, and there is no research clearly stating the specific effects of chemical fertilizers on microbial communities.

1. Diversity of soil microbe Alpha

The results of 16S high-throughput sequencing were shown in Table 5.12. Overall, there was little difference in soil bacterial diversity. The α-diversity (Ace, Chao1, Shannon and Simpson indexes) of soil bacteria under urea (UR) and biochar slow release nitrogen fertilizer (BN) treatments was significantly lower than that in the CK, but there was no significant difference between CK and biochar (BC) treatment groups. These results suggest that the relative abundance of bacterial 16S rRNA gene sequences is slightly reduced due to the application of fertilizer and fertilizer's interaction with biochar. The biochar-fertilizer interaction reduced the bacterial diversity compared to the application of biochar alone, possibly because the available substrates benefited only a few microorganisms involved in the nitrogen cycle, rather than the entire population. The relative abundance of specific populations involved in the nitrogen cycle and decomposition of organic matters had increased in some cases, leading to the dominance of a few microorganisms. Although the use of biochar has been reported to increase the microbial biomass (Herrmann et al., 2019), contrary studies have shown that biochar alone has no significant effect on the uniformity and richness of soil bacterial community. The use of fertilizer alone or in combination with biochar generally reduces the relative abundance of bacterial populations (Li et al., 2020a, 2020b). Consistent with the results of the study, Allison and Marting (2008) found that the application of NPK fertilizer laid a negative impact on soil microbial communities in 38 different ecosystems. This suggests that, due to the availability of substrates provided by fertilizers, only specific organisms involved in the nitrogen cycle thrived, and that the persistence of these bacterial populations was further dictated by the interaction between the populations growing under the appropriate conditions, which ensured their continued dominance and viability over other organisms (Figure 5.59).

Table 5.12 Soil bacterial diversity index of different treatment groups.

Sample\estimators	Shannon	Simpson	Ace	Chao1
CK	6.086	0.007	2022.177	2024.204
BC	6.176	0.006	2006.614	1994.298
UR	5.270	0.017	1770.991	1795.679
BN	5.418	0.017	1774.964	1769.111
CK(Cu200)	5.249	0.009	1902.594	1907.712
BC(Cu200)	5.592	0.019	1916.076	1897.351
UR(Cu200)	5.586	0.020	1927.231	1923.531
BN(Cu200)	5.489	0.022	1918.481	1933.485

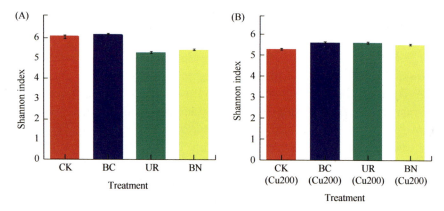

Figure 5.59 Microbial α-diversity in soil of each treatment group during the incubation. Without copper pollution (A), with copper pollution (B).

In the presence of copper pollution, there is no significant difference in the abundance of soil microbial operational taxonomic unit (Ace and Chao1 indexes). The α-diversity (Shannon and Simpson indexes) of soil bacteria in CK(Cu200) and BC(Cu200) treatment groups was lower than that in CK and BC treatment groups without copper pollution. The α-diversity (Shannon and Simpson indexes) of soil bacteria under UR(Cu200) and BN(Cu200) treatments increased slightly compared with that without copper pollution. In general, the α-diversity (Shannon and Simpson indexes) of soil bacteria in BC(Cu200), UR(Cu200) and BN(Cu200) treatment groups was higher than that in CK(Cu200) treatment group. Copper-based fungicides are the most toxic substances to soil microorganisms (Bunemann et al., 2006). Although copper is an essential trace element for the growth of soil microorganisms, an excessive content will interfere with the control mechanism of gene level, thus destroying the soil homeostasis, inhibiting the activity of microbial enzyme proteins and spoiling the metabolic pathways, usually leading to apoptosis. The number and diversity of soil microorganisms will be impaired, and the nitrification and ammonification processes as well as the activity of soil enzymes will be inhibited (Wyszkowska et al., 2013). Therefore, upon the involvement of copper pollution, the bacterial diversity of soil species in CK(Cu200) treatment group decreased, while BC(Cu200), UR(Cu200), BN(Cu200) treatment groups provided nutrients to soil microorganisms in different degrees, providing essential energy for life activities of microorganisms. The biochar in BC(Cu200) and BN(Cu200) treatment groups can also reduce copper's toxicity to microorganisms through the adsorption of copper.

2. Changes of soil microbial community structure

The microbial community structure in soil is very sensitive to the availability of chemical fertilizers and soil nutrients. As can be seen from the column chart of phylum-level microbial species abundance in soil, the top 10 phyla in the treatment groups without copper contamination are Proteobacteria, Actinobacteriota, Acidobacteriota, Chloroflexi, Firmicutes, Bacteroidota, Patescibacteria, Gemmatimonadota, Myxococcota and Planctomycetota. The relative abundance of the top 10 phyla accounted for more than 95.28% of all bacterial sequences in soil. Among them, Proteobacteria has the highest abundance, accounting for 29.03%, 27.38%, 40.15% and 43.15% of the total bacterial sequences in the control group (CK), and the biochar (BC), urea (UR) and biochar slow-release nitrogen fertilizer (BN) treatment groups, respectively. The abundance of Proteobacteria in UR and BN treatment groups was significantly higher than that in CK and BC treatment groups. Proteobacteria is a phylum of bacteria that can reflect the nutrient status of the soil. When the soil is rich in nutrients, they will become the dominant bacteria with a relatively fast growth rate (Trivedi et al., 2013). UR and BN treatments have improved the relative abundance of Proteobacteria in the soil. The results show that UR and BN treatment groups provided nutrients needed by microorganisms to promote their growth. Proteobacteria contain many bacteria that can immobilize nitrogen. For example, ammonia-oxidizing Proteobacteria (mainly β-Proteobacteria) act as an important part of the nitrogen cycle. UR and BN play an important role in the abundance and community structure of ammonia-oxidizing bacteria. Proteobacteria, Chloroflexi, Actinobacteriota and Gemmatimonadota have a dominant and symbiotic relationship, which is important for nitrogen cycle and decomposition of organic matters in soil. Acidobacteriota and Chloroflexi are oligotrophic groups, which are suitable for growing in low nutrient environment (Fierer et al., 2007). Their abundances were reduced in UR and BN treatment groups (Figure 5.60).

Upon the addition of copper, the dominant bacteria in soil changed from 10 to 7, that is, Actinobacteriota, Proteobacteria, Firmicutes, Chloroflexi, Acidobacteriota, Bacteroidota and Myxococcota. The relative abundance of the top 7 phyla accounted for more than 95.58% of all bacterial sequences in soil. In CK(Cu200) treatment group, Actinobacteriota featured the highest abundance, accounting for 29.88% of the total bacterial sequence; in BC(Cu200) and BN(Cu200) treatment groups, Proteobacteria demonstrated the highest abundance, accounting for 28.88% and 29.91% of the total bacterial sequence, respectively. Firmicutes marked the highest abundance

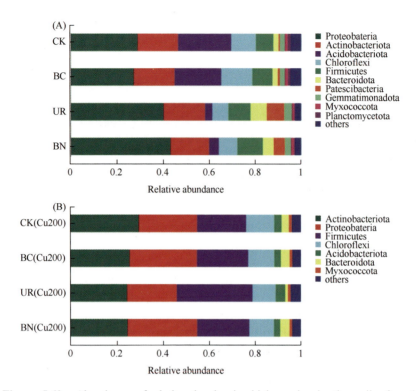

Figure 5.60 Abundance of phylum-level microbial species in the soil of each treatment group during the incubation. Without copper pollution (A), with copper pollution (B).

in UR(Cu200) treatment group, accounting for 32.97% of the total bacterial sequence. Actinobacteriota are mostly pathogenic bacteria, and an increase of this bacterium will increase the probability of soil diseases, making them a threat of soil health. Many bacteria in the Firmicutes phylum produce spores that are resistant to dehydration and extreme environment. The bacterial diversity in soil decreased in case of the copper pollution, and Actinobacteriota, Proteobacteria and Firmicutes made the dominant bacterial species with high abundance in each treatment group. The results showed that Actinobacteriota, Proteobacteria and Firmicutes were bacteria that are tolerant of copper pollution. In the BC(Cu200) and BN(Cu200) treatment groups, the Proteobacteria abundance, a sign of the livability of soil environment, is the highest, indicating that BC and BN can improve the soil environment and reduce the inhibition effect of copper on microorganisms upon the copper pollution.

5.4.4 Effects of biochar slow-release nitrogen fertilizer on plant growth and rhizosphere enzyme activity characteristics in copper-contaminated soil

5.4.4.1 Dynamic characteristics of available nitrogen in soil

1. Dynamic variation characteristics of available nitrogen in soil

As one of the most required nutrient elements of crops, nitrogen plays an important role in the growth and quality of crops. Proper application of nitrogen can improve the crop yield and quality to a certain extent. Available nitrogen in soil can be directly absorbed and utilized by crops. The content of available nitrogen in soil can reflect the nutrition status of soil and indirectly affect the growth of crops.

As can be seen from Figure 5.61A, without copper pollution, there was no significant difference in the available nitrogen content in the biochar (BC) treatment group and the CK, and the available nitrogen content in the urea (UR) and biochar slow-release nitrogen (BN) treatment groups showed a trend of increase before decrease. In the first 7 days of the culture, the available nitrogen content in UR treatment group was higher than that in BN treatment group, which was caused by the instantaneous release of simple urea applied to the soil in the early stage and the accumulation of more nitrogen. From day 14, the available nitrogen content in soil of BN treatment group was higher than that of UR treatment group, and the figure in both groups began to decline. However, the BN treatment group declined a smaller extent than the UR treatment group, because nitrogen released by pure urea was quickly lost from the soil through volatilization and leaching. Nonetheless, due to the slow release effect of biochar on urea, the

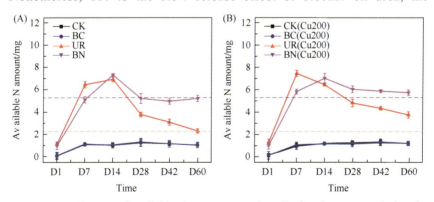

Figure 5.61 Changes of available nitrogen content in soil of each treatment during the incubation. Uncontaminated by copper (A), contaminated by copper (B). The error line is the standard error of five replicates.

content of available nitrogen in soil of BN treatment group was always maintained at a certain level. Figure 5.61B shows the change of available nitrogen content in soil of each treatment group with copper pollution. It can be seen that the available nitrogen content in the UR(Cu200) and BN(Cu200) treatment groups is higher than that in the soil without copper pollution on the whole. In the later stage, the toxicity of copper to plants inhibits the growth of plants and leads to the reduction of nitrogen uptake by plants. From day 14, the available nitrogen content in the soil of BN(Cu200) treatment group is higher than that in UR(Cu200) group, which also reflects the slow release effect of BN material.

2. Dynamic variation characteristics of cumulative content of available nitrogen in soil

In the absence of copper pollution (Figure 5.62A), the cumulative content of available nitrogen in the soil of the biochar slow-release nitrogen (BN) treatment group exceeded that of the urea (UR) group on day 28. After 60 days of culture, the cumulative content of available nitrogen in the soil of UR and BN treatment group accounts for 54.3% and 70.04% of the total nitrogen applied, respectively. It showed that BN material could reduce the loss of nitrogen in soil. In the presence of copper pollution (Figure 5.62B), the cumulative content of available nitrogen in the soil of BN(Cu200) treatment group was still higher than that of UR(Cu200) group on day 28. After 60 days of culture, the cumulative content of available nitrogen in the soil of UR(Cu200) and BN(Cu200) treatment groups came to 61.5% and 77.3% of the total application of nitrogen, respectively. This is because

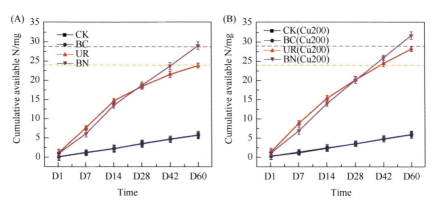

Figure 5.62 Changes of cumulative available nitrogen content in soil of each treatment group during the incubation. Uncontaminated by copper (A), contaminated by copper (B). The error line is the standard error of five replicates.

copper is toxic to plants, which inhibited the plant growth and led to a reduced nitrogen utilization by plants.

5.4.4.2 *Changes in plant growth indicators*

1. Plant height

As can be seen from Figure 5.63, with or without copper pollution, since biochar can improve the soil properties and fertility and urea can provide essential nutrients required by plants, soil applied with biochar (BC), urea (UR) and biochar slow release nitrogen fertilizer (BN) can all promote the plant growth. In the absence of copper pollution (Figure 5.63A), there was no significant difference in the plant height between the BC and UR treatment groups in the first 14 days of the culture, and plants in the two groups were higher than that of the BN group. The plant height of the BN group exceeded that of the BC and UR groups from day 14 to day 28. After 60 days of culture, the plant height was as follows: BN(66.3cm)> UR(61.5cm)>BC(54.7cm)>CK(37.1cm). It can be seen that nitrogen is the main factor influencing the plant height. In the presence of copper pollution (Figure 5.63B), the plant height was generally lower than the that without copper pollution. There was no significant difference in plant height between BC(Cu200), UR(Cu200) and BN(Cu200) treatment groups during the first 28 days of culture. No significant difference was seen in plant height between BC(Cu200) and BN(Cu200) treatment groups on day 42, and plant height of the two groups was larger than UR(Cu200) treatment group. After 60 days of culture, plant height was as follows: BN(Cu200) (50.4cm)>BC(Cu200) (43.6cm)>UR(Cu200) (40.3cm)>CK(Cu200) (27.7cm).

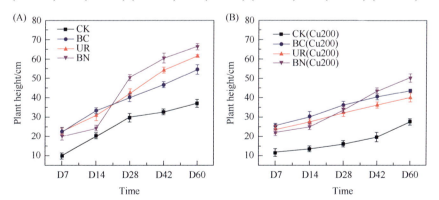

Figure 5.63 Changes of plant height in each treatment during the incubation. Uncontaminated by copper (A), contaminated by copper (B). The error line is the standard error of five replicates.

This is perhaps because the remediation effect of biochar on copper pollution in soil reduced the toxicity of copper to plants. Therefore, the plant height of BC(Cu200) treatment group was slightly higher than that of UR(Cu200) treatment group at the later stage.

2. Plant root length

Similarly, with or without copper pollution, biochar (BC), urea (UR), biochar slow-release nitrogen fertilizer (BN), BC(Cu200), UR(Cu200), and BN(Cu200) treatment groups could all promote the plant root growth. In the absence of copper pollution (Figure 5.64A), the root length of plants in the BN treatment group was significantly higher than that in the other treatment groups. During the first 42 days, there was no significant difference in the root length of plants in the BC and UR treatment groups. After 60 days of culture, the root lengths of plants were as follows: BN(25.2cm)>UR (21.7cm)>BC(18.1cm)>CK(11.3cm), indicating that root length was significantly positively correlated with plant height ($P<0.05$; Table 5.13). In the presence of copper pollution (Figure 5.64), the plant root length was lower than that without copper pollution on the whole, because copper damaged the absorption and metabolic capacity of plant roots and inhibited root vitality (Liu et al., 2015). During the first 28 days of culture, there was no significant difference in the root length of plants in BC(Cu200), UR(Cu200) and BN(Cu200) treatment groups. After 42 days, the root length of plants in BN(Cu200) treatment group was higher than that in BC(Cu200) and UR(Cu200) treatment groups. After 60 days of culture, the root lengths of plants were as follows: BN (Cu200) (19.6cm) > BC (Cu200) (16.3cm) > UR (Cu200) (15.2cm) > CK (Cu200) (9.4cm).

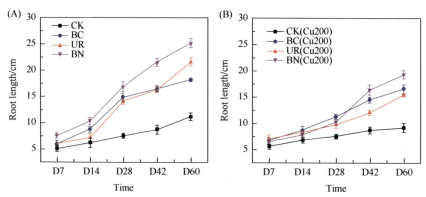

Figure 5.64 Changes of Plant Root Length in Each Treatment Group during Culture. Uncontaminated by copper (A), contaminated by copper (B). The error line is the standard error of five replicates.

Table 5.13 Analysis on correlation of all soil and plant indicators based on pearson correlation coefficient.

	Available N	Plant height	Root length	SPAD	Extent of NAG	Extent of β-glu	Extent of Phos
Available N	1	0.987**	0.982**	0.844	0.758	0.822	0.838
Plant height		1	0.930*	0.881*	0.546	0.486	0.622
Root length			1	0.935*	0.494	0.391	0.295
SPAD				1	0.378	−0.109	0.348
Extent of NAG					1	0.859*	0.954*
Extent of β-glu						1	0.993*
Extent of Phos							1

Note: SPAD represents the SPAD value of plant leaves, while NAG range, β-Glu range and Phos range respectively express the influence range of chitinase, β-glucosidase and phosphatase activity. Bold fonts indicate significant correlations. $*P<0.05$; $**P<0.01$.

3. SPAD of plant leaves

Based on chlorophyll's high absorption of red light (about 650nm) and low absorption of far infrared light (about 940nm), SPAD (Soil and Plant Analyzer Development) was first proposed in Japan. According to the SPAD value, the chlorophyll content per unit area of plant leaves was successfully predicted, which can reflect whether the plants are healthy and the soil environment is suitable for plant growth. As can be seen from Figure 5.65A, in the absence of copper pollution, the SPAD value of plants in the BC and UR treatment groups showed a gradual trend of decline during the whole culture period, while that in the BN treatment group showed a gradual trend of increase, which may be caused by the slow release of urea by BN that continuously provided nitrogen nutrition to plants. After 60 days of culture, the SPAD values of plant leaves were as follows: BN(36.2)>UR(33.0)> BC(27.1)>CK (20.9). Thus, the SPAD value of plant leaves was largely dictated by nitrogen. Upon the involvement of copper pollution, the SPAD value of plant leaves in BC(Cu200) and UR(Cu200) treatment groups showed a gradual trend of decline, while that in BN(Cu200) treatment group increased before decreased. After 60 days of culture, the SPAD values of plant leaves were as follows: BN (Cu200) (32.2) > UR (Cu200) (30.5) > BC (Cu200) (28.6) > CK (Cu200) (16.3).

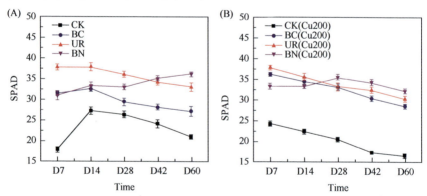

Figure 5.65 Change of SPAD value of plant leaves in each treatment during the incubation. Uncontaminated by copper (A), contaminated by copper (B). The error line is the standard error of five replicates.

4) Plant biomass

Figure 5.66 shows the aboveground and underground biomass of plants in each treatment group after 60 days of culture. In the absence of copper pollution, the aboveground biomasses of plants were as follows:

BN(1.49g)>UR(1.17g)> BC (0.87g)>CK(0.48g). There was no significant difference in the underground biomass of plants in the three treatment groups of biochar (BC), urea (UR) and biochar slow-release nitrogen fertilizer (BN), which was 0.21g, 0.19g and 0.19g, respectively, all higher than that in the (CK) (0.15g). In the presence of copper pollution, the aboveground and underground biomass of plants in each treatment group decreased. The aboveground biomass of plants in BN (Cu200) treatment group was 1.16g, significantly higher than that in BC(Cu200) (0.87g) and UR(Cu200) (0.86g) treatment groups, but there was no significant difference between BC(Cu200) and UR(Cu200) treatment groups. The three treatment groups were significantly higher than the CK(Cu200) (0.38g) control group. There was no significant difference in the underground biomass of plants in the BC(Cu200), UR(Cu200) and BN (Cu200) treatment groups, which was 0.17g, 0.14g and 0.12g, respectively, all higher than that in the CK(Cu200) control group (0.07g).

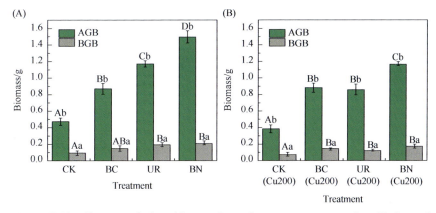

Figure 5.66 Changes of plant biomass in each treatment group after 60 days of incubation. Uncontaminated by copper (A), contaminated by copper (B). The uppercase and lowercase letters represent the significant differences between one treatment group at different sampling times and between different treatment groups at the same sampling time, respectively ($P<0.05$). The error line is the mean standard error of three parallel samples.

5.4.4.3 Influence range of soil enzyme activity in plant rhizomes

The plant roots can release not only various organic compounds, but also relevant enzymes that are used to break down rhizosphere organic matters. In addition, due to the influence of rhizosphere sediments, the microorganisms in the region became more active, and secreted enzymes to consume these unstable organic matters. As a result, the range of rhizosphere influence is broadened. The rhizosphere influence range of the enzyme is larger than its hotspot range in soil (small volume soil). The influence range of plant rhizosphere on enzyme activity usually indicates the ability of roots and microorganisms to immobilize and utilize the carbon and nutrients (Liu et al., 2020).

1. Chitinase

Due to the application of nitrogen fertilizer, the rapid growth of plants makes the roots more developed, thus enabling a high microbial population in the plant rhizosphere. Because continuous plant coverage provides varying qualities and quantities of root secretions, they do favor of the enzyme production as substrates for microorganisms. Continuous application of nitrogen fertilizer improves the substrate utilization by microorganisms and, as a result, microorganisms are activated to contribute to a higher enzyme activity (Wada and Toyota, 2007). As can be seen from Figure 5.67A, on day 7 without copper pollution, the influence range of chitinase activity in plant rhizosphere in the biochar (BC) treatment group was significantly higher than that in the CK; the influence range of activity of chitinase in plant rhizosphere in the urea (UR) and biochar slow release nitrogen fertilizer (BN) treatment groups was significantly higher than that in BC treatment group, because urea provided rich nitrogen nutrition, and the rather active microorganisms in plant roots secreted more chitinase to make the chitinase in plant rhizosphere more active. After day 14 the influence range of chitinase activity in plant rhizosphere in BN treatment group was significantly higher than that in the other three treatment groups, because the enzyme activity was heightened by the interaction between biochar and fertilizer, indicating that fertilizer increased the substrate available for nitrogen cycle enzyme to improve the enzyme activity. Forming an organic mineral layer on the biochar surface through the interaction between biochar and fertilizer may enhance the nutrient retention ability and achieve the effect of slow-release nitrogen, thus improving the microbial activity and biomass (Hagemann et al., 2017). In the presence of copper pollution, the influence range of chitinase activity in plant rhizosphere in each treatment group was still improved to a certain extent before day 14. This is perhaps because the surface of soil microbial cells in plant rhizosphere has charges capable of absorbing soil metal ions, which may be converted into nutrients required for their own growth through secretion. With the aggravation of copper pollution, in each treatment group, the influence range of chitinase activity in plant rhizosphere decreased sharply. This is because high concentration of heavy metals would destroy microbial cells, heavy metals would accumulate on the surface or in the body of microorganisms, and the number of soil microorganisms and all aspects of abilities were affected, thus leading to the decrease of soil enzyme activity and the influence range of enzyme activity, correspondingly (Maharjan et al., 2017).

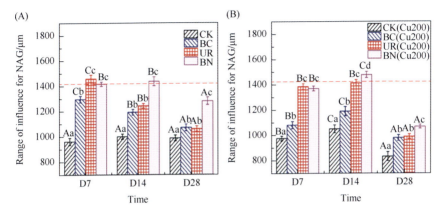

(A)　　　　　　　　　　　　　　　　(B)

Figure 5.67 Influence range of plant rhizosphere chitinase activity in each treatment group during Culture. Uncontaminated by copper (A), contaminated by copper (B). The uppercase and lowercase letters represent the significant differences between one treatment group at different sampling times and between different treatment groups at the same sampling time, respectively ($P<0.05$). The error line is the mean standard error of three parallel samples.

2. β-glucosidase

As can be seen from Figure 5.68A, in the absence of copper pollution, the influence range of β-glucosidase in plant rhizosphere of the biochar (BC), urea (UR) and biochar slow-release nitrogen fertilizer (BN) treatment groups was higher than that in the CK during the first 14 days, and that of the BC and BN treatment groups was significantly higher than that of the UR treatment group. This was because the addition of biochar could promote the carbon cycling-related enzyme activity in soil (Wang et al., 2015). On day 28, the BN treatment group had the widest influence range of β-glucosidase in plant rhizosphere, because BN boasted the promotion of biochar and the stimulation of nitrogen nutrition that made the plant rhizosphere microorganisms more active. Upon the involvement of copper contamination, the influence range of β-glucosidase in plant rhizosphere in each treatment was higher than that without copper pollution due to the stimulation of trace copper to microorganisms on day 7. With the aggravation of pollution, the influence range of β-glucosidase in plant rhizosphere sharply decreased. On day 28, BN(Cu200) treatment group had the widest influence range of β-glucosidase in plant rhizosphere.

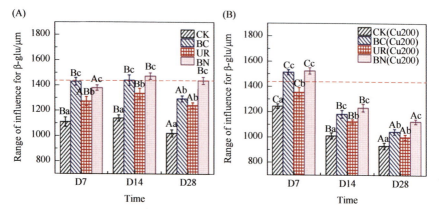

Figure 5.68 Influence range of β-glucosidase activity in plant rhizosphere of each treatment group during the incubation. Uncontaminated by copper (A), contaminated by copper (B). The uppercase and lowercase letters represent the significant differences between one treatment group at different sampling times and between different treatment groups at the same sampling time, respectively ($P<0.05$). The error line is the mean standard error of three parallel samples.

3. Phosphatase

Microorganisms in the soil convert the organophosphates into usable forms by boosting the activity of phosphatases. In the absence of copper pollution, the influence range of phosphatase activity plant rhizosphere of the biochar (BC), urea (UR) and biochar slow-release nitrogen fertilizer (BN) treatment groups was significantly higher than that in the CK (Figure 5.69A), because the limiting effect of phosphorus was lifted upon the addition of carbon and nitrogen in the soil. To obtain more phosphorus, soil microbes synthesize phosphatases to promote the release of phosphorus in organic matters. The promoting effect of plant root exudates on soil enzyme activity may be subject to the nitrogen content. Increasing the nitrogen availability tends to increase the phosphorus cycling rate, providing pathways for plants and ecosystems to adapt to changes in the nitrogen and phosphorus supply, that is, the addition of nitrogen improves the phosphorus conservation and thus promotes the nutrient balance to some extent (Marklein and Houlton,

2012). The high activity of phosphatase is mainly resulted from the production of this common enzyme by both plants and microorganisms (Allison and Vitousek, 2004). On day 28, the influence range of phosphatase activity in plant rhizosphere of the BN treatment group was the highest. In the presence of copper pollution (Figure 5.69B), on day 7, the influence range of phosphatase activity in plant rhizosphere of the BC(Cu200) and BN(Cu200) treatment groups was significantly increased, indicating that trace copper also promoted the microbial activities to a certain extent. Beside, that of the UR(Cu200) treatment group was not significantly improved, possibly due to the absence of biochar adsorption, making the copper content higher than that of BC(Cu200) and BN(Cu200) treatment groups, and inhibiting the plant rhizosphere microorganisms to a certain extent. From day 14, due to the toxicity of copper, the influence range of phosphatase activity in plant rhizosphere declined sharply. On day 28, that of the BN(Cu200) treatment group was higher than BC(Cu200) treatment group, and higher than that of the UR(Cu200) treatment group and the CK, indicating that biochar could enhance the adsorption to reduce the bioavailability of heavy metals in plant rhizosphere, thus changing the enzyme activity in plant rhizosphere of contaminated soil.

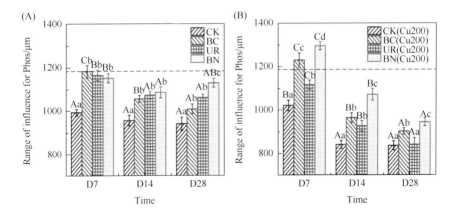

Figure 5.69 Influence range of phosphatase activity in plant rhizosphere of each treatment group during the incubation. Uncontaminated by copper (A), contaminated by copper (B). The uppercase and lowercase letters represent the significant differences between one treatment group at different sampling times and between different treatment groups at the same sampling time, respectively ($P<0.05$). The error line is the mean standard error of three parallel samples.

4. Zymogram picture of plant rhizosphere enzyme activity

The zymograms of chitinase, β-glucosidase and phosphatase in plants of each treatment group during culture are shown as follows (Figure 5.70–Figure 5.72):

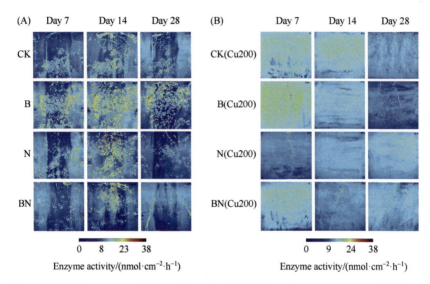

Figure 5.70 Zymogram of plant rhizosphere NAG in each treatment group during the incubation. Without copper pollution (A), with copper pollution (B).

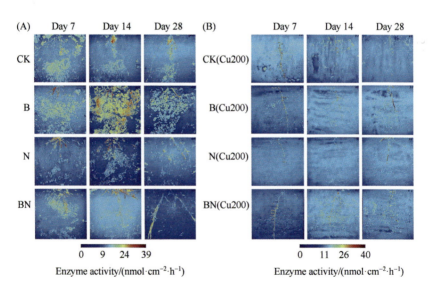

Figure 5.71 Zymogram of plant rhizosphere β-Glu in each treatment group during the incubation. Without copper pollution (A), with copper pollution (B).

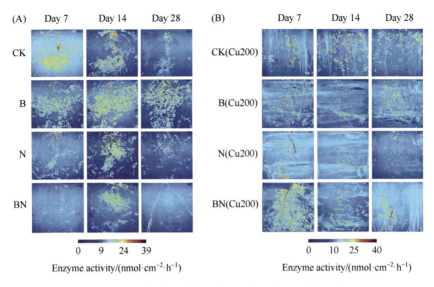

Figure 5.72 Zymogram of plant rhizosphere phosphatase in each treatment group during the incubation. Without copper pollution (A), with copper pollution (B).

5.4.4.4 Content of DTPA extractable Cu in soil

Since the different forms and effective content of heavy metals are dictated by the soil environmental conditions, the total volume of heavy metals cannot determine their environmental behaviors and ecological effects. From an ecological point of view, the effective forms of heavy metals in soil are bioavailable, that is ones that can actually be taken up and used by plants. From the perspective of soil chemistry, these include the acid-soluble, water-soluble, adsorptive, chelated and forms that can be released and absorbed by plants in a short time (Li and Chen, 2004). There was a high correlation between the heavy metal content in soil extracted by DTPA and that absorbed by plants. As can be seen from Figure 5.73, during the whole culture period, there was no significant difference in the content of copper extracted by DTPA in the soil of UR(Cu200) treatment group and that in the CK(Cu200) group, while that of BC(Cu200) and BN(Cu200) treatment groups gradually decreased. This is because biochar can convert exchangeable copper that is easily absorbed and utilized by plants into residual copper that is not easily absorbed and utilized by plants, thus reducing the content of DTPA extracted copper in soil. After 60 days of culture, DTPA-extractable copper content in BC(Cu200) and BN(Cu200) treatment groups decreased by 26.9% and 20.5%, respectively.

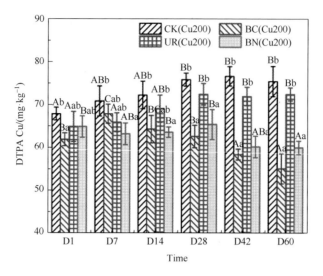

Figure 5.73 Changes of DTPA-extracted copper content in soil of each treatment group during the incubation. The uppercase and lowercase letters represent the significant differences between one treatment group at different sampling times and between different treatment groups at the same sampling time, respectively ($P<0.05$). The error line is the mean standard error of three parallel samples.

5.4.4.5 *Accumulation of Cu in plants*

The heavy metal copper in soil indirectly affects the animal and human health through its accumulation in plants. Copper is not evenly distributed in plants, with a higher content in some parts and a lower in others. The copper content is higher in underground parts than aboveground parts of plants, and lowest in fruits and seeds. As can be seen from Figure 5.74, the copper content accumulated in the underground parts of plants of each treatment group is significantly higher than that in the aboveground parts, indicating that plants growing in copper-polluted environment mainly accumulate copper in the underground parts and rarely migrate to the aboveground parts. There was no significant difference in Cu content accumulated in aboveground parts of plants in the BC(Cu200), UR(Cu200) and BN(Cu200) treatment groups, all of them lower than the CK(Cu200) group. However, the copper content accumulated in the underground parts of plants in the BC(Cu200) and BN(Cu200) treatment groups was significantly lower than that in the CK(Cu200) and UR(Cu200) treatment groups, indicating that copper in soil was partially absorbed by biochar or transformed into a form that was not easy to be absorbed by plants (residual state), thus reducing its accumulation in plants.

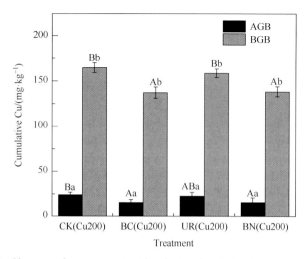

Figure 5.74 Changes of copper content in plants of each treatment group during the incubation. The uppercase and lowercase letters represent the significant differences between one treatment group at different sampling times and between different treatment groups at the same sampling time, respectively ($P<0.05$). The error line is the mean standard error of three parallel samples.

5.4.5 Summary

In this work, corn straw biochar was prepared through anoxic pyrolysis at 450℃ with corn straw as the material. Bentonite and sepiolite were used as the binders to prepare biochar slow-release nitrogen fertilizer (BN) by mixing biochar with urea. First, the microstructure and physicochemical properties of BN material were analyzed by FT-IR, SEM, XPS and XRD methods. Second, with or without copper pollution, through analysis of the changes in pH, NH_4^+-N and NO_3^--N content, morphology of copper ions, enzyme activity, microbial community structure and abundance of the soil and other indicators, the effects of BN material on the passivation of heavy metal copper in soil, on the slow release ability of urea, and on the soil microbial community were evaluated. Meanwhile, with or without copper pollution, the common crop of corn was planted. By analyzing the content of available nitrogen in soil, various growth indexes of plants, the influence range of enzyme activity in plant rhizosphere, the change of DTPA-extracted copper in soil and the content of copper in plants, the effects of BN material on plant nutrient supply, soil quality and remediation of soil copper pollution were evaluated to reveal the effects of BN material on plant growth. The main conclusions are as follows.

(1) Abundant pores were formed on the surface and inside of biochar, enabling a large specific surface area of biochar. Biochar has abundant chemical bonds and functional groups, such as C—O, C=O, —OH and —COOH, indicating that it has the structure to absorb heavy metals in soil. Urea successfully mingled with the BC and reacted with functional groups on biochar to form N-H, C-N, C-NH$_2$ and other groups.

(2) On day 60, the NO$_3^-$-N content in soil of the biochar slow-release nitrogen fertilizer (BN) treatment group was 121.37% higher than that of the urea (UR) treatment group, indicating that the BN material functioned the same with slow-release urea. With copper pollution, the NO$_3^-$-N content in soil of the BN(Cu200) treatment group was 69.98% higher than that of the UR(Cu200) treatment group.

(3) The results of BCR four-step extraction showed that the exchangeable Cu content in the soil treated by biochar (BC) and BN decreased by 26.53% and 18.37%, respectively, while the residual Cu content increased by 115% and 67.4%, respectively. These results indicate that BC and BN can convert the copper from exchangeable state which is easily absorbed and utilized by plants to stable residual state which is not easily utilized, thus reducing the toxicity of copper. The content of DTPA-extracted copper in the BN group decreased by 20.5%. The copper content accumulated in the underground parts of plants in BC and BN treatment groups was significantly lower than that in CK and UR treatment groups.

(4) With or without copper pollution, the BN treatment can boost the activity of chitinase, β-glucosidase and phosphatase in soil. BN can alleviate copper's inhibition of soil enzyme activity in the presence of copper pollution. BN treatment decreased the α-diversity of soil microorganisms. BN treatment increased the relative abundance of Proteobacteria in soil, improved the abundance of ammonia-oxidizing bacteria, and promoted the nitrogen cycle in soil. In the BN(Cu200) treatment group, the Proteobacteria abundance, a sign of the livability of soil environment, is the highest, indicating that BN can improve the soil environment and reduce the inhibition effect of copper on microorganisms upon the copper pollution.

(5) In both cases, BN treatment can promote the plant growth and significantly increase the plant height, root length, leaf SPAD value and biomass. BN can alleviate the toxicity of copper to plants due to the effect of biochar. Besides, BN treatment can expand the influence range of chitinase, β-glucoside and phosphatase activity in plant rhizosphere soil. Biochar in BN can reduce the bioavailability of heavy metals in plant rhizosphere soil

by enhancing the adsorption, so as to change the enzyme activity in contaminated plant rhizosphere soil.

Limited by the experimental conditions, the research is far from perfect and further improvement will be done in the future. Here, the following suggestions are put forward:

(1) Both the soil experiment and plant experiment in this study lasted only 60 days, with subsequent effects and changes of the biochar slow-release nitrogen fertilizer not figured out. In future experiments, the culture time can be extended appropriately to study the effects on the whole lifecycle of plants.

(2) In this study, the in situ enzyme spectrometry technique was used for nondestructive sampling, so fewer treatment groups were set up. Although a small amount was sampled each time, it may still have a slight impact on the plant growth. In future studies, multiple treatment groups can be set up for destructive sampling at the same time, so as to make the experimental results more accurate.

(3) There are some differences between laboratory experiment conditions and actual field ones. This study tries to simulate the field conditions as much as possible, but some, such as random weather changes, cannot be simulated. Therefore, the effect of biochar slow-release nitrogen fertilizer on soil and plants may be too ideal. In the future, we may consider carrying out relevant experiments on actual sites.

Main References

Abbas, A., Azeem, M., Naveed, M., et al., 2020. Synergistic use of biochar and acidified manure for improving growth of maize in chromium contaminated soil. International Journal of Phytoremediation, 22(1): 52–61. Available from: https: doi.org/10.1080/15226514.2019.1644286.

Abdu, N., Abdullahi, A.A., Abdulkadir, A., 2017. Heavy metals and soil microbes. Environmental Chemistry Letters, 15(1): 65–84. Available from: https: doi.org/10.1007/s10311-016-0587-x .

Acosta-Martinez, V., Cano, A., Johnson, J., 2018. Simultaneous determination of multiple soil enzyme activities for soil health-biogeochemical indices. Applied Soil Ecology, 126: 121–128. Available from: https: doi.org/10.1016/j.apsoil.2017.11.024.

Acostamartinez, V., Tabatabai, M.A., 2011. Phosphorus cycle enzymes. Methods of Soil Enzymology 79415 161–183. Available from: https: doi.org/10.2136/sssabookser9.c8.

Adetunji, A.T., Lewu, F.B., Mulidzi, R., et al., 2017. The biological activities of β-glucosidase, phosphatase and urease as soil quality indicators: A review. Journal of Soil Science and Plant Nutrition, 17(3): 794–807. Available from: https: doi.org/10.4067/S0718-95162017000300018.

Agegnehu, G., Srivastava, A.K., Bird, M.I., 2017. The role of biochar and biochar-compost in improving soil quality and crop performance: A review. Applied Soil Ecology, 119: 156–170. Available from: https: doi.org/10.1016/j.apsoil.2017.06.008.

Allison, S.D., Vitousek, P.M., 2005. Responses of extracellular enzymes to simple and complex nutrient inputs. Soil Biology and Biochemistry, 37(5): 937–944. Available from: https: doi.org/10.1016/j.soilbio.2004.09.014.

Allison, S.D., Martiny, J.B.H., 2008. Resistance, resilience, and redundancy in microbial communities. Proceedings of the National Academy of Sciences of the United States of America, 105(supplement_1): 11512–11519.

Antoniadis, V., Alloway, B.J., 2002. The role of dissolved organic carbon in the mobility of Cd, Ni and Zn in sewage sludge-amended soils. Environmental Pollution, 117(3): 515–521. Available from: https: doi.org/10.1016/S0269-7491(01)00172-5.

Antoniadis, V., Levizou, E., Shaheen, S.M., et al., 2017a. Trace elements in the soil-plant interface: Phytoavailability, translocation, and phytoremediation–a review. Earth-Science Reviews, 171: 621–645. Available from: https: doi.org/10.1016/j.earscirev.2017.06.005.

Antoniadis, V., Polyzois, T., Golia, E.E., et al., 2017b. Hexavalent chromium availability and phytoremediation potential of *Cichorium spinosum* as affect by manure, zeolite and soil ageing. Chemosphere, 171: 729–734. Available from: https: doi.org/10.1016/j.chemosphere.2016.11.146.

Antoniadis, V., Zanni, A.A., Levizou, E., et al., 2018. Modulation of hexavalent chromium toxicity on Origanum vulgare in an acidic soil amended with peat, lime, and zeolite. Chemosphere, 195: 291–300. Available from: https: doi.org/10.1016/j.chemosphere.2017.12.069.

Argun, M.E., Dursun, S., Karatas, M., 2009. Removal of Cd(II), Pb(II), Cu(II) and Ni(II) from water using modified pine bark. Desalination, 249(2): 519–527. Available from: https: doi.org/10.1016/j.desal.2009.01.020.

Ashraf, A., Bibi, I., Niazi, N.K., et al., 2017. Chromium(VI) sorption efficiency of acid-activated banana peel over organo-montmorillonite in aqueous solutions. International Journal of Phytoremediation, 19(7): 605–613. Available from: https: doi.org/10.1080/15226514.2016.1256372.

Auffan, M., Achouak, W., Rose, J., et al., 2008. Relation between the redox state of iron-based nanoparticles and their cytotoxicity toward *Escherichia coli*. Environmental Science & Technology, 42(17): 6730–6735. Available from: https: doi.org/10.1021/es800086f.

Awad, Y.M., Lee, S.S., Kim, K.H., et al., 2018. Carbon and nitrogen mineralization and enzyme activities in soil aggregate-size classes: Effects of biochar, oyster shells, and polymers. Chemosphere, 198: 40–48. Available from: https: doi.org/10.1016/j.chemosphere.2018.01.034.

Bailey, V.L., Fansler, S.J., Smith, J.L., et al., 2011. Reconciling apparent variability in effects of biochar amendment on soil enzyme activities by assay optimization. Soil Biology and Biochemistry, 43(2): 296–301. Available from: https: doi.org/10.1016/j.soilbio.2010.10.014.

Bais, H.P., Weir, T.L., Perry, L.G., et al., 2006. The role of root exudates in rhizosphere interactions with plants and other organisms. Annual Review of Plant Biology, 57: 233–266. Available from: https: doi.org/10.1146/annurev.arplant.57.032905.105159.

Bandara, T., Franks, A., Xu, J.M., et al., 2020. Chemical and biological immobilization mechanisms of potentially toxic elements in biochar-amended soils. Critical Reviews in Environmental Science and Technology, 50(9): 903–978. Available from: https: doi.org/10.1080/10643389.2019.1642832.

Bhaumik, M., Maity, A., Srinivasu, V.V., et al., 2011. Enhanced removal of Cr(VI) from aqueous solution using polypyrrole/Fe$_3$O$_4$ magnetic nanocomposite. Journal of Hazardous Materials, 190(1–3): 381–390. Available from: https: doi.org/10.1016/j.jhazmat.2011.03.062.

Biedrzycki, M.L., Jilany, T.A., Dudley, S.A., et al., 2010. Root exudates mediate kin recognition in plants. Communicative & Integrative Biology, 3(1): 28–35. Available from: https: doi.org/10.4161/cib.3.1.10118.

Biesinger, M.C., Payne, B.P., Grosvenor, A.P., et al., 2011. Resolving surface chemical states in XPS analysis of first row transition metals, oxides and hydroxides: Cr, Mn, Fe, Co and Ni. Applied Surface Science, 257(7): 2717–2730. Available from: https: doi.org/10.1016/j.apsusc.2010.10.051.

Bolan, N.S., Adriano, D.C., Natesan, R., et al., 2003. Effects of organic amendments on the reduction and phytoavailability of chromate in mineral soil. Journal of Environmental Quality, 32(1): 120–128. Available from: https: doi.org/10.2134/jeq2003.0120.

Bünemann, E.K., Schwenke, G.D., Van Zwieten, L., 2006. Impact of agricultural inputs on soil organisms: A review. Soil Research, 44(4): 379.

Burns, R.G., DeForest, J.L., Marxsen, J., et al., 2013. Soil enzymes in a changing environment: Current knowledge and future directions. Soil Biology and Biochemistry, 58: 216–234. Available from: https: doi.org/10.1016/j.soilbio.2012.11.009.

Carpio, I.E., Ansari, A., Rodrigues, D.F., 2018. Relationship of biodiversity with heavy metal tolerance and sorption capacity: A meta-analysis approach. Environmental Science & Technology, 52(1): 184–194. Available from: https: doi.org/10.1021/acs.est.7b04131.

Chaffai, R., Elhammadi, M.A., Seybou, T.N., et al., 2007. Altered fatty acid profile of polar lipids in maize seedlings in response to excess copper. Journal of Agronomy and Crop Science, 193(3): 207–217. Available from: https: doi.org/10.1111/j.1439-037X.2007.00252.x.

Choppala, G., Kunhikrishnan, A., Seshadri, B., et al., 2018. Comparative sorption of chromium species as influenced by pH, surface charge and organic matter content in contaminated soils. Journal of Geochemical Exploration, 184: 255–260. Available from: https: doi.org/10.1016/j.gexplo.2016.07.012.

Chu, H.Y., Lin, X.G., Fujii, T., et al., 2007. Soil microbial biomass, dehydrogenase activity, bacterial community structure in response to long-term fertilizer management. Soil Biology and Biochemistry, 39(11): 2971–2976.

Crane, R.A., Scott, T.B., 2012. Nanoscale zero-valent iron: Future prospects for an emerging water treatment technology. Journal of Hazardous Materials, 211: 112–125. Available from: https: doi.org/10.1016/j.jhazmat.2011.11.073 .

Criquet, S., Braud, A., Nèble, S., 2007. Short-term effects of sewage sludge application on phosphatase activities and available P fractions in Mediterranean soils. Soil Biology and Biochemistry, 39(4): 921–929. Available from: https: doi.org/10.1016/j.soilbio.2006.11.002.

Cui, L.Q., Noerpel, M.R., Scheckel, K.G., et al., 2019. Wheat straw biochar reduces environmental cadmium bioavailability. Environment International, 126: 69–75. Available from: https: doi.org/ 10.1016/j.envint.2019.02.022.

Dai, Z.M., Zhang, X.J., Tang, C., et al., 2017. Potential role of biochars in decreasing soil acidification: A critical review. The Science of the Total Environment, 581–582: 601–611. Available from: https: doi.org/10.1016/j.scitotenv.2016.12.169.

Daryanto, S., Wang, L.X., Gilhooly, W.P., et al., 2019. Nitrogen preference across generations under changing ammonium nitrate ratios. Journal of Plant Ecology, 12(2): 235–244. Available from: https: doi.org/10.1093/jpe/rty014.

Dinesh, R., Dubey, R.P., Prasad, G.S., 1998. Soil microbial biomass and enzyme activities as influenced by organic manure incorporation into soils of a rice-rice system. Journal of Agronomy and Crop Science, 181(3): 173–178. Available from: https: doi.org/10.1111/j.1439-037X.1998.tb00414.x.

Ding, Q.W., Qian, T.W., Liu, H.F., et al., 2011. Preparation of zero-valent iron nanoparticles and study of dispersion. Applied Mechanics and Materials, 55–57: 1748–1752.

Duan, C.J., Fang, L.C., Yang, C.L., et al., 2018. Reveal the response of enzyme activities to heavy metals through *in situ* zymography. Ecotoxicology and Environmental Safety, 156: 106–115. Available from: https: doi.org/10.1016/j.ecoenv.2018.03.015.

Eivazi, F., Tabatabai, M.A., 1977. Phosphatases in soils. Soil Biology and Biochemistry 9, 167–172. Available from: https: doi.org/10.1016/0038-0717(77)90070-0.

Ekschmitt, K., Liu, M.Q., Vetter, S., et al., 2005. Strategies used by soil biota to overcome soil organic matter stability—Why is dead organic matter left over in the soil? Geoderma, 128(1–2): 167–176. Available from: https: doi.org/ 10.1016/j.geoderma.2004.12.024.

Ertani, A., Mietto, A., Borin, M., et al., 2017. Chromium in agricultural soils and crops: A review. Water, Air, & Soil Pollution, 228(5): 190. Available from: https: doi.org/10.1007/s11270-017-3356-y.

Fajardo, C., Sánchez-Fortún, S., Costa, G., et al., 2020. Evaluation of nanoremediation strategy in a Pb, Zn and Cd contaminated soil. Science of the Total Environment, 706: 136041. . Available from: https: doi.org/10.1016/j.scitotenv.2019.136041.

Fierer, N., Bradford, M.A., Jackson, R.B., 2007. Toward an ecological classification of soil bacteria. Ecology, 88(6): 1354–1364.

Fierer, N., Leff, J.W., Adams, B.J., et al., 2012. Cross-biome metagenomic analyses of soil microbial communities and their functional attributes. Proceedings of the National Academy of Sciences of the United States of America, 109(52): 21390–21395. Available from: https: doi.org/10.1073/pnas.1215210110.

Flanagan, P.W., Van Cleve, K., 1983. Nutrient cycling in relation to decomposition and organic-matter quality in taiga ecosystems. Canadian Journal of Forest Research, 13(5): 795–817. Available from: https: doi.org/10.1139/x83-110.

Gan, Y.B., Stulen, I., Posthumus, F., et al., 2002. Effects of N management on growth, N_2 fixation and yield of soybean. Nutrient Cycling in Agroecosystems, 62(2): 163–174. Available from: https: doi.org/10.1023/A:1015528132642.

Gao, M.L., Xu, Y.L., Chang, X.P., et al., 2021. Fe–Mn oxide modified biochar decreases phthalate uptake and improves grain quality of wheat grown in phthalate-contaminated fluvo-aquic soil. Chemosphere, 270: 129428. Available from: https: doi.org/10.1016/j.chemosphere.2020.129428.

German, D.P., Weintraub, M.N., Grandy, A.S., et al., 2011. Optimization of hydrolytic and oxidative enzyme methods for ecosystem studies. Soil Biology and Biochemistry, 43(7): 1387–1397. Available from: https: doi.org/10.1016/j.soilbio.2011.11.002.

Ghosh, U., Luthy, R.G., Cornelissen, G., et al., 2011. *In-situ* sorbent amendments: A new direction in contaminated sediment management. Environmental Science & Technology, 45(4): 1163–1168. Available from: https: doi.org/10.1021/es102694h.

Gomez-Eyles, J.L., Beesley, L., Moreno-Jimenez, E., et al, 2013. The potential of biochar amendments to remediate contaminated soils. Biochar and Soil Biota，14: 100–133. Available from: https: doi.org/ 10.1201/b14585.

Gul, S., Whalen, J.K., Thomas, B.W., et al., 2015. Physico-chemical properties and microbial responses in biochar-amended soils: Mechanisms and future directions. Agriculture, Ecosystems & Environment, 206: 46–59. . Available from: https: doi.org/10.1016/j.agee.2015.03.015.

Hagemann, N., Joseph, S., Schmidt, H.P., et al., 2017. Organic coating on biochar explains its nutrient retention and stimulation of soil fertility. Nature Communications, 8(1): 1089. Available from: https: doi.org/10.1038/s41467-017-01123-0.

Han, Y.T., Cao, X., Ouyang, X., et al., 2016. Adsorption kinetics of magnetic biochar derived from peanut hull on removal of Cr (VI) from aqueous solution: Effects of production conditions and particle size. Chemosphere, 145: 336–341. Available from: https: doi.org/10.1016/j.chemosphere.2015.11.050.

He, J.Z., Hu, H.W., Zhang, L.M., 2012. Current insights into the autotrophic thaumarchaeal ammonia oxidation in acidic soils. Soil Biology and Biochemistry, 55: 146–154. Available from: https: doi.org/10.1016/j.soilbio.2012.06.006.

He, L.Z., Zhong, H., Liu, G.X., et al., 2019. Remediation of heavy metal contaminated soils by biochar: Mechanisms, potential risks and applications in China. Environmental Pollution, 252: 846–855. Available from: https: doi.org/10.1016/j.envpol.2019.05.151.

Herrmann, L., Lesueur, D., Robin, A., et al., 2019. Impact of biochar application dose on soil microbial communities associated with rubber trees in North East Thailand. Science of the Total Environment, 689: 970–979. Available from: https: doi.org/10.1016/j.scitotenv.2019.06.441.

Hinsinger, P., Plassard, C., Tang, C.X., et al., 2003. Origins of root-mediated pH changes in the rhizosphere and their responses to environmental constraints: A review. Plant and Soil, 248(1): 43–59. Available from: https: doi.org/10.1023/a:1022371130939.

Hinsinger, P., Bengough, A.G., Vetterlein, D., et al., 2009. Rhizosphere: Biophysics, biogeochemistry and ecological relevance. Plant and Soil, 321(1): 117–152.

Ho, S.H., Zhu, S.S., Chang, J.S., 2017. Recent advances in nanoscale-metal assisted biochar derived from waste biomass used for heavy metals removal. Bioresource Technology, 246: 123–134. Available from: https: doi.org/10.1016/j.biortech.2017.08.061.

Huang, G.L., 2012. Chitinase inhibitor allosamidin and its analogues: An update. Current Organic Chemistry, 16(1): 115–120. Available from: https: doi.org/10.2174/138527212798993121.

Huang, X.R., Zhu-Barker, X., Horwath, W.R., et al., 2016. Effect of iron oxide on nitrification in two agricultural soils with different pH. Biogeosciences, 13(19): 5609–5617. Available from: https: doi.org/10.5194/bg-13-5609-2016.

Huong, P.T., Lee, B.K., Kim, J., et al., 2016. Acid activation pine cone waste at differences temperature and selective removal of Pb^{2+} ions in water. Process Safety and Environmental Protection, 100: 80–90. Available from: https: doi.org/10.1016/j.psep.2015.12.002.

Husak, V., 2015. Copper and copper-containing pesticides: Metabolism, toxicity and oxidative stress. Journal of Vasyl Stefanyk Precarpathian National University, 2(1): 38–50.

ISO, 2005. Soil quality—determination of pH (ISO 10390: 2005). Available at: https://www.iso.org/standard/40879.html.

Jain, S., Mishra, D., Khare, P., et al., 2016. Impact of biochar amendment on enzymatic resilience properties of mine spoils. Science of the Total Environment, 544: 410–421. Available from: https: doi.org/10.1016/j.scitotenv.2015.11.011.

Jiao, F., Jumas, J.C., Womes, M., et al., 2006. Synthesis of ordered mesoporous Fe_3O_4 and gamma-Fe_2O_3 with crystalline walls using post-template reduction/oxidation. Journal of the American Chemical Society, 128(39): 12905–12909. Available from: https: doi.org/10.1021/ja063662i.

Kanchikerimath, M., Singh, D., 2001. Soil organic matter and biological properties after 26 years of maize–wheat–cowpea cropping as affected by manure and fertilization in a Cambisol in semiarid region of India. Agriculture, Ecosystems & Environment, 86(2): 155–162. Available from: https: doi.org/10.1016/S0167-8809(00)00280-2.

Karthik, C., Oves, M., Thangabalu, R., et al., 2016. *Cellulosimicrobium funkei*-like enhances the growth of *Phaseolus vulgaris* by modulating oxidative damage under Chromium(VI) toxicity. Journal of

Advanced Research, 7(6): 839–850. Available from: https: doi.org/10.1016/j.jare.2016.08.007.

Keller, B., Faciano, A., Tsega, A., et al., 2017. Epidemiologic characteristics of children with blood lead levels $\geqslant$ 45 μg/dL. The Journal of Pediatrics, 180: 229–234. Available from: https: doi.org/10.1016/j.jpeds.2016.09.017.

Ken, D.S., Sinha, A., 2020. Recent developments in surface modification of nano zero-valent iron (nZVI): Remediation, toxicity and environmental impacts. Environmental Nanotechnology, Monitoring and Management, 14: 100344.

Khalid, S., Shahid, M., Niazi, N.K., et al., 2017. A comparison of technologies for remediation of heavy metal contaminated soils. Journal of Geochemical Exploration, 182: 247–268. Available from: https: doi.org/10.1016/j.gexplo.2016.11.021.

Khan, K.Y., Ali, B., Cui, X.Q., et al., 2017. Impact of different feedstocks derived biochar amendment with cadmium low uptake affinity cultivar of pak choi (Brassica rapa ssb. chinensis L.) on phytoavoidation of Cd to reduce potential dietary toxicity. Ecotoxicology and Environmental Safety, 141: 129–138. Available from: https: doi.org/10.1016/j.ecoenv.2017.03.020.

Khan, M.Y., Mangrich, A.S., Schultz, J., et al., 2015. Green chemistry preparation of superparamagnetic nanoparticles containing Fe_3O_4 cores in biochar. Journal of Analytical and Applied Pyrolysis, 116: 42–48. Available from: https: doi.org/10.1016/j.jaap.2015.10.008.

Khorram, M.S., Lin, D.L., Zhang, Q., et al., 2017. Effects of aging process on adsorption–desorption and bioavailability of fomesafen in an agricultural soil amended with rice hull biochar. Journal of Environmental Sciences, 56: 180–191. Available from: https: doi.org/10.1016/j.jes.2016.09.012.

Kuzyakov, Y., Razavi, B.S., 2019. Rhizosphere size and shape: Temporal dynamics and spatial stationarity. Soil Biology and Biochemistry, 135: 343–360. Available from: https: doi.org/10.1016/j.soilbio.2019.05.011.

Lakshminarayanan, P.V., Toghiani, H., Pittman, C.U., 2004. Nitric acid oxidation of vapor grown carbon nanofibers. Carbon, 42(12–13): 2433–2442. Available from: https: doi.org/10.1016/j.carbon.2004.04.040.

Lei, C., Sun, Y.Q., Tsang, D.C.W., et al., 2018. Environmental transformations and ecological effects of iron-based nanoparticles. Environmental Pollution, 232: 10–30. Available from: https: doi.org/10.1016/j.envpol.2017.09.052.

Li, H.B., Dong, X.L., da Silva, E.B., et al., 2017. Mechanisms of metal sorption by biochars: Biochar characteristics and modifications. Chemosphere, 178: 466–478. Available from: https: doi.org/10.1016/j.chemosphere.2017.03.072.

Li, P.R., Yu, J., Huangfu, Z.X., et al., 2020. Applying modified biochar with nZVI/nFe3O4 to immobilize Pb in contaminated soil. Environmental Science and Pollution Research International, 27(19): 24495–24506. Available from: https: doi.org/10.1007/s11356-020-08458-0.

Li, S.L., Wang, S., Fan, M.C., et al., 2020. Interactions between biochar and nitrogen impact soil carbon mineralization and the microbial community. Soil and Tillage Research, 196: 104437. Available from: https: doi.org/10.1016/j.still.2019.104437.

Liang, Z., He, S., 2002. Studying on slow release effects of mineral coated compound fertilizer. Soil and environmental sciences, 11: 376–378.

Liptzin, D., Silver, W.L., 2009. Effects of carbon additions on iron reduction and phosphorus availability in a humid tropical forest soil. Soil Biology and Biochemistry, 41(8): 1696–1702. Available from: https: doi.org/10.1016/j.soilbio.2009.05.013.

Liu, C.P., Yu, H.Y., Liu, C.S., et al., 2015. Arsenic availability in rice from a mining area: Is amorphous iron oxide-bound arsenic a source or sink? Environmental Pollution, 199: 95–101. Available from: https: doi.org/10.1016/j.envpol.2015.01.025.

Liu, S.B., Zhang, X., Wang, X., 2017a. Response of soil copper speciation and bioavailability to application of biochar. Chinese Journal of Environmental Engineering, 11: 5743–5750.

Liu, S.B., Razavi, B.S., Su, X., et al., 2017b. Spatio-temporal patterns of enzyme activities after manure application reflect mechanisms of niche differentiation between plants and microorganisms. Soil Biology and Biochemistry, 112: 100–109. Available from: https: doi.org/10.1016/j.soilbio.2017.05.006.

Liu, S.B., Schleuss, P., Kuzyakov, Y., 2018. Responses of degraded tibetan kobresia pastures to n addition. Land Degradation and Development 29, 303–314. Available from: https: doi.org/10.1002/ldr.2720.

Liu, S.B., Pu, S.Y., Deng, D.L., et al., 2020. Comparable effects of manure and its biochar on reducing soil Cr bioavailability and narrowing the rhizosphere extent of enzyme activities. Environment International, 134: 105277. Available from: https: doi.org/10.1016/j.envint.2019.105277.

Liu, X.R., Liao, J.Y., Song, H.X., et al., 2019. A biochar-based route for environmentally friendly controlled release of nitrogen: Urea-loaded biochar and bentonite composite. Scientific Reports, 9(1): 9548. Available from: https: doi.org/10.1038/s41598-019-46065-3.

Lu, A.X., Zhang, S.Z., Shan, X.Q., 2005. Time effect on the fractionation of heavy metals in soils. Geoderma, 125(3–4): 225–234. Available from: https: doi.org/10.1016/j.geoderma.2004.08.002.

Lu, K.P., Yang, X., Shen, J.J., et al., 2014. Effect of bamboo and rice straw biochars on the bioavailability of Cd, Cu, Pb and Zn to *Sedum plumbizincicola*. Agriculture, Ecosystems & Environment, 191: 124–132. Available from: https: doi.org/10.1016/j.agee.2014.04.010.

Lyon, D.Y., Brunet, L., Hinkal, G.W., et al., 2008. Antibacterial activity of fullerene water suspensions (nC60) is not due to ROS-mediated damage. Nano Letters, 8(5): 1539–1543.

Ma, X.M., Gurung, A., Deng,Y., 2013. Phytotoxicity and uptake of nanoscale zero-valent iron (nZVI) by two plant species. Science of the Total Environment, 443: 844–849. Available from: https: doi.org/10.1016/j.scitotenv.2012.11.073.

Ma, X.M., Razavi, B.S., Holz, M., et al., 2017. Warming increases hotspot areas of enzyme activity and shortens the duration of hot moments in the root-detritusphere. Soil Biology and Biochemistry, 107: 226–233. Available from: https: doi.org/10.1016/j.soilbio.2017.01.009.

Mackie, K.A., Marhan, S., Ditterich, F., et al., 2015. The effects of biochar and compost amendments on copper immobilization and soil microorganisms in a temperate vineyard. Agriculture, Ecosystems & Environment, 201: 58–69. Available from: https: doi.org/10.1016/j.agee.2014.12.001.

Maharjan, M., Sanaullah, M., Razavi, B.S., et al., 2017. Effect of land use and management practices on microbial biomass and enzyme activities in subtropical top-and sub-soils. Applied Soil Ecology, 113: 22–28. Available from: https: doi.org/10.1016/j.apsoil.2017.01.008.

Maiti, S., Ghosh, N., Mandal, C., et al., 2012. Responses of the maize plant to chromium stress with reference to antioxidation activity. Brazilian Society of Plant Physiology, 24: 203–212. Available from: https: doi.org/10.1590/S1677-04202012000300007.

Malcolm, R.E., 1983. Assessment of phosphatase activity in soils. Soil Biology and Biochemistry, 15(4): 403–408. Available from: https: doi.org/10.1016/0038-0717(83)90003-2.

Mallick, S., Sinam, G., Kumar Mishra, R., et al., 2010. Interactive effects of Cr and Fe treatments on plants growth, nutrition and oxidative status in *Zea mays* L. Ecotoxicology and Environmental Safety, 73(5): 987–995. Available from: https: doi.org/10.1016/j.ecoenv.2010.03.004.

Mandal, S., Sarkar, B., Bolan, N., et al., 2017. Enhancement of chromate reduction in soils by surface modified biochar. Journal of Environmental Management, 186: 277–284. Available from: https: doi.org/10.1016/j.jenvman.2016.05.034.

Mandal, S., Pu, S.Y., He, L.L., et al., 2020a. Biochar induced modification of graphene oxide & nZVI and its impact on immobilization of toxic copper in soil. Environmental Pollution, 259: 113851. Available from: https: doi.org/10.1016/j.envpol.2019.113851.

Mandal, S., Pu, S.Y., Shangguan, L.X., et al., 2020b. Synergistic construction of green tea biochar supported nZVI for immobilization of lead in soil: A mechanistic investigation. Environment International, 135: 105374. Available from: https: doi.org/10.1016/j.envint.2019.105374

Marklein, A.R., Houlton, B.Z., 2012. Nitrogen inputs accelerate phosphorus cycling rates across a wide variety of terrestrial ecosystems. The New Phytologist, 193(3): 696–704. Available from: https: doi.org/10.1111/j.1469-8137.2011.03967.x.

Mašková, T., Herben, T., 2018. Root: Shoot ratio in developing seedlings: How seedlings change their allocation in response to seed mass and ambient nutrient supply. Ecology and Evolution, 8(14): 7143–7150. Available from: https: doi.org/10.1002/ece3.4238.

Masto, R.E., Kumar, S., Rout, T.K., et al., 2013. Biochar from water hyacinth (Eichornia crassipes) and its impact on soil biological activity. CATENA, 111: 64–71. Available from: https: doi.org/10.1002/ece3. 4238.

McConnaughay, K.D.M., Coleman, J. S., 1999. Biomass allocation in plants: Ontogeny or optimality? A test along three resource gradients. Ecology, 80(8): 2581. Available from: https: doi.org/10.2307/ 177242.

Merino, C., Godoy, R., Matus, F., 2016. Soil microorganisms and enzyme activity at different levels of organic matter stability. Journal of Soil Science and Plant Nutrition, 16: 14–30.

Merino, N.A., Barbero, B.P., Eloy, P., et al., 2006. La_{1-x} Ca $_x$ CoO$_3$ perovskite-type oxides: Identification of the surface oxygen species by XPS. Applied Surface Science, 253(3): 1489–1493. Available from: https: doi.org/10.1016/j.apsusc.2006.02.035.

Mikanova, O., 2006. Effects of heavy metals on some soil biological parameters. Journal of Geochemical Exploration, 88(1–3): 220–223. Available from: https: doi.org/10.1016/j.gexplo.2005.08.043.

Muscolo, A., Sidari, M., Mercurio, R., 2007. Influence of gap size on organic matter decomposition, microbial biomass and nutrient cycle in Calabrian pine (Pinus laricio, Poiret) stands. Forest Ecology and Management, 242(2–3): 412–418. Available from: https: doi.org/10.1016/j.foreco.2007.01.058.

Navarro-Pedreño, J., Almendro-Candel, M., Gómez Lucas, I., et al., 2018. Trace metal content and availability of essential metals in agricultural soils of Alicante (Spain). Sustainability, 10(12): 4534. Available from: https: doi.org/10.3390/su10124534.

Neeli, S.T., Ramsurn, H., 2018. Synthesis and formation mechanism of iron nanoparticles in graphitized carbon matrices using biochar from biomass model compounds as a support. Carbon, 134: 480–490. Available from: https: doi.org/10.1016/j.carbon.2018.03.079.

Němeček, J., Lhotský, O., Cajthaml, T., 2014. Nanoscale zero-valent iron application for in situ reduction of hexavalent chromium and its effects on indigenous microorganism populations. Science of the Total Environment, 485: 739–747. Available from: https: doi.org/10.1016/j.scitotenv.2013.11.105.

Nigam, N., Yadav, V., Mishra, D., et al., 2019. Biochar amendment alters the relation between the Pb distribution and biological activities in soil. International Journal of Environmental Science and Technology, 16(12): 8595–8606.

O'Connor, D., Hou, D.Y., Ye, J., et al., 2018. Lead-based paint remains a major public health concern: A critical review of global production, trade, use, exposure, health risk, and implications. Environment International, 121: 85–101. Available from: https: doi.org/10.1016/j.envint.2018.08.052.

Oburger, E., Gruber, B., Schindlegger, Y., et al., 2014. Root exudation of phytosiderophores from soil-grown wheat. The New Phytologist, 203(4): 1161–1174. Available from: https: doi.org/10.1111/nph.12868.

Oleszczuk, P., Kołtowski, M., 2017. Effect of co-application of nano-zero valent iron and biochar on the total and freely dissolved polycyclic aromatic hydrocarbons removal and toxicity of contaminated soils. Chemosphere, 168: 1467–1476. Available from: https: doi.org/10.1016/j.chemosphere.2016.11.100.

Park, J.H., Lamb, D., Paneerselvam, P., et al., 2011. Role of organic amendments on enhanced bioremediation of heavy metal(loid) contaminated soils. Journal of Hazardous Materials, 185(2–3): 549–574. Available from: https: doi.org/10.1016/j.jhazmat.2010.09.082.

Pei, G.P., Zhu, Y.E., Wen, J.G., et al., 2020. Vinegar residue supported nanoscale zero-valent iron: Remediation of hexavalent chromium in soil. Environmental Pollution, 256: 113407. Available from: https: doi.org/10.1016/j.envpol.2019.113407.

Pietrzykowski, M., Socha, J., van Doorn, N.S., 2014. Linking heavy metal bioavailability (Cd, Cu, Zn and Pb) in Scots pine needles to soil properties in reclaimed mine areas. Science of the Total Environment, 470: 501–510. Available from: https: doi.org/10.1016/j.scitotenv.2013.10.008.

Pu, S.Y., Yan, C., Huang, H.Y., et al., 2019. Toxicity of nano-CuO particles to maize and microbial community largely depends on its bioavailable fractions. Environmental Pollution, 255: 113248. Available from: https: doi.org/10.1016/j.envpol.2019.113248.

Punz, W.F., Sieghardt, H., 1993. The response of roots of herbaceous plant species to heavy metals. Environmental and Experimental Botany, 33(1): 85–98.

Qiao, J.T., Liu, T.X., Wang, X.Q., et al., 2018. Simultaneous alleviation of cadmium and arsenic accumulation in rice by applying zero-valent iron and biochar to contaminated paddy soils. Chemosphere, 195: 260–271. Available from: https: doi.org/10.1016/j.chemosphere.2017.12.081.

Qin, W.X., Fang, G.D., Wang, Y.J., et al., 2018. Mechanistic understanding of polychlorinated biphenyls degradation by peroxymonosulfate activated with $CuFe_2O_4$ nanoparticles: Key role of superoxide radicals. Chemical Engineering Journal, 348: 526–534. Available from: https: doi.org/10.1016/j.cej.2018.04.215.

Rajajayavel, S.R.C., Ghoshal, S., 2015. Enhanced reductive dechlorination of trichloroethylene by sulfidated nanoscale zerovalent iron. Water Research, 78: 144–153. Available from: https: doi.org/10.1016/j.watres.2015.04.009.

Razavi, B.S., Zarebanadkouki, M., Blagodatskaya, E., et al., 2016. Rhizosphere shape of lentil and maize: Spatial distribution of enzyme activities. Soil Biology and Biochemistry, 96: 229–237. Available from: https: doi.org/10.1016/j.soilbio.2016.02.020.

Reyhanitabar, A., Alidokht, L., Khataee, A.R., et al., 2012. Application of stabilized Fe^0 nanoparticles for remediation of Cr(VI)-spiked soil. European Journal of Soil Science, 63(5): 724–732. Available from: https: doi.org/10.1111/j.1365-2389.2012.01447.x.

Rinklebe, J., Shaheen, S.M., 2017. Geochemical distribution of Co, Cu, Ni, and Zn in soil profiles of Fluvisols, Luvisols, Gleysols, and Calcisols originating from Germany and Egypt. Geoderma, 307: 122–138. Available from: https: doi.org/10.1016/j.geoderma.2017.08.005.

Rizvi, A., Khan, M.S., 2019. Heavy metal-mediated toxicity to maize: Oxidative damage, antioxidant defence response and metal distribution in plant organs. International Journal of Environmental Science and

Technology, 16(8): 4873–4886. Available from: https: doi.org/10.1007/s13762-018-1916-3.

Rumpel, C., 2011. Biochar for environmental management: Science and technology. Natures Sciences Societes, 19(4): 457–458.

Schimmelpfennig, S., Glaser, B., 2012. One step forward toward characterization: Some important material properties to distinguish biochars. Journal of Environmental Quality, 41(4): 1001–1013. Available from: https: doi.org/10.2134/jeq2011.0146.

Shahid, M., Shamshad, S., Rafiq, M., et al., 2017. Chromium speciation, bioavailability, uptake, toxicity and detoxification in soil-plant system: A review. Chemosphere, 178: 513–533. Available from: https: doi.org/10.1016/j.chemosphere.2017.03.074.

Shakoor, M.B., Niazi, N.K., Bibi, I., et al., 2019. Exploring the arsenic removal potential of various biosorbents from water. Environment International, 123: 567–579. Available from: https: doi.org/10.1016/j.envint.2018.12.049.

Shan, A.L., Idrees, A., Zaman, W.Q., et al., 2021. Synthesis of nZVI-Ni@BC composite as a stable catalyst to activate persulfate: Trichloroethylene degradation and insight mechanism. Journal of Environmental Chemical Engineering, 9(1): 104808. Available from: https: doi.org/10.1016/j.jece.2020.104808.

Shen, T.T., Tang, Y.Y., Lu, X.Y., et al., 2018. Mechanisms of copper stabilization by mineral constituents in sewage sludge biochar. Journal of Cleaner Production, 193: 185–193. Available from: https: doi.org/10.1016/j.jclepro.2018.05.071.

Shi, W., Ju, Y.Y., Bian, R.J., et al., 2020. Biochar bound urea boosts plant growth and reduces nitrogen leaching. The Science of the Total Environment, 701: 134424. Available from: https: doi.org/10.1016/j.scitotenv.2019.134424.

Shimoi, Y., Honma, D., Kurematsu, A., et al., 2020. Effects of chitin degradation products N-acetylglucosamine and N, N'-diacetylchitobiose on chitinase activity and bacterial community structure in an incubated upland soil. Soil Science and Plant Nutrition, 66(3): 429–437. Available from: https: doi.org/10.1080/00380768.2020.1767488.

Shin, K.H., Cha, D.K., 2008. Microbial reduction of nitrate in the presence of nanoscale zero-valent iron. Chemosphere, 72(2): 257–262. Available from: https: doi.org/10.1016/j.chemosphere.2008.01.043.

Shipley, B., Meziane, D., 2002. The balanced-growth hypothesis and the allometry of leaf and root biomass allocation. Functional Ecology, 16(3): 326–331. Available from: https: doi.org/10.1046/j.1365-2435.2002.00626.x.

Singh, R., Misra, V., Singh, R.P., 2011. Synthesis, characterization and role of zero-valent iron nanoparticle in removal of hexavalent chromium from chromium-spiked soil. Journal of Nanoparticle Research, 13(9): 4063–4073. Available from: https: doi.org/10.1007/s11051-011-0350-y.

Sinsabaugh, R.L., Lauber, C.L., Weintraub, M.N., et al., 2008. Stoichiometry of soil enzyme activity at global scale. Ecology Letters, 11(11): 1252–1264. Available from: https: doi.org/10.1111/j.1461-0248.2008.01245.x.

Smith, J.L., Collins, H.P., Bailey, V.L., 2010. The effect of young biochar on soil respiration. Soil Biology and Biochemistry, 42(12): 2345–2347. Available from: https: doi.org/10.1016/j.soilbio.2010.09.013.

Stefaniuk, M., Oleszczuk, P., Ok, Y.S., 2016. Review on nano zerovalent iron (nZVI): From synthesis to environmental applications. Chemical Engineering Journal, 287: 618–632. Available from: https: doi.org/10.1016/j.cej.2015.11.046.

Su, H.J., Fang, Z.Q., Tsang, P.E., et al., 2016. Remediation of hexavalent chromium contaminated soil by biochar-supported zero-valent iron nanoparticles. Journal of Hazardous Materials, 318: 533–540. Available from: https: doi.org/10.1016/j.jhazmat.2016.07.039.

Sun, J., Wang, H.M., 2016. Soil nitrogen and carbon determine the trade-off of the above- and below-ground biomass across alpine grasslands, Tibetan Plateau. Ecological Indicators, 60: 1070–1076. Available from: https: doi.org/10.1016/j.ecolind.2015.08.038.

Sun, P.F., Hui, C., Azim Khan, R., et al., 2015. Efficient removal of crystal violet using Fe_3O_4-coated biochar: The role of the Fe_3O_4 nanoparticles and modeling study their adsorption behavior. Scientific Reports, 5: 12638. Available from: https: doi.org/10.1038/srep12638.

Sun, Y.H., Zheng, F.Y., Wang, W.J., et al., 2020. Remediation of Cr(VI)-contaminated soil by nano-zero-valent iron in combination with biochar or humic acid and the consequences for plant performance. Toxics, 8(2): 26. Available from: https: doi.org/10.3390/toxics8020026.

Tan, X.F., Liu, Y.G., Zeng, G.M., et al., 2015. Application of biochar for the removal of pollutants from aqueous solutions. Chemosphere, 125: 70–85. Available from: https: doi.org/10.1016/j.chemosphere. 2014.12.058.

Tang, J.C., Lv, H.H., Gong, Y.Y., et al., 2015. Preparation and characterization of a novel graphene/biochar composite for aqueous phenanthrene and mercury removal. Bioresource Technology, 196: 355–363. Available from: https: doi.org/10.1016/j.biortech.2015.07.047.

Tang, J.Y., Zhang, L.H., Zhang, J.C., et al., 2020. Physicochemical features, metal availability and enzyme activity in heavy metal-polluted soil remediated by biochar and compost. Science of the Total Environment, 701: 134751. Available from: https: doi.org/10.1016/j.scitotenv.2019.134751.

Tang, S.C.N., Lo, I.M.C., 2013. Magnetic nanoparticles: Essential factors for sustainable environmental applications. Water Research, 47(8): 2613–2632. Available from: https: doi.org/10.1016/j.watres.2013. 02.039.

Tilman, D., Cassman, K.G., Matson, P.A., et al., 2002. Agricultural sustainability and intensive production practices. Nature, 418(6898): 671–677.

Tiwari, J.N., Mahesh, K., Le, N.H., et al., 2013. Reduced graphene oxide-based hydrogels for the efficient capture of dye pollutants from aqueous solutions. Carbon, 56: 173–182. vailable from: https: doi.org/10.1016/j.carbon.2013.01.001.

Trivedi, P., Anderson, I.C., Singh, B.K., 2013. Microbial modulators of soil carbon storage: Integrating genomic and metabolic knowledge for global prediction. Trends in Microbiology, 21(12): 641–651. Available from: https: doi.org/10.1016/j.tim.2013.09.005.

Turner, B.L., 2010. Variation in pH optima of hydrolytic enzyme activities in tropical rain forest soils. Applied and Environmental Microbiology, 76(19): 6485–6493. Available from: https: doi.org/10.1128/ aem.00560-10.

Turu, Y., Peter, H., 2008. Analysis of XPS spectra of Fe^{2+} and Fe^{3+} ions in oxide materials. Applied Surface Science, 254(8): 2441–2449. Available from: https: doi.org/10.1016/j.apsusc.2009.04.153.

Uksa, M., Schloter, M., Endesfelder, D., et al., 2015. Prokaryotes in subsoil-evidence for a strong spatial separation of different *Phyla* by analysing co-occurrence networks. Frontiers in Microbiology, 6: 1269. Available from: https: doi.org/10.3389/fmicb.2015.01269.

Várhegyi, G., Bobály, B., Jakab, E., et al., 2011. Thermogravimetric study of biomass pyrolysis kinetics. A distributed activation energy model with prediction tests. Energy & Fuels, 25(1): 24–32. Available from: https: doi.org/10.1021/ef101079r.

Ventura, M., Canchaya, C., Tauch, A., et al., 2007. Genomics of Actinobacteria: Tracing the evolutionary history of an ancient Phylum. Microbiology and Molecular Biology Reviews: MMBR, 71(3): 495–548. Available from: https: doi.org/10.1128/MMBR.00005-07.

Waa, B., Sma, B., Nd, C., et al., 2021. Utilization of *Citrullus lanatus* L. seeds to synthesize a novel MnFe$_2$O$_4$-biochar adsorbent for the removal of U(VI) from wastewater: Insights and comparison between modified and raw biochar. The Science of the Total Environment, 771: 144955. Available from: https: doi.org/10.1016/j.scitotenv.2021.144955.

Wada, S., Toyota, K., 2007. Repeated applications of farmyard manure enhance resistance and resilience of soil biological functions against soil disinfection. Biology and Fertility of Soils, 43(3): 349–356. Available from: https: doi.org/10.1007/s00374-006-0116-3.

Wang, J.L., Wang, S.Z., 2019. Preparation, modification and environmental application of biochar: A review. Journal of Cleaner Production, 227: 1002–1022. Available from: https: doi.org/10.1016/j.jclepro.2019.04.282.

Wang, X.B., Song, D.L., Liang GQ, et al., 2015. Maize biochar addition rate influences soil enzyme activity and microbial community composition in a fluvo-aquic soil. Applied Soil Ecology, 96: 265–272. Available from: https: doi.org/10.1016/j.apsoil.2015.08.018.

Weber, K., Quicker, P., 2018. Properties of biochar. Fuel, 217: 240–261. Available from: https: doi.org/10.1016/j.fuel.2017.12.054.

Wei, X.M., Ge, T.D., Zhu, Z.K., et al., 2019. Expansion of rice enzymatic rhizosphere: Temporal dynamics in response to phosphorus and cellulose application. Plant and Soil, 445(1): 169–181. Available from: https: doi.org/10.1007/s11104-018-03902-0.

Weil, R.R., Brady, N.C., 2016. Soil acidity. In: Brady, N.C., and Weil, R.R. (Eds.), The Nature and Properties of Soils. Pearson Education Inc., USA, pp. 374–417.

Wen, E.G., Yang, X., Chen, H.B., et al., 2021. Iron-modified biochar and water management regime-induced changes in plant growth, enzyme activities, and phytoavailability of arsenic, cadmium and lead in a paddy soil. Journal of Hazardous Materials, 407: 124344. Available from: https: doi.org/10.1016/j.jhazmat.2020.124344.

Wen, P., Wu, Z.S., Han, Y.J., et al., 2017. Microwave-assisted synthesis of a novel biochar-based slow-release nitrogen fertilizer with enhanced water-retention capacity. ACS Sustainable Chemistry & Engineering, 5(8): 7374–7382. Available from: https: doi.org/10.1021/acssuschemeng.7b01721.

Wu, J.Z., Li, Z.T., Wang, L., et al., 2020. A novel calcium-based magnetic biochar reduces the accumulation of As in grains of rice (*Oryza sativa* L.) in As-contaminated paddy soils. Journal of Hazardous Materials, 394: 122507. Available from: https: doi.org/10.1016/j.jhazmat.2020.122507.

Wyszkowska, J., Borowik, A., Kucharski, M., et al., 2013. Effect of cadmium, copper and zinc on plants, soil microorganisms and soil enzymes. Journal of Elementology, 18: 769–796. Available from: https: doi.org/10.5601/jelem.2013.18.4.455.

Xiong, Z., He, F., Zhao, D.Y., et al., 2009. Immobilization of mercury in sediment using stabilized iron sulfide nanoparticles. Water Research, 43(20): 5171–5179. Available from: https: doi.org/10.1016/j.watres.2009.08.018.

Xu, C., Peng, C., Sun, L.J., et al., 2015. Distinctive effects of TiO$_2$ and CuO nanoparticles on soil microbes and their community structures in flooded paddy soil. Soil Biology and Biochemistry, 86: 24–33. Available from: https: doi.org/10.1016/j.soilbio.2015.03.011.

Xu, J.M., Tang, C.X., He, J.Z., 2010. Molecular environmental soil science at the interfaces in the Earth's critical zone. Journal of Soils and Sediments, 10(5): 797–798. Available from: https:

doi.org/10.1007/s11368-010-0236-0.

Xu, X.Y., Cao, X.D., Zhao, L., 2013. Comparison of rice husk- and dairy manure-derived biochars for simultaneously removing heavy metals from aqueous solutions: Role of mineral components in biochars. Chemosphere, 92(8): 955–961. Available from: https: doi.org/10.1016/j.chemosphere.2013.03.009.

Yanardağ, I.H., Zornoza, R., Bastida, F., et al., 2017. Native soil organic matter conditions the response of microbial communities to organic inputs with different stability. Geoderma, 295: 1–9. Available from: https: doi.org/10.1016/j.geoderma.2017.02.008.

Yang, G.X., Jiang, H., 2014. Amino modification of biochar for enhanced adsorption of copper ions from synthetic wastewater. Water Research, 48: 396–405. Available from: https: doi.org/10.1016/j.watres.2013.09.050.

Yang, L., Bian, X.G., Yang, R.P., et al., 2018. Assessment of organic amendments for improving coastal saline soil. Land Degradation & Development, 29(9): 3204–3211. Available from: https: doi.org/10.1002/ldr.3027.

Yang, Y.Q., Chen, N., Feng, C.P., et al., 2018. Chromium removal using a magnetic corncob biochar/polypyrrole composite by adsorption combined with reduction: Reaction pathway and contribution degree. Colloids and Surfaces A: Physicochemical and Engineering Aspects, 556: 201–209. Available from: https: doi.org/10.1016/j.colsurfa.2018.08.035.

Yao, Y., Zhou, H., Yan, X.L., et al., 2021. The Fe_3O_4-modified biochar reduces arsenic availability in soil and arsenic accumulation in indica rice (*Oryza sativa* L.). Environmental Science and Pollution Research International, 28(14): 18050–18061. Available from: https: doi.org/10.1007/s11356-020-11812-x.

Yin, D.X., Wang, X., Peng, B., et al., 2017. Effect of biochar and Fe-biochar on Cd and As mobility and transfer in soil-rice system. Chemosphere, 186: 928–937. Available from: https: doi.org/10.1016/j.chemosphere.2017.07.126.

Yirsaw, B.D., Megharaj, M., Chen, Z.L., et al., 2016. Environmental application and ecological significance of nano-zero valent iron. Journal of Environmental Sciences, 44: 88–98. Available from: https: doi.org/10.1016/j.jes.2015.07.016.

Yu, B.Z., Li, D.Q., Wang, Y.H., et al., 2020. The compound effects of biochar and iron on watercress in a Cd/Pb-contaminated soil. Environmental Science and Pollution Research International, 27(6): 6312–6325. Available from: https: doi.org/10.1007/s11356-019-07353-7.

Yuan, J.H., Xu, R.K., Qian, W., et al., 2011. Comparison of the ameliorating effects on an acidic ultisol between four crop straws and their biochars. Journal of Soils and Sediments, 11(5): 741–750. Available from: https: doi.org/10.1007/s11368-011-0365-0.

Zand, A.D., Tabrizi, A.M., Heir, A.V., 2020. Incorporation of biochar and nanomaterials to assist remediation of heavy metals in soil using plant species. Environmental Technology & Innovation, 20: 101134. Available from: https: doi.org/10.1016/j.eti.2020.101134 .

Zhang, J.Y., Zhou, H., Zeng, P., et al., 2021. Nano-Fe_3O_4-modified biochar promotes the formation of iron plaque and cadmium immobilization in rice root. Chemosphere, 276: 130212. Available from: https: doi.org/10.1016/j.chemosphere.2021.130212.

Zhang, W.K., Song, S.Q., Nath, M., et al., 2021. Inhibition of Cr^{6+} by the formation of *in situ* Cr^{3+} containing solid-solution in Al_2O_3–CaO–Cr_2O_3–SiO_2 system. Ceramics International, 47(7): 9578–9584. Available from: https: doi.org/10.1016/j.ceramint.2020.12.092.

Zhou, X., Qiao, M., Su, J.Q., et al., 2019. Turning pig manure into biochar can effectively mitigate antibiotic resistance genes as organic fertilizer. Science of the Total Environment, 649: 902-908.

Available from: https: doi.org/10.1016/j.scitotenv.2018.08.368.

Zhu, X.M., Chen, B.L., Zhu, L.Z., et al., 2017. Effects and mechanisms of biochar-microbe interactions in soil improvement and pollution remediation: A review. Environmental Pollution, 227: 98–115. Available from: https: doi.org/10.1016/j.envpol.2017.04.032.

Zimmerman, A.R., Gao, B., Ahn, M.Y., 2011. Positive and negative carbon mineralization priming effects among a variety of biochar-amended soils. Soil Biology and Biochemistry, 43(6): 1169–1179. Available from: https: doi.org/10.1016/j.soilbio.2011.02.005.

Zornoza, R., Moreno-Barriga, F., Acosta, J.A., et al., 2016. Stability, nutrient availability and hydrophobicity of biochars derived from manure, crop residues, and municipal solid waste for their use as soil amendments. Chemosphere, 144: 122–130. Available from: https: doi.org/10.1016/j.chemosphere.2015.08.046.

Chapter 6

Enzyme activities in the rhizosphere of soil and groundwater

6.1 Nutrient availability and relationships with enzyme activities in the rhizosphere

6.1.1 Introduction

The global agricultural productivity must increase by up to 70% to meet the food demand of nearly 10 billion people expected to populate Earth by 2050 (Hunter et al., 2017). This onerous challenge must be met without the expansion of arable land and by using lower inputs of fertilizers and pesticides, or we may risk increasing natural ecosystem loss and causing the depletion of rock phosphate reserves (Abelson, 1999; Chowdhury et al., 2017; Sattari et al., 2012; Tilman et al., 2011). Nutrient deficiency, extensively confirmed in most terrestrial ecosystems, threatens plant production and ecosystem stability (Elser et al., 2007; Hou et al., 2020). One of the ways by which plants attempt to overcome this stress is by reinforcing nutrient availability in the rhizosphere (Hinsinger et al., 2003; Kuzyakov and Razavi, 2019). The rhizosphere is one of the most pivotal microbial hotspots determining the dynamics and cycling of nutrients, therefore the plant-microbial interactions are very intense in this critical zone (Kuzyakov and Blagodatskaya, 2015). The interplay between roots and microbes determines the available nutrient content by regulating three process groups: ①the rate of nutrient uptake by plants, immobilization by microorganisms, or release from organic matter decomposition; ②nutrient mobility in soils; ③the conversion between unavailable and available forms of nutrients (Darrah, 1993).

Many interrelated factors, such as original soil properties, plant characteristics and climate factors, influence the three process groups and their interplay, and modulate the rhizosphere nutrient status (Jones et al., 2004). For instance, in low-fertile environments, a range of mechanisms (e.g., exudation of organic compounds and protons) is activated to increase the solubilization of nutrients in the rhizosphere (Rengel and Marschner, 2005). In acidic soil, plants elevate pH and optimize rhizosphere nutrient availability by releasing various root exudates, enhancing the uptake of anions (e.g., nitrate) or excreting OH^- from nitrate reduction (Sugihara et al., 2016). In contrast, in alkaline soils, roots secrete organic anions (e.g., citrate and malate) and protons, leading to rhizosphere acidification (Sun et al., 2020). In addition, compared with plants with fibrous root systems, taproots are more strongly colonized by arbuscular mycorrhizal fungi, contributing to greater nutrient acquisition by accessing soil micropores and transporting nutrients [e.g., phosphorus (P) and nitrogen (N)] to plants (Hodge et al., 2001; Hodge and Storer, 2014; Yang et al., 2015). The formation of root nodules in legumes allows the conversion of N_2 to ammonia and amino acids and induces a divergent nutrient status in the rhizosphere, as compared to nonlegumes (Day et al., 2001; Moreau et al., 2019). Furthermore, moisture and temperature also strongly regulate nutrient acquisition by plants by influencing the diffusion rate, cell membrane permeability, root physiology, and interactions with microorganisms (Praeg et al., 2019; Warren, 2009). For instance, the root uptake of NO_3^- is more preferential relative to NH_4^+ or glycine under warmer temperatures (Warren, 2009). The colonization of arbuscular mycorrhizal fungi is higher under drier conditions, inducing positive mycorrhizal P responses in plants (Cavagnaro, 2016). In summary, individual studies have identified differences between the rhizosphere and bulk soil in terms of their nutrient content and availability (Li et al., 2020; Marschner et al., 2002; Massaccesi et al., 2015; Steer and Harris, 2000). Similarly, the mechanism by which rhizosphere nutrient availability is modulated by only one or several interrelated factors has long been unraveled. However, a thorough and quantitative understanding of the effects of these factors on rhizosphere nutrient availability is still lacking and strongly recommended for further clarification (Cheng et al., 2014; Jones et al., 2004; Marschner et al., 2004; Pii et al., 2016).

Bacteria and mycorrhizal/saprotrophic fungi are the dominant components of rhizobiomes. They play important roles in improving nutrient acquisition, facilitating plant growth by exuding enzymes and organic molecules, and inhibiting fungal pathogens (Artursson et al., 2006; Philippot et al., 2013). For instance, in P-depleted soil, P-mobilizing bacteria release available P from organic compounds, which provides a P pool for plant growth and arbuscular fungal hyphae to extend to roots (Kucey et al., 1989). Even though the mobility and dispersal of bacteria and fungi are different, specific bacteria can migrate via fungal hyphal structures and colonize novel microhabitats (e.g., the rhizosphere). Many bacteria also feed on living saprotrophic and pathogenic fungi, as well as fungal remains (Ballhausen and de Boer, 2016).

These relationships suggest that bacteria and fungi have specific survival and colonizing strategies, which may contribute to diverse changes in their densities in the rhizosphere as compared to bulk soil. Microbial densities in the rhizosphere, are two-to-three orders of magnitude greater than those in bulk soil (Ramos et al., 2000). The classification of microorganisms based on their life strategies (i.e., copiotrophs and oligotrophs) suggests that copiotrophs (grow faster and rely on resource availability; r-strategists) are more abundant in the rhizosphere, whereas oligotrophs (efficiently exploit resources at the expense of growth rate; K-strategists) are abundant in resource-poor environments (such as bulk soil; Ho et al., 2017; Liu et al., 2022; Ling et al., 2022). For instance, the abundances of the copiotrophic phyla *Bacteroidetes* and *β-Proteobacteria* are generally greater in the rhizosphere than those in the surrounding bulk soil (Fan et al., 2017; Kavamura et al., 2018). The fungal phylum *Ascomycota*, a copiotrophic taxon, is more abundant in the rhizosphere by utilizing a higher number of resources compared to *Basidiomycota* (Egidi et al., 2019; Pascual et al., 2018). While the shift of microbial communities in the rhizosphere has been acknowledged, a quantitative elucidation of the relative abundance of bacterial and fungal communities in the rhizosphere is still seldom (Ling et al., 2022).

We collected studies on nutrient content and availability in the rhizosphere, depending on plant characteristics, soil properties, and climatic factors. These studies mainly reported nutrients (e.g., N, P and K) content, organic carbon (C) content, microbial C, activity of enzymes, bacterial and fungal numbers, and other original soil properties (e.g., pH, total N and P). Plant species information (e.g., legumes/non-legumes, fibrous roots/taproots, trees/herbaceous plants/ forbs) and climate factors [mean annual precipitation (MAP) and mean annual temperature (MAT)] were recorded. A meta-analysis was performed to elucidate the rhizosphere effects on nutrients [i.e., N, P and potassium (K)] content. Our main aims were to: ①quantify nutrient availability in the rhizosphere compared to bulk soil; ②identify how plant characteristics, original soil properties, and climatic factors influence nutrient availability in the rhizosphere relative to bulk soil; ③quantitatively elucidate the relative abundance of bacterial and fungal communities in the rhizosphere compared to bulk soil. We hypothesized that: ①available pools of nutrients are more depleted in the rhizosphere than in bulk soil due to intensive root uptake and microbial immobilization; ② plants with taproots accumulate more organic C and nutrients as compared to plants with fibrous roots owing to stronger interactions with arbuscular mycorrhizal fungi during their longer life; ③copiotrophic groups are more abundant than oligotrophs in the rhizosphere than in bulk soil.

6.1.2 Soil biochemical and biological factors in the rhizosphere vs. in bulk soil

The increase in pools with fast turnover, for example, MBC and MBN, was approximately twice stronger in the rhizosphere (19% and 32%, respectively) compared with SOC (9%) and TN (14%) (Figure 6.1).

The activities of all enzymes were greater in the rhizosphere than in the bulk soil (Figure 6.2). Remarkably higher bacterial (205%) and fungal (254%) numbers in the rhizosphere were also confirmed (Figure 6.1G and Figure 6.1H), empha-sizing the importance of the rhizosphere as a microbial hotspot. Positive relationships between MBC and bacterial numbers, MBC and fungal numbers, and effect sizes of bacterial and fungal numbers were also confirmed (Figure 6.3).

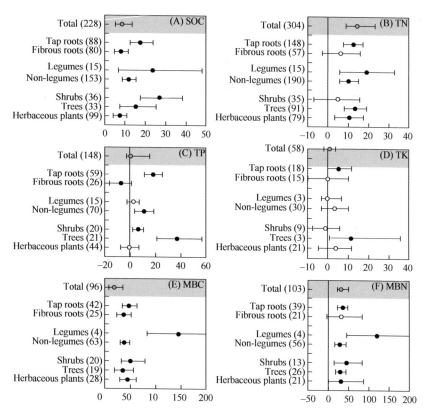

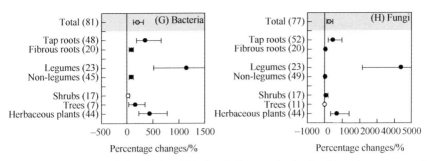

Figure 6.1 Changes of chemical and biochemical properties in the rhizosphere compared with bulk soil. The negative and positive values represent the depletion and increase, respectively, of the nutrients in the rhizosphere as compared to bulk soil. The open symbols indicate that the value is not significantly different from zero. Error bars represent bootstrapped 95% confidence intervals. Note that the total effect size of each soil property does not reflect the means from individual groups, because approximately 5%–42% of the observations cannot be subdivided to one of these groups, owing to the absence of sufficient information in the original articles. SOC, soil organic carbon; TN, total nitrogen; TP, total phosphorus; TK, total potassium; MBC, microbial biomass carbon; MBN, microbial biomass nitrogen. Credit: Reprinted with permission from Liu et al., 2022.

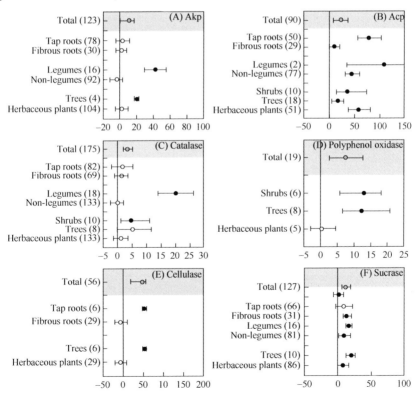

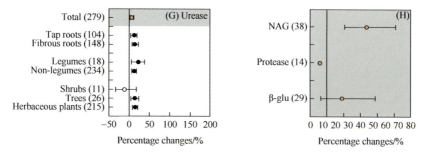

Figure 6.2 Changes of enzyme activities in the rhizosphere compared with bulk soil. The negative and positive values represent the decrease and increase, respectively, of the enzyme activities in the rhizosphere compared to bulk soil. The open symbols indicate that the value is not significantly different with zero. The number in the parentheses beside each enzyme is the sample size. Error bars represent bootstrapped 95% confidence intervals. Note that the total effect size of the activity of each enzyme does not reflect the means from individual groups, because approximately 9%—36% of the observations cannot be subdivided to one of these groups, owing to the absence of sufficient information in the original articles. Acp: acid phosphatase; Akp: alkaline phosphatase; NAG: N-acetyl-β-D-glucosaminidase; Dehydro: dehydrogenase; β-glu: β-glucosidase. Credit: Reprinted with permission from Liu et al., 2022.

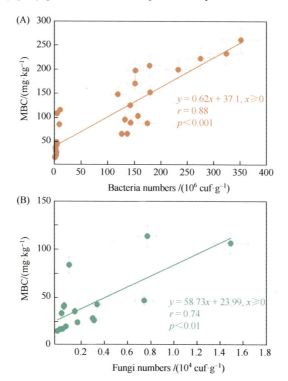

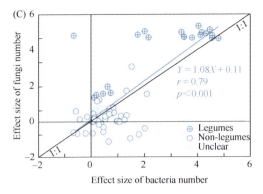

Figure 6.3 Relationship between bacterial numbers and MBC (A), fungal numbers and MBC (B), and effect sizes of bacterial and fungal numbers (C). Error bars represent standard deviations. Credit: Reprinted with permission from Liu et al., 2022.

The relative abundances of bacteria and archaea, such as Proteobacteria, Bacteroidetes, and Firmicutes, were greater in the rhizosphere than in the bulk soil (Figure 6.4A). In contrast, the relative abundance of Crenarchaeota, Actinobacteria, and Acidobacteria in the rhizosphere decreased.

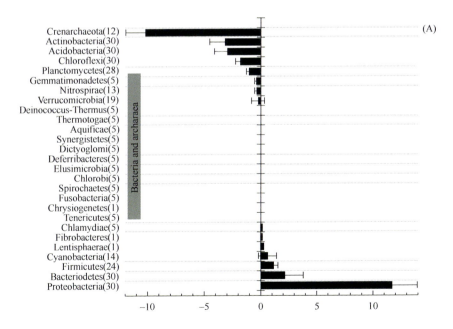

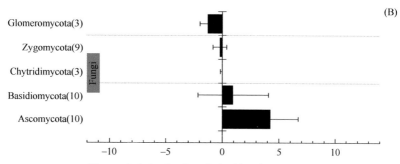

Change of relative abundance in the rhizospher compared to bulk soil/%

Figure 6.4 Relative abundance of bacteria (A) and fungi (B) presented as a difference between the rhizosphere and bulk soil. The data were extracted from seven articles (Bakker et al. 2015; Fonseca et al. 2018; García-Salamanca et al. 2013; Mendes et al. 2014; Pascual et al. 2018; Ren et al. 2021; Sugiyama et al. 2014). Credit: Reprinted with permission from Liu et al., 2022.

6.1.3 Effects of climatic factors and original soil properties on available nutrients in the rhizosphere

The available N (AN) was 10% higher in the rhizosphere than in the bulk soil (Figure 6.5A). The NH_4^+ (11%), NO_3^- (24%), TN (14%), and MBN (32%) content in the rhizosphere increased (Figure 6.1 and Figure 6.5). Higher enrichment of NH_4^+ and NO_3^- was observed in the rhizosphere when either MAT or MAP was relatively low (Figure 6.6, $P<0.05$), which generally represents boreal or relatively dry conditions. The effect size of AN increased at high original soil pH or when the original SOC, TN and TP were low, while the opposite trends were found for the effect size of NO_3^- (Figure 6.6, $P<0.05$).

Unlike AN, AP decreased by 12% and was more depleted in the rhizosphere than in the bulk soil (Figure 6.5B). The effect size of AP increased at high original soil pH, but decreased with an increase in the original TP and AP (Figure 6.6, $P<0.05$).

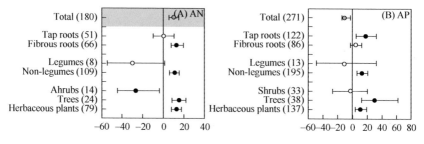

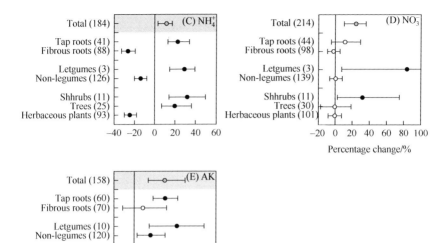

Figure 6.5 Changes of available nutrients in the rhizosphere compared with bulk soil. The negative and positive values represent the depletion and accumulation, respectively, of the nutrients in the rhizosphere as compared to bulk soil. The open symbols indicate that the value is not significantly different from zero. The number in the parentheses beside each available nutrient is the sample size. Error bars represent bootstrapped 95% confidence intervals. Note that the total effect size of each available nutrient does not reflect the means from individual groups, because approximately 17%–34% of the observations cannot be subdivided to one of these groups, owing to the absence of sufficient information in the original articles. AN: available nitrogen; AP: available phosphorus; AK: available potassium. Credit: Reprinted with permission from Liu et al., 2022. Copyright 2022 Elsevier.

The AK increased by 15% in the rhizosphere compared to the bulk soil (Figure 6.5E), whereas the TK content was similar in the rhizosphere and bulk soil (Figure 6.1D). The effect size of AK decreased at high precipitation (Figure 6.6, $P<0.05$).

Climatic factors, such as precipitation and temperature, have specific effects on nutrient availability within the rhizosphere. While NH_4^+ and NO_3^- enrichments were strongly regulated by precipitation and temperature, these did not influence AP in the rhizosphere (Figure 6.6). NH_4^+ and NO_3^- are the two most important N forms for root uptake, the content of which is mediated by soil moisture and temperature. For instance, the transformation between NH_4^+ and NO_3^-, as well as the plant preference for

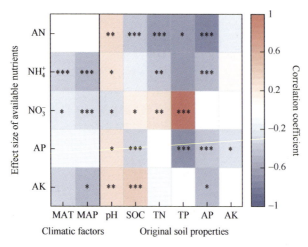

Figure 6.6 Spearman's rank correlation coefficients between effect sizes of available nutrients and climatic factors and original soil properties. SOC: soil organic carbon; TN: total nitrogen; TP: total phosphorus; AN: available nitrogen; AP: available phosphorus; AK: available potassium; MAT: mean annual temperature; MAP: mean annual precipitation. Cells colored with pink or blue indicate that the relationship is positively or negatively significant, respectively. * = $P<0.05$, **= $P<0.01$, *** = $P<0.001$. Credit: Reprinted with permission from Liu et al., 2022.

N uptake, strongly increases with temperature (Clarkson, 1986; Ganmore-Neumann and Kafkafi, 1980; Kafkafi, 1990). Warming increases soil organic matter decomposition (Kuzyakov and Bol, 2006), thereby also N mineralization, but also stimulates N uptake by plants and immobilization by microorganisms (Dong et al., 2001), which depletes NH_4^+ and NO_3^- in the rhizosphere. The effects of temperature on the root uptake of either NH_4^+ or NO_3^- differs because of N diffusion, root physiology, membrane permeability, presence of mycorrhizal symbionts and competition with microorganisms for N (Clarke and Barley, 1968; Warren, 2009). Clarkson (1986) suggested that 85% of N uptake is in the form of NH_4^+ rather than NO_3^- at temperatures below 9°C. High NO_3^- uptake is more common under warmer conditions (Warren, 2009). NH_4^+ can be adsorbed on and in clay particles, which strongly decreases its diffusivity as compared to NO_3^- and influences the effect of temperature change on NH_4^+ uptake. N availability in the rhizosphere is greater under drought conditions due to decreased plant and/or microbial N uptake (Cregger et al., 2014; Deng et al., 2021). This explains the negative relationship between MAP and the effect sizes of NO_3^- (Figure 6.3). Similar to NO_3^-, drought-induced reduction in K uptake by roots also contributes to K enrichment in the rhizosphere (Ge et al., 2012), leading to a negative relationship between MAP and the effect size of AK.

Compared to the effects of climatic factors, the original soil properties (pH, SOC, TN, TP, AP, and AK) were more important for the effect sizes of AN, AP, and AK (Figure 6.6). Soil with originally high fertility (high content of SOC, TN, TP, AP and AK) provides sufficient available nutrients for both plant and microbial communities (Tang et al., 2021; Tian et al., 2020). Nutrient delivery from bulk soil to the roots by diffusion and mass flow meets the plant and microbial demand. In contrast, when soil is less fertile (lower SOC, TN and TP content), the mass flow from the bulk soil to the rhizosphere is low. Microorganisms and roots release more enzymes and carboxylates to increase nutrient mineralization from organic pools to maintain their growth (Rengel and Marschner, 2005).

6.1.4 Acidity and alkalinity neutralization in the rhizosphere modulates nutrient availability

ΔpH decreased with increasing original soil pH (Figure 6.7A), indicating a neutralizing effect and buffering capacity of plant roots in both acidic and alkaline soils. On average, the soil pH in the rhizosphere was 0.16 ± 0.02 units lower than in bulk soil (Figure 6.7B). Plant characteristics strongly influenced the rhizosphere soil pH (Figure 6.7C—Figure 6.7E). Leguminous species induced a much higher reduction in rhizosphere pH (-0.44 units) than non-legumes (-0.11 units).

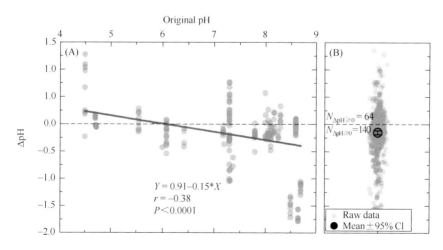

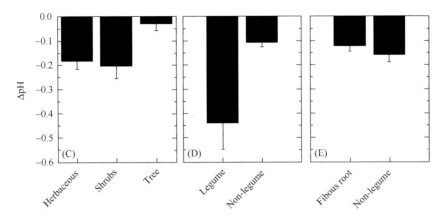

Figure 6.7 Relationships between ΔpH and original pH (A), average ΔpH (B), and influence of plant characteristics on ΔpH [(C), (D) and (E)]. The ΔpH was calculated by subtracting the pH value of the rhizosphere from the value of the bulk soil: the negative values show acidification (more common in neutral and alkaline soils). Error bars represent standard errors. $N_{\Delta pH} \geqslant 0$ and $N_{\Delta pH} < 0$ represent the number of observations with ΔpH above (or equal to) and below 0, respectively. Credit: Reprinted with permission from Liu et al., 2022.

Although most plant species acidify the rhizosphere, the ability of plants to influence the rhizosphere pH depends strongly on the original soil pH (Figure 6.7A). Specifically, the plant increases or decreases the rhizosphere pH to maintain the equilibrium between cations and anions at the root-soil interface (Figure 6.7 and Figure 6.8). Roots acidify rhizosphere soil through the following processes: ①uptake of cations coupled with proton release; ②release of organic acids (Jones, 1998); ③CO_2 release via root respiration (Rao et al., 2002); ④release of chelates removing cations from exchange sites. Microorganisms acidify the rhizosphere through similar processes. In particular, legumes strongly acidify the rhizosphere (−0.44 unit, Figure 6.7D) because of the release of protons following the excess uptake of cations over anions during biological N_2 fixation (Haynes, 1983; Israel and Jackson, 1982; Liu et al., 1989), as well as nitrification of NH_4^+ to NO_3^- (Ma et al., 2019; Raza et al., 2020). These processes help overcome the deficiencies of P and Fe as well as micronutrients in soil (Custos et al., 2020).

In contrast, in acidic soil, plants generally increase the rhizosphere soil pH to tackle Al stress (Bagayoko et al., 2000) and increase the availability of all macro- and most micro-nutrients. The plant uptake of anions (e.g., NO_3^-) in excess of cations often causes the roots to secrete HCO_3^-. The excretion of

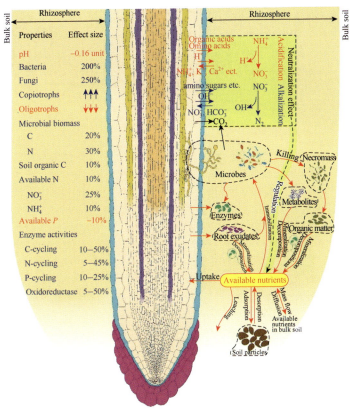

Figure 6.8 Changes of soil biochemical properties in the rhizosphere compared to bulk soil (left), and nutrient depletion and accumulation processes (right). Note that the values in this conceptual figure are rounded up. Most soil properties (evaluated in this meta-analysis) increase in the rhizosphere as compared to the surrounding bulk soil. Exceptionally, soil pH decreases by 0.16 units. The available P decreased roughly by 10%, indicating extreme P demand for plants and microorganisms in the rhizosphere and slow diffusion from the surrounding soil. The relative abundance of oligotrophic microorganisms (K-strategists) decreased in the rhizosphere with highly competitive copiotrophic microorganisms (r-strategists) under greater C and nutrient availability. The available nutrients in the rhizosphere are mainly derived from the decomposition and mineralization of various organic compounds (organic matter, root exudates, metabolites, etc.) and microbial cells killed by phages and soil animals. Enzymes released by microorganisms and roots accelerate nutrient cycling. Mass flow, diffusion, and desorption from soil clay particles also contribute to nutrient accumulation in the rhizosphere. In contrast, root uptake, immobilization by microorganisms, leaching, and adsorption by soil particles lead to nutrient depletion in the rhizosphere. Acidity or alkalinity neutralization by roots and microorganisms regulates the nutrient availability in the rhizosphere. Credit: Reprinted with permission from Liu et al., 2022.

OH^- from NO_3^- reduction is also the main mechanism to increase rhizosphere pH (Sugihara et al., 2016). For legumes, the reduction of protons to H_2 by nitrogenase during biological N_2 fixation also contributes to an increase in rhizosphere pH (Mahon, 1979). In summary, these neutralizing effects of the rhizosphere contribute to higher content of AN, NH_4^+, AP and AK in the rhizosphere when the original soil pH is low (Figure 6.6).

6.1.5 Nutrient depletion and accumulation in the rhizosphere

The rhizosphere volume is crucial for plant nutrition, as it defines the total amount of each nutrient immediately available to the plant. The rhizosphere is expected to be depleted with nutrients compared to bulk soil (Joner et al., 1995; Wang et al., 2007). However, the global meta-analysis confirmed this trend only for AP and AN in the rhizosphere of shrubs (Figure 6.5B), which contradicts our first general hypothesis. Low P availability in the rhizosphere coupled with high acid and alkaline phosphatase activity (Figure 6.2A and Figure 6.2B) compared to bulk soil, shows extreme P demand for plants and microorganisms. P availability in the rhizosphere is limited at low and high soil pH (Haynes, 1982). For instance, calcium phosphate precipitates in an alkaline environment (i.e., pH>8.0). At pH values lower than 5.0, iron or aluminum phosphates are formed, and P is adsorbed onto Fe or Al (oxyhydr)oxides. Neutralization of acidity and alkalinity in the rhizosphere increases P availability. It should be noted that when we calculated the total effect size of AP (Figure 6.5B), all cases (including monoculture and polyculture) were included, but only monocultures were considered when analyzing the influence of plant characteristics (e.g., taproots and fibrous roots). The negative total effect size of AP, compared to the positive influence of most plant characteristics (Figure 6.5B), indicates that increased plant species richness may exacerbate P depletion in the rhizosphere.

In contrast to AP, high nutrient content (i.e., AN, AK, NH_4^+ and NO_3^-, Figure 6.5) in the rhizosphere could be because mass flow (the movement of nutrients with water) is faster than the nutrient uptake by roots (Barber, 1962). If the nutrient supply from mass flow exceeds their uptake, then those nutrients increase around the root and create accumulation zones (Barber and Ozanne, 1970; York et al., 2016). This is common especially for Ca^{2+} uptake from calcareous soils (Zamanian et al., 2016), or by Fe accumulation around the roots in iron-rich water-saturated soils (Hoefer et al., 2017; Jeewani et al., 2020; Williams et al., 2014). However, if nutrients are more enriched in the rhizosphere than in bulk soil, diffusion may occur opposite to the mass flow because of the concentration gradient. The diffusion rate of a nutrient is influenced by the exchange of nutrients between soil particle surfaces and surrounding water (Kuzyakov and Razavi, 2019). Organic compounds excreted by roots (i.e., low and high molecular weight organic compounds) also affect the diffusion rate through acidification, reduction/complexation, or ligand exchange reactions (Hoefer et al., 2017; Kramer-Walter et al., 2016; Terzano et al., 2014), but are mainly defined by microbial utilization (Kuzyakov et al., 2003). These effects on diffusion can reduce the nutrient uptake of roots by 13%–61% compared to roots in a stirred nutrient solution (Drew et al., 1969). Therefore, even though nutrients are enriched in the rhizosphere, the diffusion from the rhizosphere to the soil is diminished. On the other hand, the stronger soil organic matter decomposition and mineralization rate[the rhizosphere priming effects (Huo et al., 2017; Pausch et al., 2013) as indicated by increased enzyme activities (Sanaullah et al., 2016), MBC and microbial abundance] increase nutrient mineralization and their availability (Figure 6.1 and Figure 6.2).

6.1.6 Nutrient availability within the rhizosphere is mediated by plant and root characteristics

Most available nutrients (AP, AK, NH_4^+ and NO_3^-) were more enriched in the rhizosphere of taproots than in fibrous roots, as indicated by their larger change around taproots (Figure 6.5). Similar strong enrichments were also found for the SOC, TN, TP, TK, MBC and MBN content around taproots (Figure 6.1). Further, bacteria and fungi were more abundant in the rhizosphere of taproots than in fibrous roots (Figure 6.1G and Figure 6.1H).

AN and AP were more exhausted in the rhizosphere of legumes than in non-legumes (Figure 6.5A and Figure 6.5B). In contrast, the MBC and MBN content is higher in the legume than in the non-legume rhizospheres (Figure 6.1E and Figure 6.1F). Similarly, the effect size of bacterial and fungal numbers was also higher in the rhizosphere of legumes than in that of non-legumes (Figure 6.3C).

Plant nutrient exploitation depends on root morphology (e.g., taproots or fibrous roots), interaction with symbionts (e.g., mycorrhiza and rhizobia), and release of root exudates and enzymes (Dinkelaker and Marschner, 1992; Moreau et al., 2019). Taproots acquire more water and nutrients from deeper horizons (Kautz et al., 2013). Lateral roots growing from taproots extend the nutrient exploring capacity of the plants and enhance their interactions with beneficial microorganisms. A larger increase in bacterial (+352%) and fungal (+483%) numbers around taproots also implies stronger microbial activities in their rhizosphere than plants with fibrous roots (Figure 6.1G and Figure 6.1H), contributing to the slightly higher AP, AK, NH_4^+ and NO_3^- in the rhizosphere of taproots than fibrous roots. The stronger increase in the rhizosphere SOC of taproots, in accordance with our second hypothesis, demonstrates the stronger ability of taproots to store C and energy sources during their longer life, which provides abundant substrates for enzyme synthesis, thus a larger release of available nutrients (Figure 6.1A). It should be noted that plants with fibrous root systems have much larger absorptive surface areas than taproots because of more highly branched fine roots and active root hairs which facilitate direct nutrient acquisition (Holz et al., 2018). In contrast, the ability of taproots to absorb nutrients is limited because of their coarser system; therefore, assistance from microbial symbionts (e.g., arbuscular mycorrhizal fungi) is required, explaining the larger increase in bacterial and fungal numbers. This indicates that both root systems have different absorptive strategies for nutrients from soils (Yang et al., 2015).

Legumes create symbiotic relationships with N-fixing bacteria to increase N availability. However, to effectively form symbiotic relationships with bacteria and fix N_2, leguminous plants acquire more P (Magadlela et al., 2016; Stevens et al., 2019), which explains why AP is more exhausted in the rhizosphere of legumes than of other species (Figure 6.5B). This was supported by the increased alkaline (42%) and acid (108%) phosphatase activities in the rhizosphere of legumes (Figure 6.2A and Figure 6.2B) and also suggested stronger mineralization of organic P in legume plants (Maltais-Landry, 2015). Contrary to our expectation, the effect size of AN in the rhizosphere of legumes was not different from zero (Figure 6.5A). Even though the number of observations (8 vs. 109) for the AN of legumes and non-legumes is imbalanced, individual studies have shown either increased (Dai et al., 2016; Qiu et al., 2014) or decreased AN (Ding et al., 2017; Xu et al., 2020; Zhou, 2014) in the rhizosphere of legumes. It has been argued that legumes could be either "altruistic" or "self-serving" (Xu et al., 2020). These divergent results can be attributed to the shift in the N-acquisition strategy of plants between N_2-fixation and mineral N uptake pathways, depending on whether the canopy accumulates enough N to support photosynthesis and rhizodeposition (Henneron et al., 2020).

While AN decreased in the rhizosphere of legumes, NH_4^+ and NO_3^-

increased, as was the case in non-legumes. This inconsistent trend could be explained by the low proportion of NH_4^+ and NO_3^- in AN (Figure 6.9). The AN content was approximately equal to 5.5%–6.0% of the TN, while NH_4^+ and NO_3^- accounted for only approximately 1% and 0.5% of the TN, respectively. This indicates that the remainder of AN, (e.g., peptides and amino acids) accounts for approximately 4%–4.5% of the TN, which equals approximately 73%–75% of AN. This is consistent with the findings of previous studies (Jämtgård et al., 2010; Young and Aldag, 2015), which demonstrated that the content of soluble organic N forms is higher than that of mineral N, especially in natural ecosystems. The significantly higher proportion of soluble organic N content in AN suggests that the change in organic N content, not mineral N, mainly reflects the variation in rhizosphere AN.

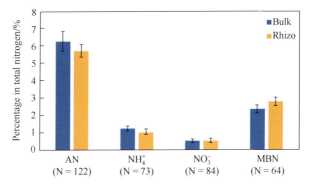

Figure 6.9 Percentage change of N forms in total N. Credit: Reprinted with permission from Liu et al., 2022.

6.1.7 Microbial activities and community shift in the rhizosphere

The enzyme activity in the rhizosphere increased by 3%–52% as compared to that in the bulk soil (Figure 6.2) and demonstrates an enzyme-specific pattern. Catalase and polyphenol oxidase are two defense-related antioxidant enzymes (Babu et al., 2015) and their activities increased by 3.4% and 7.6%, respectively, in the rhizosphere. These enzymes can be expressed by plant growth-promoting bacteria (e.g., *Pseudomonas*) or other beneficial biota, such as endophytes, which in turn not only strengthen the resistance of plants to diseases induced by pathogenic fungi (Hayden et al., 2018; Lazcano et al., 2021), but also facilitate the transfer of nutrients to the host plant by colonizing root surfaces. For instance, endophytic bacteria (e.g.,

Bacillus tequilensis belonging to the Firmicutes phylum) stimulate root hair development and N uptake by roots (Beltran-Garcia et al., 2014). Increased polyphenol oxidase facilitates the degradation of phenolic compounds and induces the breakdown of polyphenol-protein complexes, which enables the availability of a large portion of N contained in the complexes (Hättenschwiler and Vitousek, 2000). Catalase is related to the redox ability of soil and is associated with aerobic microbial activity (Nowak et al., 2004). In particular, catalase activity in the rhizosphere of legumes increased by 20% because of the stronger increase in bacterial and fungal numbers (Figure 6.1 and Figure 6.3), indicating that leguminous species have a stronger ability to resist environmental stress and protect nodules from pathogens.

The increase in bacterial numbers in the rhizosphere was similar to that of fungi (Figure 6.1G and Figure 6.1H), consistent with the close correlation between the effect size of bacterial and fungal numbers (Figure 6.3). A much larger increase in bacterial and fungal numbers (205%– 254%) in the rhizosphere than in microbial biomass (indicated as MBC, +19%) (Figure 6.1 and Figure 6.3A and Figure 6.3B) suggests that the average biomass of individual microbial cells decreased. This could be attributed to the greater dominance of copiotrophic microorganisms (e.g., Proteobacteria, Bacteroidetes and Firmicutes) in the rhizosphere (Figure 6.4A), which have much faster growth rates in available C-rich environment than oligotrophs (Fierer et al., 2007; Gharechahi et al., 2021). Copiotrophs invest more resources to proliferate instead of increasing the biomass of each microbial cell. In contrast, Acidobacteria and Actinobacteria, mostly belonging to oligotrophic microorganisms (Lebeau, 2011), showed decreased relative abundance of approximately 3.0% (Figure 6.4A). The shift to copiotrophic microorganisms is important for plant-microbial interactions and nutrient availability because some members of the phylum (e.g., *Mucilaginibacter* belonging to Bacteroidetes and *Pseudomonas* belonging to Proteobacteria) can produce large quantities of polysaccharides to optimize the rhizosphere microenvironment, for example, aggregate soil particles (Agnihotri et al., 2022), provide an energy source for microorganisms, and promote the uptake of trace elements (Danhorn and Fuqua, 2007; Naseem et al., 2018). Ascomycota and Basidiomycota were the dominant fungi in the rhizosphere soil (Qin et al., 2017). Basidiomycota can degrade lignocellulose organic matter more easily (Yelle et al., 2008), whereas Ascomycota are more prone to use easily degradable organic compounds for fast-growing fungi (Lundell et al., 2010). The higher relative abundance of Ascomycota (+4%, Figure

6.4B) in the rhizosphere ensures greater production of organic matter-degrading enzymes and the facilitation of symbiotic associations (Challacombe et al., 2019). Furthermore, Crenarchaeota is an ammonia-oxidation archaeon that colonize roots (Simon et al., 2000). Its relative abundance in this meta-analysis decreased by 10% in the rhizosphere as compared to that in bulk soil (Figure 6.4), corroborating the results of Xu et al. (2018). Strong competition from plants for N may constrain the diversification and growth of Crenarchaeota (Liu et al., 2019; Treusch et al., 2005). It should be noted that even though relative abundance of some phyla in the rhizosphere as compared to bulk soil was revealed based on the data extracted from the seven studies, a more comprehensive systematic review is required to elucidate the shift of relative abundance and functions of microbial communities.

6.1.8 Summary

The findings of our review and meta-analysis have important implications for the elucidation of rhizosphere nutrient regulation by plant-microbial interactions. First, the available N and K are enriched in the rhizosphere, while the available P is depleted as compared to the bulk soil. Plants trigger a variety of mechanisms to acquire nutrients, including intensive root exudation and neutralization of the acid or alkaline environment. Second, plants have various levels of interactions with microorganisms to acquire nutrients. For instance, plants with taproot systems will accumulate more organic compounds in the rhizosphere and induce greater microbial density relative to plants with fibrous roots, which provide microorganisms with enough substrates to synthesize enzymes and produce polysaccharides beneficial for the microenvironment. Legumes create symbiotic relationships with N-fixing bacteria and facilitate the release of root exudates and H^+, leading to greater microbial number in the rhizosphere and alleviating serious P deficiency. Bacterial and fungal numbers increase much more than the total microbial biomass. Together with the more abundant copiotrophs in the rhizosphere, these findings suggest that microorganisms in the rhizosphere invest more resources to proliferate instead of increasing their size. Even though small cells are presumed to be favored in oligotrophic settings, specific rhizosphere environments (e.g., stronger microbial activities and higher resource availability) may induce specific microbial growth strategies in this critical zone. For more comprehensive analysis of the drivers determining nutrient content and availability in the rhizosphere, the effects of various types and concentration of root exudates and other soil properties (e.g., moisture, texture and types) on rhizosphere nutrient availability require generalization. Furthermore, examination of processes controlling nutrient content and availability from the perspective of molecular genetics of plants and microorganisms is necessary.

6.2 Response mechanism of soil enzymes in plant rhizosphere to heavy metal pollution

6.2.1 Introduction

With the rapid development of modern industry, the soil pollution by heavy metals is becoming more and more serious. In 2014, the National Soil Pollution Survey Communique jointly issued by the former Ministry of Environmental Protection (PRC) and the former Ministry of Land and Resources (PRC) showed that the comprehensive over standard rate of soil pollution sites was 16.1%, and inorganic pollutant sites accounted for 82.8% of all. Soil pollution by heavy metals is widely seen in southern provinces of China, and the background concentration of heavy metal in southwest China is obviously higher than that in other provinces. The average concentrations of Cd and As, two main pollutant elements, in each province ranged from $36.5mg \cdot kg^{-1}$ to $112.9mg \cdot kg^{-1}$ and $5.8mg \cdot kg^{-1}$ to $31.1mg \cdot kg^{-1}$ (Cheng et al., 2015). About 10 million hm^2 of cultivated land soil is polluted by heavy metals, resulting in a loss of about 12 million tons of grain yield every year (Teng et al., 2010). Heavy metal pollution has a widespread and serious harm, characterized by being concealed and lagging. Therefore, it is of great significance to work out appropriate characterization means to quickly identify and evaluate the toxicity and ecological effects of heavy metals in soil.

Soil enzyme is a kind of macromolecular active protein, which can catalyze a variety of biochemical reactions, as closely related to the material cycle and energy flow in soil (Yao et al., 2006). As an important biochemical index to characterize the soil quality, enzyme activity is very sensitive to environmental stress. Soil enzymes can not only reflect the biogeochemical cycle of soil but also the diversity of microorganisms and other biological characteristics in soil (Nannipieri et al., 2018). Current researches on soil enzyme activity mainly focus on the rhizosphere soil enzymes. The term "rhizosphere" was first proposed by Hiltner, which mainly refers to the soil area around plant roots that is strongly influenced by root functions. To date, there is no clear definition of the rhizosphere coverage, which is generally admitted by researchers as the range of 0.5—4mm on the surface of plant roots (Kuzyakov and Bahar, 2019). As the main interface of plant-microbe-soil interaction, plant rhizosphere is one of the most active sites for energy flow and material metabolism, where the enzyme activity is higher than in non-rhizosphere soils. Rhizosphere enzyme activity is a sensitive index to monitor the microbial activity, community composition and functional changes, which reflects the interaction between plants and microorganisms (Liu et al., 2017). Due to the sensitivity of enzyme activity to environmental changes, the relationship between heavy metals and soil enzyme activity has become a hot topic in recent years. In particular, it is of great significance to study the response mechanism of soil enzymes in plant rhizosphere to heavy metal pollution for revealing the response mechanism of soil and plants to heavy metals.

Based on a large number of literature researches, this section systematically reviews and summarizes the research status quo related to the response mechanism of soil enzymes in plant rhizosphere to heavy metal pollution. On this basis, the existing problems are discussed, and the future research is prospected to provide references for the research on the ecological effects of soil pollution by heavy metals.

6.2.2 Enzyme activity in rhizosphere soil and its main influencing factors

Rhizosphere enzymes are proteins released from the roots of microorganisms and plants into the rhizosphere soil environment, and their activity is distributed in a gradient in the rhizosphere. The enzyme activity in rhizosphere soil is 1.3—2 times that of nonrhizosphere soil. The rhizosphere range of most enzymes is 1—3mm, and the enzyme activity decreases with the increase of root distance. The spatial distribution of enzyme activity in rhizosphere soil is typically studied by in situ enzyme spectrometry or soil slice method. Compared with the destructive sampling, the visualization technology can be used for in situ quantitative analysis of soil molecular substances. With less disturbance to soil, it has attracted extensive attention in recent years. Rhizosphere microorganisms refer to ones that grow and reproduce in the rhizosphere soil, mainly rhizosphere bacteria, rhizosphere fungi and archaea. The quantity and activity of microorganisms in plant rhizosphere are higher than those in non-rhizosphere soil, which is called the "rhizosphere effect" (Mapelli et al., 2018). The species of soil organisms can affect the activity of rhizosphere soil enzymes. In addition, the physiochemical properties of soil and the addition of nutrients are also important factors.

6.2.2.1 Plant types

The difference of enzyme activity in plant rhizosphere varies to different plant species. For example, the activity of protease, phosphatase, invertase and catalase in the rhizosphere soil of wheat (*Triticum aestivum* L.) is lower than that of rhizosphere soil of *Trifolium repens* L. Urease activity in rhizosphere soil of legumes is higher than that of other crops. Different rhizosphere microorganisms can secrete different kinds of enzymes. As an important part of the rhizosphere environment, plant roots can also secrete H^+, enzymes, amino acids and other organic substances (Marschner, 1986). Therefore, rhizosphere microorganisms and plant root exudates are also important factors affecting the enzyme activity in rhizosphere soil.

A large number of studies have shown that the main sources of rhizosphere soil enzymes include: ①Microbial secretion. Microbial cells are one of the main sources of soil enzymes, and some enzymes can be synthesized and secreted by rhizosphere bacteria or fungi inside the cells. Studies have shown that 90%—95% of enzymes in soil are directly secreted by microorganisms. ②Secretion by plant roots. A small part of enzymes in rhizosphere soil is secreted by plant roots. ③Animal excrement. Enzymes released by animal excretion can be transferred into the rhizosphere soil. Table 6.1 lists the main sources and some types of enzymes in rhizosphere soil.

6.2.2.2 Physical and chemical properties of soil

Temperature, humidity and pH are important physical and chemical properties of the soil, whose changes may affect the enzyme activity in rhizosphere soil. The change of soil pH is related to H^+ and OH^- secreted by roots. Plant roots secrete H^+ and OH^- to stimulate nutrients and maintain the electrochemical potential balance on root surfaces. H^+ and OH^- secreted by plant roots are distributed within 2—3mm of the root surface. This process changes the pH around plant roots, enabling a gradient distribution of pH with the increase of distance. Protonation is an important process in plant root activities. The hydrogen ion adenosine triphosphate (H^+-ATP) produced by the protonation can provide the counter ions required for the exudation of organic anions and amino acids in the root medium. The

Table 6.1 Main sources and types of enzymes in rhizosphere soil.

Main source		Major enzymes	Rhizosphere soil sampling method	Enzyme activity assay	References
Rhizosphere microbe	Most bacteria and fungi	Phosphatase, urease, cellulose, pectase	—	—	Aon and Colaneri, 2001
	Streptomyces viridosporus Craw.	Catalase, esterase	—	—	Magnuson and Crawford, 1992
	Lysobacter capsici P.	Lysozyme, gelatinase	—	Enzyme spectrum method	Lee et al., 2015
Plant root system	Lupinus micranthus Guss	Cellulase, chitinase, phosphatase	Nondestructive	Enzyme spectrum method	Spohn and Kuzyakov, 2014
	Hordeum vulgare Linn.	Phosphatase	Nondestructive	Enzyme spectrum method	Giles et al., 2018
	Taraxacum mongolicum Hand.-Mazz.	Hosphomonoesterase, β-glucosidase, β-cellobiohydrolase	Destructive	Fluorescence	Tischer et al., 2019
	Trifolium repens L.	N-acetylglucosaminidase, β-glucosidase, xylanase	Destructive	Fluorescence	Tischer et al., 2019
	Arabidopsis thaliana L. Heynh.	Phytate aspergillus enzyme	Destructive	Molybdenum blue method	Richardson et al., 2001
Animal excrements	Lumbricus terrestris L. feces	Cholinesterase, carboxylesterase	Destructive	—	Sanchez-Hernandez et al., 2018
	Acromyrmex echinatior Johan. feces	Xyloglucanase	Destructive	—	Kooij et al., 2016

change of pH affects the protonation process and thus the secretion of H^+-ATPase (Shahsavari et al., 2019). Studies have shown that the increase of soil temperature or the continuous wetting and drying of soil environment will change the composition of soil microbial community, and may increase the microbial biomass and enzyme activity. Low soil moisture will significantly inhibit the activity of β-glucosidase, and with the increase of humidity, the enzyme activity will also increase. Ge et al. (2017) found that the sensitivity of enzyme activity to temperature was low, and high temperature usually improved the enzyme activity.

It can be seen that soil enzyme activity is sensitive to temperature and moisture, and too high or too low a temperature and humidity will affect the enzyme activity. The change of pH, one of the major physiochemical properties of soil, also affects the activity of corresponding enzymes.

6.2.2.3 Nutrients

C, P, N and S are important nutrient elements required for the physiological and metabolic activities of plants and animals in soil. Due to their different content or ratios, they will impose different effects on the enzyme activity of rhizosphere soil.

The content of P and the value of C/P can change the rhizosphere range of plants, and then affect the enzyme activity in rhizosphere soil. Phosphatase is related to the P cycle. Razavi et al. (2017) analyzed the effects of nematodes on enzyme activity in the rhizosphere of *Lupinus micranthus* Guss through zymography. The results showed that compared with normal plants, the demand for phosphorus in infected plants was significantly increased, thus expanding the rhizosphere range of phosphatase around the root system by 1mm. Wei et al. (2019) found that the application of cellulose increased the rhizosphere influence range of acid and alkaline phosphatase, because it increased the C/P value, thus affecting the phosphorus mineralization process of soil organic matters. β-glucosidase is closely related to the C cycle, and the addition of C element can improve its activity. Urease is related to the N cycle. Studies have shown that fertilization increases storage capacity of C, N, P and S elements of soil, and enhances the catalytic activity of urease.

In summary, the lack or increase of nutrient elements is an important factor affecting the rhizosphere range, and the ratio of different nutrient elements also affects the enzyme activity of rhizosphere soil, triggering corresponding changes in enzyme activity.

6.2.3 Research progress in response of soil enzymes in plant rhizosphere to heavy metal pollution

The concentration and pollution type of heavy metals will affect the enzyme activity in rhizosphere soil. Excessive accumulation of heavy metals in rhizosphere soil will cause environmental pollution, do serious harm to plants and microorganisms, and even threaten human health. By far, types of heavy metal pollution mainly include: single heavy metal pollution, composite heavy metal pollution, heavy metal-organic compound pollution, and nano heavy metal and its oxide pollution. The changes in enzyme activity also vary from one pollution type to other.

6.2.3.1 Single heavy metal pollution

The effect of a single heavy metal on physicochemical properties of the plants or microbes is the key subject in laboratory researches. Different kinds of heavy metals have different ecotoxicological effects, and the responses of enzyme activity in rhizosphere soil to heavy metal pollution also vary according to the concentrations and types of heavy metals.

The response of enzyme activity in rhizosphere soil to heavy metal concentration is characterized by the "promotion at low concentration and inhibition at high concentration." Studies have shown that the addition of a low concentration of zinc can enhance the hydrolysis and transport activity of H^+-ATPase (Shahsavari et al., 2019). Yang et al. found through the root-bag pot experiment that a low concentration of Cd significantly promoted the activity of sucrase and urease in rhizosphere soil, and the enzyme activity gradually decreased with the increase of Cd concentration. When Cd concentration was $30mg \cdot kg^{-1}$, the inhibition rate of sucrase activity reached 7.2%. Deng et al. studied the effects of exogenous lead pollution on microbial enzyme activity in purple soil, and found that the activity of β-glucosidase and acid phosphatase in forest and cultivated soil was significantly correlated with the amount of lead added externally and the content of available lead. This indicates that within a certain range, the higher the concentration of heavy metals, the stronger the inhibitory effect on the enzyme activity of rhizosphere soil.

In nature, although the environmental pollution caused by a single pollutant occurs from time to time, the single heavy metal pollution factually does not exist in an absolute sense, because the pollution is normally concomitant and comprehensive. Therefore, it is not realistic to consider the pollution of only one heavy metal; we need to study the pollution of multiple heavy metals.

6.2.3.2 Compound pollution of heavy metals

The interactions between heavy metals mainly include antagonism and synergy. Antagonism refers to the case where the total toxicity of heavy metals interacting with each other is smaller than the sum of toxicity of each heavy metal. The synergistic effect of multiple heavy metals shows higher toxicity than that of a single heavy metal. Site competition is the cause of antagonism between heavy metals, and the sites mainly include active sites of the metabolic system or adsorption sites in the media, such as the binding sites on metallothionein, and the adsorption sites in soil. The synergistic effect is triggered by different heavy metals acting on different enzymes or proteins, which increases the damage to enzyme molecules.

The results show that $Cr_2O_7^{2-}$ and Pb^{2+} are easy to precipitate, weakening the combined toxicity of Cr^{6+} and Pb^{2+}, which indicates that the combined pollution of Cr and Pb may have produced the antagonistic effect. Compared with the effect of Zn added alone on the growth of rice (*Oryza sativa* L.), the synergistic effect of Cd and Zn is more toxic to rice. Bielińska et al. (2009) planted *Taraxacum mongolicum* Hand.-Mazz in the soil polluted by Zn, Pb and Cu. Analysis on the correlation basis suggests that the lower the content of Zn, Pb and Cu in the soil, the weaker the dehydrogenase activity; meanwhile, the activity of dehydrogenase in rhizosphere soil is higher than that in non-rhizosphere soil. This indicates that the deactivation rate of enzyme activity in rhizosphere soil is smaller than that in nonrhizosphere soil under the stress of compound heavy metals. Jia et al. studied the redox enzyme activity in rhizosphere soil of winter wheat seedlings to find that a low Pb content could reduce the influence of Cd on the activity of polyphenol oxidase and aggravate the inhibitory effect of Cd on the activity of dehydrogenase.

Aside from the compound pollution of heavy metals, there are often organic pollution phenomena in nature, such as soil pollution caused by lead and pyrene. As to the study on soil ecotoxicology, it is of great significance to discuss the heavy metal-organic compound pollution.

6.2.3.3 Heavy metal-organic compound pollution

Organic pollution is the environmental problem caused by many industrial and agricultural activities, including oil exploitation, smelting and pesticide spraying. With the aggravation of heavy metal pollution, the heavy metal-organic compound pollution has attracted increasingly attention. Generally, organic pollutants are absorbed in plant rhizosphere in two ways: ①Plant rhizosphere microorganisms convert the organic matters into small molecules. Amino acids in soil can be absorbed by microorganisms, and finally mineralized into inorganic nitrogen, ammonia (NH_4^+) and nitrate (NO_3^-) through ammoniation, nitrification and denitrification (Kieloaho et al., 2016). ②Plant root exudates can degrade the organic pollutants. For example, the roots of Sorghum sudanense (Piper) Stapf are rich in sphinsosinol, which can effectively degrade PAHs and other pollutants in the soil (Dominguez et al., 2019).

Studies have shown that the pyrene-lead compound pollution increases the secretion of related redoxases in the roots of *Scirpus triqueter* Linn, which are closely related to the pollutant degradation. Gao et al. (2019) found that in the soil contaminated by cesium and dibutyl phthalate (DBP), the activity of dehydrogenase, urease and acid phosphatase in the rhizosphere soil of wheat at seedling stage increased, while the activity of phenol oxidase and β-glucosidase was significantly inhibited, and the inhibition was apparently enhanced with the increase of DBP concentration. With the growth of wheat seedlings in the later stage, the enzyme activity in rhizosphere soil was restored.

The study of compound pollution is one of the major directions in the current development of environmental science. Therefore, as to solving the ecological security problems and fully evaluating the migration/transformation behavior of various pollutants, it is of great significance to reveal the response mechanism of plant rhizosphere processes, such as matter circulation and energy flow, to the compound pollution.

6.2.3.4 *Nano heavy metal and its oxide pollution*

With the extensive application of nano-metals in industry and agriculture, the environmental problems caused by nano-metals and their oxides are increasing. Soil is the main sink of artificial nanomaterials in the environment. Nanomaterials may lay potential impacts on the ecological environment due to their unique physical and chemical properties, thus inducing hazards to human health. A single nanomaterial has different effects on enzyme activity in different soils. For example, with nano Fe_3O_4 particles added to the black soil and saline-alkali soil of corn field (*Zea mays* L.), it is found that the catalase activity in salinized soil is significantly reduced, while the phosphatase activity in black soil has no significant changes (You et al.,

2018). Nano-ZnO can improve the utilization efficiency of zinc in plant rhizosphere and the activity of related enzymes in rhizosphere soil. Raliya and Tarafdar (2013) sprayed nano-ZnO solution with a concentration of 10 $mg \cdot L^{-1}$ on the 14-day-old *Legume* L. leaves, and 6 weeks later found the activity of rhizosphere acid phosphatase, alkaline phosphatase and phytase increased by 73.5%, 48.7% and 72.4%, respectively, indicating that low concentration of ZnO nanoparticles (NPs) promoted the activity of rhizosphere enzymes in pod plants. Sillen et al. (2015) found that the addition of low-concentration Ag NPs increased the biomass of corn, and the rhizosphere bacterial community significantly, ultimately leading to changes in the enzyme activity.

Nano heavy metals and their oxides are a new type of pollutants, which are difficult to monitor and control because of their small particle size. By far, there have been limited studies on the influence of nanomaterials on enzyme activity in rhizosphere soil, with improvement to be made in many aspects.

6.2.4 Main influencing mechanism of heavy metals on enzyme activity in rhizosphere soil

Heavy metals' inhibition of enzyme activity is temporary. Studies have shown that the enzyme activity of wheat field contaminated by cesium at seedling stage gradually recovers after the growth period. The influencing mechanism of heavy metals on enzyme activity in rhizosphere soil is mainly studied from three aspects: enzyme molecules, rhizosphere microorganisms and root secretions.

The action mechanisms of heavy metals on enzyme molecules mainly include: ①Heavy metals bind to the hydrophobic, amine and carboxyl groups of the enzyme molecules, or occupy the active center of enzymes, thus inhibiting the catalytic function of enzymes and destroying their structural integrity. Wang et al. (2019) found that long-term As pollution would change the functional stability of soil enzymes, and soil lightly polluted by As had higher enzyme activity than one heavily polluted by As. ②Heavy metal ions, as the auxiliary groups of enzyme molecules, promote the coordinated bonding between enzyme active sites and substrates, so that the enzyme molecules and their active sites maintain a certain specific structure, thus changing the equilibrium property of enzyme catalytic reaction and the surface charge of enzyme protein, and enhancing the enzyme activity. The results showed that under the stress of Cu, the catalase activity in rhizosphere soil of *Pinus sylvestris* Linn. infected by ectomycorrhizal fungi increased before decreased with the deepening of stress. This is perhaps because the presence of low concentration of Cd promoted the coordination between the active sites of enzymes and the substrates, which increased the surface charge of the enzyme protein and the enzyme activity, while high concentration of Cd inhibited these reactions and reduced the enzyme activity. ③Under the stress of heavy metals, the enzyme activity was not affected, that is, there was no specific correlation between the two. Cd (concentration of 0.5 mg·kg^{-1}) can promote the urease activity in rhizosphere soil of TPRC2001-1 Stylosanthes guianensias SW., but the same concentration of Cd has no significant effect on urease activity in rhizosphere soil of Stylosanthes guianensias SW.

The gas exchange between the rhizosphere environment and the outside world is triggered by the physiological metabolism of the root system or the respiration of microorganisms. There are a large number of microorganisms in the rhizosphere environment. The stress of heavy metals can induce the oxidative stress mechanism of microorganisms, and the occurrence of oxidative stress will inhibit the cytoplasmic enzyme activity of microorganisms and destroy the cell structure. This will affect the activity of enzymes in rhizosphere soil. Root exudates (such as amino acids, organic acids, clays and surfactants) can interact with heavy metals to reduce their toxicity. This is because amino acids combine with metal cations through ester group ($—COO$) and amine group ($—NH$) to form complexes, thus reducing the migration and transformation ability of metal ions. Under extremely acidic or alkaline conditions, H^+ and OH^- secreted by roots can also reduce the toxicity of some heavy metal ions, such as that of Al^{3+} caused by iron and manganese deficiency under acidic or alkaline pH conditions (Jaillard et al., 2003). In addition, the stress of heavy metals can also damage the structure of enzymes. In summary, Figure 6.10 shows the response mechanism of plant rhizosphere to heavy metal pollution.

During evaluation of the effects of heavy metals on enzyme activity in rhizosphere soil, the combined effects of various root exudes should also be considered. Specific species of plants or microorganisms can reduce the toxicity of heavy metals to the enzyme activity in rhizosphere soil to a certain extent. For areas with high heavy metal pollution, phytoremediation or plant-microbe combined remediation can be considered.

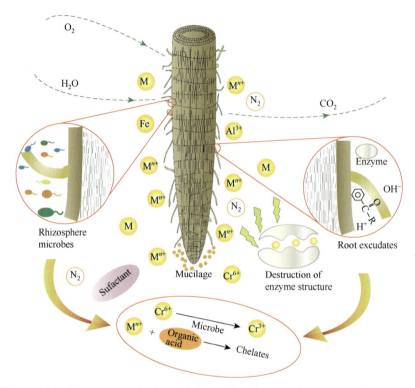

Figure 6.10 Response mechanism of plant rhizosphere environment to heavy metal pollution.

6.2.5 Summary

There have been a mass of studies on the effects of heavy metals on soil enzyme activity, but relatively few on the effects of heavy metals on enzyme activity in plant rhizosphere soil. A large number of studies have shown that the response of enzyme activity in rhizosphere soil to heavy metal concentration is characterized by "promotion at low concentration and inhibition at high concentration." Heavy metal-organic compound pollution is a common type in nature, whose effects on the rhizosphere soil enzymes vary to the growth cycle of plants. Nanomaterials are a new type of pollutants, whose effects on rhizosphere enzyme activity are closely related to the pollutant and soil types. Future studies should focus on the following aspects.

(1) Visualization technology should be further optimized and developed. The current in situ enzyme spectrometry method has significant limitations

in sensitivity, spatial and temporal resolution, and applicability, making accurate quantitative analysis unavailable. In the future, we should focus on the quantitative study of enzyme activity in rhizosphere soil, and use site-specific sampling to analyze the influence of various new factors that are crucial to the root development and rhizosphere characteristics.

(2) We will consider more realistic factors, extend the experiment period, and carry out in situ experimental researches. The laboratory conditions cannot be exact equivalent of nature, with a short experiment period, and unclear cycle for full recovery of enzyme activity. In the future, the response mechanism of enzyme activity in rhizosphere soil to heavy metal pollution in real environment should be studied to minimize the deviation caused by experimental simulation data.

(3) We will link the enzyme activity with microorganisms to establish more microbial indicators. Microorganisms are the main sources of soil enzymes, and heavy metal pollution can also affect the activity of rhizosphere microorganisms. The soil-plant-microbe interactions should be explored to find out the sensitive microorganisms corresponding to various enzymes, so as to provide more biological indicators for the evaluation of soil quality.

Main References

Abelson, P., 1999. A potential phosphate crisis. Science, 283: 2015–2015. Available from: https: doi.org/10.1126/science.283.5410.2015.

Agnihotri, R., Sharma, M.P., Prakash, A., et al., 2022. Glycoproteins of arbuscular mycorrhiza for soil carbon sequestration: Review of mechanisms and controls. Science of the Total Environment, 806: 150571. Available from: https: doi.org/10.1016/j.scitotenv.2021.150571.

Aon, M.A., Colaneri, A.C., 2001. II. Temporal and spatial evolution of enzymatic activities and physico-chemical properties in an agricultural soil. Applied Soil Ecology, 18(3): 255–270. Available from: https: doi.org/10.1016/S0929-1393(01)00161-5.

Artursson, V., Finlay, R.D., Jansson, J.K., 2006. Interactions between arbuscular mycorrhizal fungi and bacteria and their potential for stimulating plant growth. Environmental Microbiology, 8(1): 1–10. Available from: https: doi.org/10.1111/j.1462-2920.2005.00942.x.

Babu, A.N., Jogaiah, S., Ito, S.I., et al., 2015. Improvement of growth, fruit weight and early blight disease protection of tomato plants by rhizosphere bacteria is correlated with their beneficial traits and induced biosynthesis of antioxidant peroxidase and polyphenol oxidase. Plant Science, 231: 62–73. Available from: https: doi.org/10.1016/j.plantsci.2014.11.006.

Bagayoko, M., Alvey, S., Neumann, G., et al., 2000. Root-induced increases in soil pH and nutrient availability to field-grown cereals and legumes on acid sandy soils of Sudano-Sahelian West Africa. Plant and Soil, 225(1): 117–127. Available from: https: doi.org/10.1023/A:1026570406777.

Bakker, M.G., Chaparro, J.M., Manter, D.K., et al., 2015. Impacts of bulk soil microbial community structure on rhizosphere microbiomes of Zea mays. Plant and Soil, 392(1): 115–126. Available from:

https: doi.org/10.1007/s11104-015-2446-0.

Ballhausen, M. B., de Boer, W., 2016. The sapro-rhizosphere: Carbon flow from saprotrophic fungi into fungus-feeding bacteria. Soil Biology and Biochemistry, 102: 14–17. Available from: https: doi.org/10.1016/j.soilbio.2016.06.014.

Barber, S.A., 1962. A diffusion and mass-flow concept of soil nutrient availability. Science, 93(1): 39–49. Available from: https: doi.org/10.1097/00010694-196201000-00007.

Barber, S.A., Ozanne, P.G., 1970. Autoradiographic evidence for the differential effect of four plant species in altering the calcium content of the rhizosphere soil. Soil Science Society of America Journal, 34(4): 635–637. Available from: https: doi.org/10.2136/sssaj1970.03615995003400040027x.

Beltran-Garcia, M.J., White, J., Prado, F.M., et al., 2014. Nitrogen acquisition in *Agave tequilana* from degradation of endophytic bacteria. Scientific Reports, 4: 6938. Available from: https: doi.org/10.1038/srep06938.

Bielińska, E.J., Kołodziej, B., 2009. The effect of common dandelion (taraxacum officinale web.) rhizosphere on heavy metal content and enzymatic activity of soil. Acta Horticulturae, (826): 245–250.

Cavagnaro, T.R., 2016. Soil moisture legacy effects: Impacts on soil nutrients, plants and mycorrhizal responsiveness. Soil Biology and Biochemistry, 95: 173–179. Available from: https: doi.org/10.1016/j.soilbio.2015.12.016.

Challacombe, J.F., Hesse, C.N., Bramer, L.M., et al., 2019. Genomes and secretomes of Ascomycota fungi reveal diverse functions in plant biomass decomposition and pathogenesis. BMC Genomics, 20(1): 976. Available from: https: doi.org/10.1186/s12864-019-6358-x.

Chen, H.Y., Teng, Y.G., Lu, S.J., et al., 2015. Contamination features and health risk of soil heavy metals in China. Science of the Total Environment, 512: 143–153.

Cheng, W.X., Parton, W.J., Gonzalez-Meler, M.A., et al., 2014. Synthesis and modeling perspectives of rhizosphere priming. The New Phytologist, 201(1): 31–44. Available from: https: doi.org/10.1111/nph.12440.

Chowdhury, R.B., Moore, G.A., Weatherley, A.J., et al., 2017. Key sustainability challenges for the global phosphorus resource, their implications for global food security, and options for mitigation. Journal of Cleaner Production, 140: 945–963. Available from: https: doi.org/10.1016/j.jclepro.2016.07.012.

Clarke, A.L., Barley, K.P., 1968. The uptake of nitrogen from soils in relation to solute diffusion. Soil Research, 6(1): 75. Available from: https: doi.org/10.1071/SR9680075.

Clarkson, D.T., 1986. Regulation of the absorption and release of nitrate by plant cells: A review of current ideas and methodology. In: Lambers, H., Neeteson, J.J., Stulen, I., Fundamental, Ecological and Agricultural Aspects of Nitrogen Metabolism in Higher Plants. Dordrecht: Springer: 3–27. Available from: https: doi.org/10.1007/978-94-009-4356-8_1.

Cregger, M.A., McDowell, N.G., Pangle, R.E., et al., 2014. The impact of precipitation change on nitrogen cycling in a semi-arid ecosystem. Functional Ecology, 28(6): 1534–1544. Available from: https: doi.org/10.1111/1365-2435.12282.

Custos, J.M., Moyne, C., Sterckeman, T., 2020. How root nutrient uptake affects rhizosphere pH: A modelling study. Geoderma, 369: 114314. Available from: https: doi.org/10.1016/j.geoderma.2020.114314.

Dai, Y.T., Hou, X.Y., Yan, Z.J., et al., 2016. Soil microbes and the chemical properties of the rhizosphere and non-rhizosphere soil under two types of vegetation restoration in the hobq sandy land of inner Mongolia, China. Acta Ecologica Sinica, 36(20): 6353–6364. Available from: https: doi.org/10.5846/

stxb201504190802.

Danhorn, T., Fuqua, C., 2007. Biofilm formation by plant-associated bacteria. Annual Review of Microbiology, 61: 401–422. Available from: https: doi.org/10.1146/annurev.micro.61.080706.093316.

Darrah, P.R., 1993. The rhizosphere and plant nutrition: A quantitative approach. Plant and Soil, 155(1): 1–20.

Day, D.A., Poole, P.S., Tyerman, S.D., et al., 2001. Ammonia and amino acid transport across symbiotic membranes in nitrogen-fixing legume nodules. Cellular and Molecular Life Sciences: CMLS, 58(1): 61–71. Available from: https: doi.org/10.1007/PL00000778.

Deng, L., Peng, C.H., Kim, D.G., et al., 2021. Drought effects on soil carbon and nitrogen dynamics in global natural ecosystems. Earth-Science Reviews, 214: 103501. Available from: https: doi.org/ 10.1016/j.earscirev.2020.103501.

Ding, X.J., Jing, R.Y., Huang, Y.L., et al., 2017. Bacterial structure and diversity of rhizosphere and bulk soil of robinia pseudoacacia forests in Yellow River Delta. Acta Pedologica Sinica , 54(5): 1293–1302.

Dinkelaker, B., Marschner, H., 1992. *In vivo* demonstration of acid phosphatase activity in the rhizosphere of soil-grown plants. Plant and Soil, 144(2): 199–205. Available from: https: doi.org/10.1007/ BF00012876.

Dominguez, J.J.A., Bacosa, H.P., Chien, M.F., et al., 2019. Enhanced degradation of polycyclic aromatic hydrocarbons (PAHs) in the rhizosphere of sudangrass (*Sorghum* × drummondii). Chemosphere, 234: 789–795. Available from: https: doi.org/10.1016/j.chemosphere.2019.05.290.

Dong, S., Scagel, C.F., Cheng, L., et al., 2001. Soil temperature and plant growth stage influence nitrogen uptake and amino acid concentration of apple during early spring growth. Tree Physiology, 21(8): 541–547. Available from: https: doi.org/10.1093/treephys/21.8.541.

Drew, M.C., Nye, P.H., Vaidyanathan, L.V., 1969. The supply of nutrient ions by diffusion to plant roots in soil - I. Absorption of potassium by cylindrical roots of onion and leek. Plant and Soil, 30: 252–270. Available from: https: doi.org/10.1007/BF01349514.

Egidi, E., Delgado-Baquerizo, M., Plett, J.M., et al., 2019. A few Ascomycota taxa dominate soil fungal communities worldwide. Nature Communications, 10(1): 2369. Available from: https: doi.org/ 10.1038/s41467-019-10373-z.

Elser, J.J., Bracken, M.E.S., Cleland, E.E., et al., 2007. Global analysis of nitrogen and phosphorus limitation of primary producers in freshwater, marine and terrestrial ecosystems. Ecology Letters, 10(12): 1135–1142. Available from: https: doi.org/10.1111/j.1461-0248.2007.01113.x.

Fan, K.K., Cardona, C., Li, Y.T., et al., 2017. Rhizosphere-associated bacterial network structure and spatial distribution differ significantly from bulk soil in wheat crop fields. Soil Biology and Biochemistry, 113: 275–284. Available from: https: doi.org/10.1016/j.soilbio.2017.06.020.

Fierer, N., Bradford, M.A., Jackson, R.B., 2007. Toward an ecological classification of soil bacteria. Ecology, 88(6): 1354–1364.

Fonseca, J.P., Hoffmann, L., Cabral, B.C.A., et al., 2018. Contrasting the microbiomes from forest rhizosphere and deeper bulk soil from an Amazon rainforest reserve. Gene, 642: 389–397.Available from: https: doi.org/10.1016/j.gene.2017.11.039.

Ganmore-Neumann, R., Kafkafi, U., 1980. Root temperature and percentage NO_3^-/NH_4^+ effect on tomato development II. nutrients composition of tomato Plants[1]. Agronomy Journal, 72(5): 762–766. Available from: https: doi.org/10.2134/agronj1980.00021962007200050017x.

Gao, M.L., Zhang, Z., Song, Z.G., 2019. Effects of di-n-butyl phthalate on rhizosphere and non-rhizosphere soil microbial communities at different growing stages of wheat. Ecotoxicology and Environmental Safety, 174: 658–666. Available from: https: doi.org/10.1016/j.ecoenv.2019.01.125.

García-Salamanca, A., Molina-Henares, M. A., van Dillewijn, P., et al., 2013. Bacterial diversity in the rhizosphere of maize and the surrounding carbonate-rich bulk soil. Microbial Biotechnology, 6(1): 36–44. Available from: https: doi.org/10.1111/j.1751-7915.2012.00358.x.

Ge, T.D., Sun, N.B., Bai, L.P., et al., 2012. Effects of drought stress on phosphorus and potassium uptake dynamics in summer maize (*Zea mays*) throughout the growth cycle. Acta Physiologiae Plantarum, 34(6): 2179–2186. Available from: https: doi.org/10.1007/s11738-012-1018-7.

Ge, T.D., Wei, X.M., Razavi, B.S., et al., 2017. Stability and dynamics of enzyme activity patterns in the rice rhizosphere: Effects of plant growth and temperature. Soil Biology and Biochemistry, 113: 108–115. Available from: https: doi.org/10.1016/j.soilbio.2017.06.005.

Gharechahi, J., Vahidi, M.F., Bahram, M., et al., 2021. Metagenomic analysis reveals a dynamic microbiome with diversified adaptive functions to utilize high lignocellulosic forages in the cattle rumen. The ISME Journal, 15(4): 1108–1120. Available from: https: doi.org/10.1038/s41396-020-00837-2.

Giles, C.D., Dupuy, L., Boitt, G., et al., 2018. Root development impacts on the distribution of phosphatase activity: Improvements in quantification using soil zymography. Soil Biology and Biochemistry, 116: 158–166. Available from: https: doi.org/10.1016/j.soilbio.2017.08.011.

Hättenschwiler, S., Vitousek, P.M., 2000. Hattenschwiler and Vitousek 2000 Polyphenols and nutrient cycling. Tree, 15: 238–243.

Hayden, H.L., Savin, K.W., Wadeson, J., et al., 2018. Comparative metatranscriptomics of wheat rhizosphere microbiomes in disease suppressive and Non-suppressive soils for *Rhizoctonia solani* Ag8. Frontiers in Microbiology, 9: 859. Available from: https: doi.org/10.3389/fmicb.2018.00859.

Haynes, R.J., 1982. Effects of liming on phosphate availability in acid soils. Plant and Soil, 68(3): 289–308. Available from: https: doi.org/10.1007/BF02197935.

Haynes, R.J., 1983. Soil acidification induced by leguminous crops. Grass and Forage Science, 38(1): 1–11. Available from: https: doi.org/10.1111/j.1365-2494.1983.tb01614.x.

Henneron, L., Kardol, P., Wardle, D.A., et al., 2020. Rhizosphere control of soil nitrogen cycling: A key component of plant economic strategies. The New Phytologist, 228(4): 1269–1282. Available from: https: doi.org/10.1111/nph.16760.

Hinsinger, P., Plassard, C., Jaillard, B., 2003. Measurements of H+ fluxes and concentrations in the rhizosphere. In Rengel Z, ed. Handbook of Soil Acidity. Boca Raton: CRC Press.

Hinsinger, P., Plassard, C., Tang, C.X., et al., 2003. Origins of root-mediated pH changes in the rhizosphere and their responses to environmental constraints: A review. Plant and Soil, 248(1): 43–59. Available from: https: doi.org/10.1023/A.

Ho, A., Di Lonardo, D.P., Bodelier, P.L.E., 2017. Revisiting life strategy concepts in environmental microbial ecology. FEMS Microbiology Ecology, 93(3): fix006. Available from: https: doi.org/10.1093/femsec/fix006.

Hodge, A., Storer, K., 2015. Arbuscular mycorrhiza and nitrogen: Implications for individual plants through to ecosystems. Plant and Soil, 386(1): 1–19. Available from: https: doi.org/10.1007/s11104-014-2162-1.

Hodge, A., Campbell, C.D., Fitter, A.H., 2001. An arbuscular mycorrhizal fungus accelerates decomposition and acquires nitrogen directly from organic material. Nature, 413(6853): 297–299.

Available from: https: doi.org/10.1038/35095041.

Hoefer, C., Santner, J., Borisov, S.M., et al., 2017. Integrating chemical imaging of cationic trace metal solutes and pH into a single hydrogel layer. Analytica Chimica Acta, 950: 88–97. Available from: https: doi.org/10.1016/j.aca.2016.11.004.

Holz, M., Zarebanadkouki, M., Kuzyakov, Y., et al., 2018. Root hairs increase rhizosphere extension and carbon input to soil. Annals of Botany, 121(1): 61–69. Available from: https: doi.org/10.1093/aob/mcx127.

Hou, E.Q., Luo, Y.Q., Kuang, Y.W., et al., 2020. Global meta-analysis shows pervasive phosphorus limitation of aboveground plant production in natural terrestrial ecosystems. Nature Communications, 11(1): 637. Available from: https: doi.org/10.1038/s41467-020-14492-w.

Hunter, M.C., Smith, R.G., Schipanski, M.E., et al., 2017. Agriculture in 2050: Recalibrating targets for sustainable intensification. BioScience, 67(4): 386–391. Available from: https: doi.org/10.1093/biosci/bix010.

Huo, C.F., Luo, Y.Q., Cheng, W.X., 2017. Rhizosphere priming effect: A meta-analysis. Soil Biology and Biochemistry, 111: 78–84. Available from: https: doi.org/10.1016/j.soilbio.2017.04.003.

Israel, D.W., Jackson, W.A., 1982. Ion balance, uptake, and transport processes in n(2)-fixing and nitrate- and urea-dependent soybean plants. Plant Physiology, 69(1): 171–178. Available from: https: doi.org/10.1104/pp.69.1.171.

Jämtgård, S., Näsholm, T., Huss-Danell, K., 2010. Nitrogen compounds in soil solutions of agricultural land. Soil Biology and Biochemistry, 42(12): 2325–2330. Available from: https: doi.org/10.1016/j.soilbio.2010.09.011.

Jeewani, P.H., Gunina, A., Tao, L., et al., 2020. Rusty sink of rhizodeposits and associated keystone microbiomes. Soil Biology and Biochemistry, 147: 107840. Available from: https: doi.org/10.1016/j.soilbio.2020.107840.

Joner, E.J., Magid, J., Gahoonia, T.S., et al., 1995. P depletion and activity of phosphatases in the rhizosphere of mycorrhizal and non-mycorrhizal cucumber (cucumis sativus L.). Soil Biology and Biochemistry, 27(9): 1145–1151. Available from: https: doi.org/10.1016/0038-0717(95)00046-H.

Jones, D.L., 1998. Organic acids in the rhizosphere–a critical review. Plant and Soil, 205(1): 25–44. Available from: https: doi.org/10.1023/A:1004356007312.

Jones, D.L., Hodge, A., Kuzyakov, Y., 2004. Plant and mycorrhizal regulation of rhizodeposition. The New Phytologist, 163(3): 459–480. Available from: https: doi.org/10.1111/j.1469-8137.2004.01130.x.

Kafkafi, U., 2008. Root temperature, concentration and the ratio NO_3^- / NH_4^+ effect on plant development. Journal of Plant Nutrition, 13: 1291–1306. Available from: https: doi.org/10.1080/01904169009364152.

Kautz, T., Amelung, W., Ewert, F., et al., 2013. Nutrient acquisition from arable subsoils in temperate climates: A review. Soil Biology and Biochemistry, 57: 1003–1022. Available from: https: doi.org/10.1016/j.soilbio.2012.09.014.

Kavamura, V.N., Hayat, R., Clark, I.M., et al., 2018. Inorganic nitrogen application affects both taxonomical and predicted functional structure of wheat rhizosphere bacterial communities. Frontiers in Microbiology, 9: 1074. Available from: https: doi.org/10.3389/fmicb.2018.01074.

Kieloaho, A.J., Pihlatie, M., Dominguez Carrasco, M., et al., 2016. Stimulation of soil organic nitrogen pool: The effect of plant and soil organic matter degrading enzymes. Soil Biology and Biochemistry, 96: 97–106. Available from: https: doi.org/10.1016/j.soilbio.2016.01.013.

Kooij, P.W., Pullens, J.W.M., Boomsma, J.J., et al., 2016. Ant mediated redistribution of a xyloglucanase enzyme in fungus gardens of *Acromyrmex echinatior*. BMC Microbiology, 16: 81.

Kramer-Walter, K.R., Bellingham, P.J., Millar, T.R., et al., 2016. Root traits are multidimensional: Specific root length is independent from root tissue density and the plant economic spectrum. Journal of Ecology, 104(5): 1299–1310. Available from: https: doi.org/10.1111/1365-2745.12562.

Kucey, R.M.N., Janzen, H.H., Leggett, M.E., 1989. Microbially mediated increases in plant-available phosphorus. Advances in Agronomy, 42: 199–228. Available from: https: doi.org/10.1016/S0065-2113(08)60525-8.

Kuzyakov, Y., Bol, R., 2006. Sources and mechanisms of priming effect induced in two grassland soils amended with slurry and sugar. Soil Biology and Biochemistry, 38(4): 747–758. Available from: https: doi.org/10.1016/j.soilbio.2005.06.025.

Kuzyakov, Y., Blagodatskaya, E., 2015. Microbial hotspots and hot moments in soil: Concept & review. Soil Biology and Biochemistry, 83: 184–199. Available from: https: doi.org/10.1016/j.soilbio.2015.01.02.

Kuzyakov, Y., Razavi, B.S., 2019. Rhizosphere size and shape: Temporal dynamics and spatial stationarity. Soil Biology and Biochemistry, 135: 343–360. Available from: https: doi.org/10.1016/j.soilbio.2019.05.011.

Kuzyakov, Y., Raskatov, A., Kaupenjohann, M., 2003. Turnover and distribution of root exudates of *Zea mays*. Plant and Soil, 254(2): 317–327. Available from: https: doi.org/10.1023/A:1025515708093.

Lazcano, C., Boyd, E., Holmes, G., et al., 2021. The rhizosphere microbiome plays a role in the resistance to soil-borne pathogens and nutrient uptake of strawberry cultivars under field conditions. Scientific Reports, 11(1): 3188. Available from: https: doi.org/10.1038/s41598-021-82768-2.

Lebeau, T., 2011. Bioaugmentation for in situ soil remediation: How to ensure the success of such a process. In:Soil Biology. Berlin, Heidelberg: Springer Berlin Heidelberg: 129–186. Available from: https: doi.org/10.1007/978-3-642-19769-7_7.

Lee ,Y.S., Nguyen, X.H., Naing, K.W., et al., 2015. Role of lytic enzymes secreted by *Lysobacter capsici* YS1215 in the control of root-knot nematode of tomato plants. Indian Journal of Microbiology, 55(1): 74–80. Available from: https: doi.org/10.1007/s12088-014-0499-z.

Li, Q.W., Liu, Y., Gu, Y.F., et al., 2020. Ecoenzymatic stoichiometry and microbial nutrient limitations in rhizosphere soil along the Hailuogou Glacier forefield chronosequence. Science of the Total Environment, 704: 135413. Available from: https: doi.org/10.1016/j.scitotenv.2019.135413.

Ling, N., Wang, T., Kuzyakov, Y., 2022. Bacterial microbiome of the rhizosphere: from Structure to Functions. Nature Communications (accepted). Available from: https: doi.org/10.1038/s41467-022-28448-9.

Liu, N.N., Hu, H.F., Ma, W.H., et al., 2019. Contrasting biogeographic patterns of bacterial and archaeal diversity in the top- and subsoils of temperate grasslands. mSystems, 4(5): 519–566. Available from: https: doi.org/10.1128/msystems.00566-19.

Liu, S.B., Razavi, B.S., Su, X., et al., 2017. Spatio-temporal patterns of enzyme activities after manure application reflect mechanisms of niche differentiation between plants and microorganisms. Soil Biology and Biochemistry, 112: 100–109. Available from: https: doi.org/10.1016/j.soilbio.2017.05.006.

Liu, S.B., He, F.K., Kuzyakov, Y., et al., 2022. Nutrients in the rhizosphere: A meta-analysis of content, availability, and influencing factors. Science of the Total Environment, 826: 153908.

Liu, W.C., Lund, L.J., Page, A.L., 1989. Acidity produced by leguminous plants through symbiotic dinitrogen fixation. Journal of Environmental Quality, 18(4): 529–534.

Liu, X., Lu, X., Zhao, W.Q., et al., 2022. The rhizosphere effect of native legume Albizzia julibrissin on coastal saline soil nutrient availability, microbial modulation, and aggregate formation. Science of the Total Environment, 806: 150705. Available from: https: doi.org/10.1016/j.scitotenv.2021.150705.

Lundell, T.K., Mäkelä, M.R., Hildén, K., 2010. Lignin-modifying enzymes in filamentous basidiomycetes: Ecological, functional and phylogenetic review. Journal of Basic Microbiology, 50(1): 5–20. Available from: https: doi.org/10.1002/jobm.200900338.

Ma, X.M., Mason-Jones, K., Liu, Y., et al., 2019. Coupling zymography with pH mapping reveals a shift in lupine phosphorus acquisition strategy driven by cluster roots. Soil Biology and Biochemistry, 135: 420–428. Available from: https: doi.org/10.1016/j.soilbio.2019.06.001.

Magadlela, A., Pérez-Fernández, M.A., Kleinert, A., et al., 2016. Source of inorganic N affects the cost of growth in a legume tree species (*Virgilia divaricata*) from the Mediterrean-type Fynbos ecosystem. Journal of Plant Ecology, 9(6): 752–761. Available from: https: doi.org/10.1093/jpe/rtw015.

Magnuson, T.S., Crawford, D.L., 1992. Comparison of extracellular peroxidase- and esterase-deficient mutants of *Streptomyces viridosporus* T7A. Applied and Environmental Microbiology, 58(3): 1070–1072. Available from: https: doi.org/10.1128/AEM.58.3.1070-1072.1992.

Mahon, J.D., 1979. Environmental and genotypic effects on the respiration associated with symbiotic nitrogen fixation in peas. Plant Physiology, 63(5): 892–897. Available from: https: doi.org/10.1104/pp. 63.5.892.

Maltais-Landry, G., 2015. Legumes have a greater effect on rhizosphere properties (pH, organic acids and enzyme activity) but a smaller impact on soil P compared to other cover crops. Plant and Soil, 394(1): 139–154. Available from: https: doi.org/10.1007/s11104-015-2518-1.

Mapelli, F., Marasco, R., Fusi, M., et al., 2018. The stage of soil development modulates rhizosphere effect along a High Arctic Desert chronosequence. The ISME Journal, 12(5): 1188–1198. Available from: https: doi.org/10.1038/s41396-017-0026-4.

Marschner, P., Marino, W., Lieberei, R., 2002. Seasonal effects on microorganisms in the rhizosphere of two tropical plants in a polyculture agroforestry system in Central *Amazonia*, Brazil. Biology and Fertility of Soils, 35(1): 68–71. Available from: https: doi.org/10.1007/s00374-001-0435-3.

Marschner, P., Crowley, D., Yang, C.H., 2004. Development of specific rhizosphere bacterial communities in relation to plant species, nutrition and soil type. Plant and Soil, 261(1): 199–208. Available from: https: doi.org/10.1023/B:PLSO.0000035569.80747.c5.

Martin, M.H., Marschner, H., 1988. The mineral nutrition of higher plants. The Journal of Ecology, 76(4): 1250.

Massaccesi, L., Benucci, G.M.N., Gigliotti, G., et al., 2015. Rhizosphere effect of three plant species of environment under periglacial conditions (*Majella* Massif, central Italy). Soil Biology and Biochemistry, 89: 184–195. Available from: https: doi.org/10.1016/j.soilbio.2015.07.010.

Mendes, L.W., Kuramae, E.E., Navarrete, A.A., et al., 2014. Taxonomical and functional microbial community selection in soybean rhizosphere. The ISME Journal, 8(8): 1577–1587. Available from: https: doi.org/10.1038/ismej.2014.17.

Moreau, D., Bardgett, R.D., Finlay, R.D., et al., 2019. A plant perspective on nitrogen cycling in the rhizosphere. Functional Ecology, 33(4): 540–552. Available from: https: doi.org/10.1111/1365-2435. 13303.

Nannipieri, P., Trasar-Cepeda, C., Dick, R.P., 2018. Soil enzyme activity: A brief history and biochemistry as a basis for appropriate interpretations and meta-analysis. Biology and Fertility of Soils, 54(1): 11–19. Available from: https: doi.org/10.1007/s00374-017-1245-6.

Naseem, H., Ahsan, M., Shahid, M.A., et al., 2018. Exopolysaccharides producing rhizobacteria and their role in plant growth and drought tolerance. Journal of Basic Microbiology, 58(12): 1009–1022. Available from: https: doi.org/10.1002/jobm.201800309.

Nowak, J., Kaklewski, K., Ligocki, M., 2004. Influence of selenium on oxidoreductive enzymes activity in soil and in plants. Soil Biology and Biochemistry, 36(10): 1553–1558. Available from: https: doi.org/10.1016/j.soilbio.2004.07.002.

Pascual, J., Blanco, S., Ramos, J.L., et al., 2018. Responses of bulk and rhizosphere soil microbial communities to thermoclimatic changes in a Mediterranean ecosystem. Soil Biology and Biochemistry, 118: 130–144. Available from: https: doi.org/10.1016/j.soilbio.2017.12.013.

Pausch, J., Zhu, B., Kuzyakov, Y., et al., 2013. Plant inter-species effects on rhizosphere priming of soil organic matter decomposition. Soil Biology and Biochemistry, 57: 91–99. Available from: https: doi.org/10.1016/j.soilbio.2012.08.029.

Philippot, L., Raaijmakers, J.M., Lemanceau, P., et al., 2013. Going back to the roots: The microbial ecology of the rhizosphere. Nature Reviews Microbiology, 11(11): 789–799. Available from: https: doi.org/10.1038/nrmicro3109.

Pii, Y., Borruso, L., Brusetti, L., et al., 2016. The interaction between iron nutrition, plant species and soil type shapes the rhizosphere microbiome. Plant Physiology and Biochemistry, 99: 39–48. Available from: https: doi.org/10.1016/j.plaphy.2015.12.002.

Praeg, N., Pauli, H., Illmer, P., 2019. Microbial diversity in bulk and rhizosphere soil of *Ranunculus glacialis* along a high-alpine altitudinal gradient. Frontiers in Microbiology, 10: 1429. Available from: https: doi.org/10.3389/fmicb.2019.01429.

Qin, S.H., Yeboah, S., Xu, X.X., et al., 2017. Analysis on fungal diversity in rhizosphere soil of continuous cropping potato subjected to different furrow-ridge mulching managements. Frontiers in Microbiology, 8: 845. Available from: https: doi.org/10.3389/fmicb.2017.00845.

Qiu, Q., Li, J.Y., Wang, J.H., et al., 2014. Microbes, enzyme activities and nutrient characteristics of rhizosphere and non-rhizosphere soils under four shrubs in Xining Nanshan, Prefecture, China. Acta Ecologica Sinica, 34(24): 7411–7420.

Raliya, R., Tarafdar, J.C., 2013. ZnO nanoparticle biosynthesis and its effect on phosphorous-mobilizing enzyme secretion and gum contents in clusterbean (*Cyamopsis tetragonoloba* L.). Agricultural Research, 2(1): 48–57.

Ramos, C., Mølbak, L., Molin, S., 2000. Bacterial activity in the rhizosphere analyzed at the single-cell level by monitoring ribosome contents and synthesis rates. Applied and Environmental Microbiology, 66(2): 801–809. Available from: https: doi.org/10.1128/AEM.66.2.801-809.2000.

Rao, T.P., Yano, K., Iijima, M., et al., 2002. Regulation of rhizosphere acidification by photosynthetic activity in cowpea (*Vigna unguiculata* L. walp.) seedlings. Annals of Botany, 89(2): 213–220. Available from: https: doi.org/10.1093/aob/mcf030.

Raza, S., Miao, N., Wang, P.Z., et al., 2020. Dramatic loss of inorganic carbon by nitrogen-induced soil acidification in Chinese croplands. Global Change Biology, 26(6): 3738–3751. Available from: https: doi.org/10.1111/gcb.15101.

Razavi, B.S., Hoang, D.T., Blagodatskaya. E., et al., 2017. Mapping the footprint of nematodes in the rhizosphere: Cluster root formation and spatial distribution of enzyme activities. Soil Biology and Biochemistry, 115: 213–220. Available from: https: doi.org/10.1016/j.soilbio.2017.08.027.

Ren, C.J., Zhou, Z.H., Guo, Y.X., et al., 2021. Contrasting patterns of microbial community and enzyme activity between rhizosphere and bulk soil along an elevation gradient. CATENA, 196: 104921. Available from: https: doi.org/10.1016/j.catena.2020.104921.

Rengel, Z., Marschner, P., 2005. Nutrient availability and management in the rhizosphere: Exploiting genotypic differences. The New Phytologist, 168(2): 305–312. Available from: https: doi.org/ 10.1111/j.1469-8137.2005.01558.x.

Richardson, A.E., Hadobas, P.A., Hayes, J.E., 2001. Extracellular secretion of *Aspergillus* phytase from *Arabidopsis* roots enables plants to obtain phosphorus from phytate. The Plant Journal: for Cell and Molecular Biology, 25(6): 641–649. Available from: https: doi.org/10.1046/j.1365-313x.2001.00998.x.

Sanaullah, M., Razavi, B.S., Blagodatskaya, E., et al., 2016. Spatial distribution and catalytic mechanisms of β-glucosidase activity at the root-soil interface. Biology and Fertility of Soils, 52(4): 505–514. Available from: https: doi.org/10.1007/s00374-016-1094-8.

Sanchez-Hernandez, J.C., Notario del Pino, J., Capowiez, Y., et al., 2018. Soil enzyme dynamics in chlorpyrifos-treated soils under the influence of earthworms. Science of the Total Environment, 612: 1407–1416. Available from: https: doi.org/10.1016/j.scitotenv.2017.09.043.

Sattari, S.Z., Bouwman, A.F., Giller, K.E., et al., 2012. Residual soil phosphorus as the missing piece in the global phosphorus crisis puzzle. Proceedings of the National Academy of Sciences of the United States of America, 109(16): 6348–6353. Available from: https: doi.org/10.1073/pnas.1113675109.

Shahsavari, F., Khoshgoftarmanesh, A.H., Ali Mohammad Mirmohammady Maibody, S., et al., 2019. The role of root plasma membrane ATPase and rhizosphere acidification in zinc uptake by two different Zn-deficiency-tolerant wheat cultivars in response to zinc and histidine availability. Archives of Agronomy and Soil Science, 65(12): 1646–1658. Available from: https: doi.org/10.1080/03650340.2019.1572881.

Sillen, W.M.A., Thijs, S., Abbamondi, G.R., et al., 2015. Effects of silver nanoparticles on soil microorganisms and maize biomass are linked in the rhizosphere. Soil Biology and Biochemistry, 91: 14–22. Available from: https: doi.org/10.1016/j.soilbio.2015.08.019.

Simon, H.M., Dodsworth, J.A., Goodman, R.M., 2000. Crenarchaeota colonize terrestrial plant roots. Environmental Microbiology, 2(5): 495–505. Available from: https: doi.org/10.1046/j.1462-2920.2000.00131.x.

Spohn, M., Kuzyakov, Y., 2014. Spatial and temporal dynamics of hotspots of enzyme activity in soil as affected by living and dead roots—a soil zymography analysis. Plant and Soil, 379(1): 67–77. Available from: https: doi.org/10.1007/s11104-014-2041-9.

Steer, J., Harris, J.A., 2000. Shifts in the microbial community in rhizosphere and non-rhizosphere soils during the growth of *Agrostis stolonifera*. Soil Biology and Biochemistry, 32(6): 869–878. Available from: https: doi.org/10.1016/S0038-0717(99)00219-9.

Stevens, G.G., Pérez-Fernández, M.A., Morcillo, R.J.L., et al., 2019. Roots and nodules response differently to P starvation in the Mediterranean-type legume *Virgilia divaricata*. Frontiers in Plant Science, 10: 73. Available from: https: doi.org/10.3389/fpls.2019.00073.

Sugihara, S., Tomita, Y., Nishigaki, T., et al., 2016. Effects of different phosphorus-efficient legumes and soil texture on fractionated rhizosphere soil phosphorus of strongly weathered soils. Biology and Fertility of Soils, 52(3): 367–376. Available from: https: doi.org/10.1007/s00374-015-1082-4.

Sugiyama, A., Ueda, Y., Zushi, T., et al., 2014. Changes in the bacterial community of soybean rhizospheres during growth in the field. PLoS One, 9(6): e100709. Available from: https: doi.org/10.1371/journal.pone.0100709.

Sun, B.R., Gao, Y.Z., Wu, X., et al., 2020. The relative contributions of pH, organic anions, and phosphatase to rhizosphere soil phosphorus mobilization and crop phosphorus uptake in maize/alfalfa polyculture. Plant and Soil, 447(1): 117–133. Available from: https: doi.org/10.1007/s11104-019-04110-0.

Tang, Y.Q., Tian, J., Li, X.Z., et al., 2021. Higher free-living N_2 fixation at rock-soil interfaces than topsoils during vegetation recovery in Karst soils. Soil Biology and Biochemistry, 159: 108286. Available from: https: doi.org/10.1016/j.soilbio.2021.108286.

Teng, Y.G., Ni, S.J., Wang, J.S., et al., 2010. A geochemical survey of trace elements in agricultural and non-agricultural topsoil in Dexing Area, China. Journal of Geochemical Exploration, 104(3): 118–127. Available from: https: doi.org/10.1016/j.gexplo.2010.01.006.

Terzano, R., Cesco, S., Mimmo, T., 2015. Dynamics, thermodynamics and kinetics of exudates: Crucial issues in understanding rhizosphere processes. Plant and Soil, 386(1): 399–406. Available from: https: doi.org/10.1007/s11104-014-2308-1.

Tian, P., Razavi, B.S., Zhang, X.C., et al., 2020. Microbial growth and enzyme kinetics in rhizosphere hotspots are modulated by soil organics and nutrient availability. Soil Biology and Biochemistry, 141: 107662. Available from: https: doi.org/10.1016/j.soilbio.2019.107662.

Tilman, D., Balzer, C., Hill, J., et al., 2011. Global food demand and the sustainable intensification of agriculture. Proceedings of the National Academy of Sciences of the United States of America, 108(50): 20260–20264. Available from: https: doi.org/10.1073/pnas.1116437108.

Tischer, A., Sehl, L., Meyer, U.N., et al., 2019. Land-use intensity shapes kinetics of extracellular enzymes in rhizosphere soil of agricultural grassland plant species. Plant and Soil, 437(1): 215–239. Available from: https: doi.org/10.1007/s11104-019-03970-w.

Treusch, A.H., Leininger, S., Kletzin, A., et al., 2005. Novel genes for nitrite reductase and Amo-related proteins indicate a role of uncultivated mesophilic crenarchaeota in nitrogen cycling. Environmental Microbiology, 7(12): 1985–1995. Available from: https: doi.org/10.1111/j.1462-2920.2005.00906.x.

Wang, Z.Q., Tian, H.X., Tan, X.P., et al., 2019. Long-term As contamination alters soil enzyme functional stability in response to additional heat disturbance. Chemosphere, 229: 471–480. Available from: https: doi.org/10.1016/j.chemosphere.2019.05.055.

Wang, Z.Y., Kelly, J.M., Kovar, J.L., 2007. Depletion of macro-nutrients from rhizosphere soil solution by juvenile corn, cottonwood, and switchgrass plants. Plant and Soil, 270(1): 213–221. Available from: https: doi.org/10.1007/s11104-004-1538-z.

Warren, C.R., 2009. Why does temperature affect relative uptake rates of nitrate, ammonium and glycine: A test with *Eucalyptus pauciflora*. Soil Biology and Biochemistry, 41(4): 778–784. Available from: https: doi.org/10.1016/j.soilbio.2009.01.012.

Wei, X.M., Hu, Y.J., Razavi, B.S., et al., 2019. Rare taxa of alkaline phosphomonoesterase-harboring microorganisms mediate soil phosphorus mineralization. Soil Biology and Biochemistry, 131: 62–70. Available from: https: doi.org/10.1016/j.soilbio.2018.12.025.

Williams, P.N., Santner, J., Larsen, M., et al., 2014. Localized flux maxima of arsenic, lead, and iron around root apices in flooded lowland rice. Environmental Science & Technology, 48(15): 8498–8506. Available from: https: doi.org/10.1021/es501127k.

Xu, H., Detto, M., Fang, S.Q., et al., 2020. Soil nitrogen concentration mediates the relationship between leguminous trees and neighbor diversity in tropical forests. Communications Biology, 3(1): 317. Available from: https: doi.org/10.1038/s42003-020-1041-y.

Xu, J., Zhang, Y.Z., Zhang, P.F., et al., 2018. The structure and function of the global citrus rhizosphere microbiome. Nature Communications, 9(1): 4894. Available from: https: doi.org/10.1038/s41467-018-07343-2.

Yang, H.S., Zhang, Q., Dai, Y.J., et al., 2015. Effects of arbuscular mycorrhizal fungi on plant growth depend on root system: A meta-analysis. Plant and Soil, 389(1): 361–374. Available from: https: doi.org/10.1007/s11104-014-2370-8.

Yao, X.H., Min, H., Lü, Z.H., et al., 2006. Influence of acetamiprid on soil enzymatic activities and respiration. European Journal of Soil Biology, 42(2): 120–126. Available from: https: doi.org/10.1016/j.ejsobi.2005.12.001 .

Yelle, D.J., Ralph, J., Lu, F.C., et al., 2008. Evidence for cleavage of lignin by a brown rot basidiomycete. Environmental Microbiology, 10(7): 1844–1849. Available from: https: doi.org/10.1111/j.1462-2920.2008.01605.x.

York, L.M., Carminati, A., Mooney, S.J., et al., 2016. The holistic rhizosphere: Integrating zones, processes, and semantics in the soil influenced by roots. Journal of Experimental Botany, 67(12): 3629–3643. Available from: https: doi.org/10.1093/jxb/erw108.

You, T.T., Liu, D.D., Chen, J., et al., 2018. Effects of metal oxide nanoparticles on soil enzyme activities and bacterial communities in two different soil types. Journal of Soils and Sediments, 18(1): 211–221. Available from: https: doi.org/10.1007/s11368-017-1716-2.

Young, J.L., Aldag, R. W., 2015. Inorganic forms of nitrogen in soil. In Nitrogen in Agricultural Soils. Madison, WI, USA: American Society of Agronomy, Crop Science Society of America, Soil Science Society of America: 43–66. Available from: https: doi.org/10.2134/agronmonogr22.c2.

Zamanian, K., Pustovoytov, K., Kuzyakov, Y., 2016. Pedogenic carbonates: Forms and formation processes. Earth-Science Reviews, 157: 1–17. Available from: https: doi.org/10.1016/j.earscirev.2016.03.003.

Zhou, Q., 2014. Research on crop rhizosphere effects of winter crops in paddy fields. Nanchang: Jiangxi Agricultural University.

Final remarks

This section is intended to summarize the main ideas of this book and give a perspective into the future of researches on enzyme activities in soil and groundwater remediation.

Soil enzymes, as the biological activity index of soil quality and the evaluation index of soil fertility, play an important role in soil ecosystem. During the agricultural production, with the aim of activating the soil nutrients, improving the soil fertility and enhancing the nutrient utilization efficiency, the biochemical characteristics of soil enzymes should be made full use to give full play to the advantages of biological activity of soil enzymes. With the development of science and the introduction of new technology, the study on soil enzymes has made remarkable progress. However, different studies come to different conclusions. In recent years, a large number of studies using bibliometric analysis or meta-analysis methods to study the qualitative or quantitative changes in enzyme activity or groundwater have found that there are still many problems to be solved regarding the soil enzyme activity. As key contents of the soil science and microbial science, studying the biochemical dynamic characteristics of soil enzymes and the influencing factors of enzyme activity, combining the soil enzymology knowledge with agricultural production practices, soil environmental ecological protection and pollution control, and using the soil enzymology knowledge to deal with the practical problems of agricultural and forestry ecological environment are the development prospect and trend of soil enzymology.

First of all, the method of extracting enzymes from soil is not quite mature, because the enzyme compound with soil matrix often loses its integrity during the extraction process. Therefore soil enzymes are often studied by measuring their activity, which makes it difficult to distinguish the activity of non biological enzymes from that of soil-living organisms. Second, it is not known how non biological enzymes in soil catalyze the natural substrates in situ. In addition, the proposed conceptual model of enzymes is an experimental humus/clay-enzyme one that assumes a close relationship between non biological enzymes and organisms in the soil. Burns hypothesizes that humus-enzyme complexes play an important role in substrate catalysis. Some organisms, substrates, or their decomposition products may be beneficial to organisms, but unavailable to microorganisms for their large molecular weight or insolubility. The microbes cannot sense the presence of the substrate, thus they cannot be induced to synthesize or secrete extracellular enzymes to hydrolyze the substrate. In turn, products released by complex soil enzymes can be quickly broken down by other enzymes or taken up by soil microorganisms. It may be beneficial for soil microbial cells to be distributed on the colloidal surface of soil humus with numerous enzymes. In fact, some kinds of soil microorganisms can successfully live in unfavorable soil environment, possibly because of their connection with the humus-colloid complexes.

When soil enzyme is taken as an index of the soil quality, the determination method of soil enzymes must be considered. For existing researches on soil enzymes, further improvement should be made in the soil pre-treatment, soil sterilization, analytical conditions, and even the expression of active units of soil enzyme. It is urgent now to standardize the research methods of soil enzyme activity.

The enzymes in the soil are derived from living microorganisms, plants or soil animals, as well as from the excrement or even dead cells of living cells, which could combine with organic mineral colloids. Enzymes can catalyze various reactions in the soil that are necessary for life activities of soil microbes, including the degradation of organic residues, nutrient cycling, and formation of organic matters and soil structure. The role of non biological enzyme activity is made unclear by the difficulty in distinguishing between the biological and non biological enzyme activity.

Previous studies have shown that there is no direct correlation between soil enzyme activity and other soil biological properties. This is partly due to the incomplete measurement of soil biological properties (microbial counts and respiration) and the specificity of enzymatic catalysis, or the presence of non biological enzyme activity. With the study of the new soil biological parameters, it was found that there was a relatively stable relationship between the microorganisms and the enzyme activity of soil. However, the primary purpose of studying soil enzyme activity is not to measure the biological activity itself, but to find a comprehensive index of enzyme activity to characterize the changes of soil chemistry and biochemistry under external management or environmental conditions. In particular, certain enzymes are well suited to be used as such indicators. Numerous studies have shown that soil enzyme activity is sensitive to a variety of land management practices, including plant decomposition, soil compaction, tillage and crop rotation. After a few months or a year of soil treatment, some soil enzymes will change. Therefore, soil enzyme activity indicates the advantages and disadvantages of soil management measures in a relatively short period of time, while changes in organic matter are perceived in a longer period of time.

With the continuous growth of the global population, the ecological environment in some regions is deteriorating. In this situation, the soil quality has aroused widespread attention. Thus, it turns out a chief task in soil enzymology to find a sensitive and universal comprehensive index, thus eliminating the need to measure multiple soil parameters and compare multiple treatments. In the future, the research of enzyme activity should focus on solving practical problems in agriculture, forestry, ecological environment and other fields.